Magnetic Properties of Josephson Junction Networks

An Introduction

Magnetic Properties of Josephson Junction Networks
An Introduction

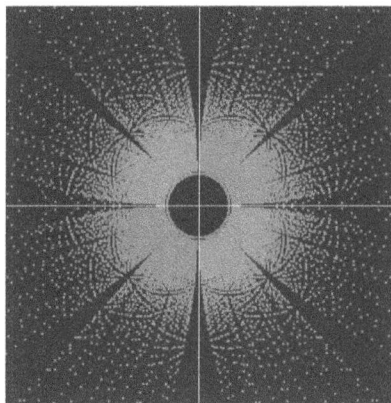

Roberto De Luca
Università degli Studi di Salerno, Italy

World Scientific

NEW JERSEY · LONDON · SINGAPORE · BEIJING · SHANGHAI · HONG KONG · TAIPEI · CHENNAI · TOKYO

Published by

World Scientific Publishing Co. Pte. Ltd.

5 Toh Tuck Link, Singapore 596224

USA office: 27 Warren Street, Suite 401-402, Hackensack, NJ 07601

UK office: 57 Shelton Street, Covent Garden, London WC2H 9HE

British Library Cataloguing-in-Publication Data
A catalogue record for this book is available from the British Library.

MAGNETIC PROPERTIES OF JOSEPHSON JUNCTION NETWORKS
An Introduction

ISBN 978-981-120-925-3 (hardcover)
ISBN 978-981-120-926-0 (ebook for institutions)
ISBN 978-981-120-927-7 (ebook for individuals)

For any available supplementary material, please visit
https://www.worldscientific.com/worldscibooks/10.1142/11525#t=suppl

Desk Editor: Nur Syarfeena Binte Mohd Fauzi

Dedication

To my two daughters, Francesca and Mariateresa, to whom I shall always be grateful.

Preface

This book has been conceived as a study guide for advanced physics students wishing to get acquainted with the still largely unexplored realm of three-dimensional Josephson junction networks. Part of this work is taken and organized by well-known textbooks, part is contained in my Ph.D. thesis, and part has been object of studies in successive years. The idea of the present book arose when teaching a Ph.D. course at University of Salerno (Italy). At that time the overwhelming teaching duties and other circumstances did not favor full development of this fascinating subject. However, I felt that students could still benefit from these lectures for two reasons. First of all, the material was mostly composed by first-hand information, so that they could attend lectures given by the same Author who elaborated part of the original work included in the course. Secondly, students could acquire some knowledge on this interesting subject, in order to try, with the right dose of enthusiasm, to make some progress in the field.

In this way, the first four chapters of the present book are dedicated to the essential analytic instruments to face the study of Josephson junction networks. One-dimensional and two-dimensional networks are studied in Chapter 5 and 6, respectively. A possible application of Josephson junction network models in determining the low-field magnetic properties of granular superconducting systems is illustrated in Chapter 7. In Chapter 8, the theoretical bases for studying three-dimensional Josephson junction networks are laid and some examples of simple networks are given. In Chapter 9, finally, the essential properties of a promising device based on superconducting quantum interference in a three-dimensional Josephson junction network are investigated. This system, already named 3D-SQUID in the literature, could be used as an ultra-sensitive flux-to-voltage transducer able to detect all components of an extremely weak local magnetic field, in much the same way as a SQUID detects magnetic field components orthogonal to the plane of the device. Will Mankind acquire this tiny accessory to continue its walk on the road of scientific progress? It is certainly hard to answer to this question, which some scientists may probably retain not even pertinent. From my point of view, however, the question is not only pertinent, but also important. In fact, the effort made in writing this book has been supported by a clear perspective: sooner or later the topic of three-dimensional Josephson junction networks will become an important issue for applications in many different fields, in such a way that young scientists could give their contribution to the development of novel superconducting devices. In this way, the essential knowledge on this subject could be readily found in the pages of the present work. This book also provides useful material to those scientists interested in getting a deeper insight on the low-field magnetic response of granular superconductors.

The work has been accomplished with the awareness that a single person cannot easily achieve, all by himself, relevant goals. Therefore, I need to acknowledge the patience of my family and the support of many colleagues: S. Pace for having introduced me, long ago, to these studies, A. Naddeo and F. Romeo for critical reading of the manuscript; M. Di Mauro, O. Faella, G. Monetti, L. Orza, I. Rabuffo, A. Stabile for their constant attention to my present activities and for their support.

Roberto De Luca

Contents

Chapter 1

The Josephson Junction

1.1. Feynman's and Ohta's Models

B. D. Josephson was the first to describe the quantum properties of superconducting junctions (Josephson, 1962) by deriving the dynamic equations that are now named after him. For the "theoretical predictions of the properties of a supercurrent through a tunnel barrier, in particular those phenomena which are generally known as the Josephson effects", the scientist received the Nobel prize in 1973. A Josephson junction (JJ) is a system made of two weakly coupled superconductors, S_1 and S_2. In particular, one can consider a tunnel junction, in which S_1 and S_2 are separated by a very thin insulating barrier, as shown in Fig. 1.1. By denoting the superconducting wave-functions in the superconductors S_1 and S_2 as ψ_1 and ψ_2 respectively, we can set (Schmidt, 1997)

$$\psi_1 = \sqrt{N_1}e^{-i\theta_1}, \quad \psi_2 = \sqrt{N_2}e^{-i\theta_2}, \qquad (1.1)$$

where N_k is the number density of Cooper pairs and θ_k is the superconducting phase of the k–th electrode ($k = 1, 2$). When both wave-functions, ψ_1 and ψ_2, do not depend on position, we can take N_k to be the number of Cooper pairs, provided we redefine the scalar product through the following bilinear application:

$$< \Psi|\Psi > = |\psi_1|^2 + |\psi_2|^2 = N_1 + N_2 = N. \qquad (1.2)$$

In this way, the vector wave-function

$$\Psi = \begin{pmatrix} \psi_1 \\ \psi_2 \end{pmatrix} \qquad (1.3)$$

could be normalized with respect to the total number of electron pairs N.

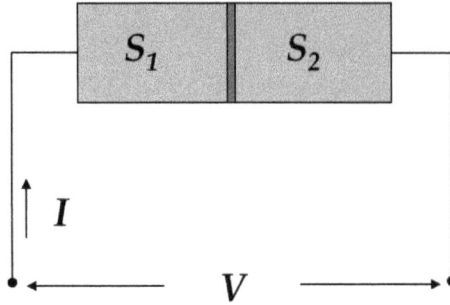

Fig. 1.1. Schematic representation of a Josephson junction consisting of two superconductors, S_1 and S_2, separated by a very thin insulator. A voltage V is applied to the junction and the same current I flows in the two electrodes, which are both connected to the external "classical" world.

We shall not proceed in normalizing the two wave-functions for reasons which will be clear in the following discussion. Furthermore, we shall take N_k to represent the number of Cooper pairs in the k-th electrode of the Josephson junction. Having defined the main characteristics of the physical system, Josephson equations can be written as follows (Barone & Paternò, 1982):

$$I = I_{J0} \sin \phi \,; \tag{1.4a}$$

$$V = \frac{\Phi_0}{2\pi} \frac{d\phi}{dt}, \tag{1.4b}$$

where I is the superconducting current flowing through the insulating barrier, I_{J0} being its maximum value, $\phi = \theta_2 - \theta_1$ is the superconducting phase difference across the Josephson junction and $\Phi_0 = h/2e$ is the elementary flux quantum, expressed as the ratio of Planck's constant h and of the absolute value of the Cooper pair charge $2e$.

R. P. Feynman was able to capture the main physical properties of the JJ, by means of a rather immediate analytical approach (Feynman, Leighton, & Sands, 1965). Feynman's model, which can be considered as a specific description of a weakly coupled two-level quantum system, does not provide, however, a consistent account of the external bias circuit. As a consequence, the voltage-frequency relation given by (1.4b) does not appear in its "strict" form. This not trivial inconsistency in Feynman's model was carefully considered by H. Ohta, who introduced a semi-classical model (Ohta, 1977) based on a rigorous quantum derivation. Ohta noticed that the weakly coupled two-level quantum system adopted by Feynman did not include an additional term due to energy contribution of the external classical circuit biasing the Josephson junction. By adding this missing term and by making use of a semi-classical approach, the latter scientist was able to recover the strict form of Eq. (1.4b). Ohta's model has been proven useful in deriving the correct Josephson equations for a tri-layer system (double-barrier Josephson junction), in which the two external electrodes of the JJ are analyzed using Ohta's approach and the sandwiched superconductor is considered to be a pure quantum system (De Luca & Romeo, 2009). Let us now take a close look to both models.

1.1.1. *Feynman's model*

Consider two weakly coupled superconductors forming a Josephson junction, as shown in Fig. 1.1. Following Feynman's model (Feynman, Leighton, & Sands, 1965), we would like to obtain the current-phase and the voltage-frequency relations for the Josephson junction as reported in Eq. (1.4a) and (1.4b). We then start by saying that the dynamics of the two weakly coupled quantum systems can be described by means of Schrödinger equation (Sakurai, 2004):

$$i\hbar \frac{\partial \Psi}{\partial t} = H_0 \Psi, \qquad (1.5)$$

where H_0 is the full Hamiltonian of the quantum system, given by the following matrix

$$H_0 = \begin{pmatrix} E_1 & -K \\ -K & E_2 \end{pmatrix}. \qquad (1.6)$$

In the previous equation, K is the coupling energy constant between the superconductors S_1 and S_2. By now substituting Eq. (1.6) into Eq. (1.5), we may write:

$$i\hbar \frac{\partial \psi_1}{\partial t} = E_1 \psi_1 - K \psi_2, \tag{1.7a}$$

$$i\hbar \frac{\partial \psi_2}{\partial t} = E_2 \psi_2 - K \psi_1. \tag{1.7b}$$

By making use of Eq. (1.1), the latter two relations can be rewritten as follows:

$$\hbar \frac{d\theta_1}{dt} + \frac{i\hbar}{2N_1} \frac{dN_1}{dt} = E_1 - K \sqrt{\frac{N_2}{N_1}} e^{-i(\theta_2 - \theta_1)}, \tag{1.8a}$$

$$\hbar \frac{d\theta_2}{dt} + \frac{i\hbar}{2N_2} \frac{dN_2}{dt} = E_2 - K \sqrt{\frac{N_1}{N_2}} e^{i(\theta_2 - \theta_1)}. \tag{1.8b}$$

Let us now define the superconducting phase difference as $\phi = \theta_2 - \theta_1$. Let us also separate the real and imaginary parts in Eq. (1.8a) and Eq. (1.8b), to get

$$\frac{dN_2}{dt} = -\frac{dN_1}{dt} = -\frac{2K}{\hbar} \sqrt{N_1 N_2} \sin \phi, \tag{1.9a}$$

$$\hbar \frac{d\phi}{dt} = E_2 - E_1 - K \left(\sqrt{\frac{N_1}{N_2}} - \sqrt{\frac{N_2}{N_1}} \right) \cos \phi. \tag{1.9b}$$

From Eq. (1.9a), we may notice that the system obeys charge conservation, since $\dot{N}_1 + \dot{N}_2 = 0$, where the dot on top of a variable is taken to stand for the derivative with respect to time of the same quantity. The external voltage applied to the quantum system in Fig. 1.1 means that the Cooper pairs in the two condensates occupy distinct energy levels, so that we may set $E_2 - E_1 = 2eV$. Furthermore, by defining

$$I = -2e \dot{N}_1, \quad I_{J0} = -\frac{4eK}{\hbar} \sqrt{N_1 N_2}, \tag{1.10}$$

we may rewrite Eq. (1.9a) and Eq. (1.9b), respectively, as follows:

$$I = I_{J0} \sin \phi, \tag{1.11a}$$

$$\frac{d\phi}{dt} = \frac{2e}{\hbar}V - \frac{K}{\hbar}\left(\sqrt{\frac{N_1}{N_2}} - \sqrt{\frac{N_2}{N_1}}\right)\cos\phi. \qquad (1.11b)$$

Therefore, Eq. (1.11a) completely agrees with Eq. (1.4a), denoted as the "current-phase relation" (CPR) of a Josephson junction. On the other hand, Eq. (1.11b) differs from the strict "voltage-frequency relation" in Eq. (1.4b) by the additional cosine term, which pertains to the properties of isolated quantum systems. Because of this discrepancy, Ohta successfully introduced a semi-classical analysis by which he was able to reproduce both the current-phase relation and the strict voltage-frequency relation given in Eq. (1.4a) and in Eq. (1.4b), respectively. We shall illustrate Ohta's approach in the following section.

1.1.2. *Ohta's model*

Let us again consider the two weakly coupled superconductors forming a Josephson junction in Fig. 1. This time we agree that the physical system needs to be described semi-classically, by retaining Feynman approach for the quantum system, but by also introducing the energy contribution due to the external circuit. In his important work (Ohta, 1977), Ohta stated that he had long been puzzled by the fact that one could not achieve a strict frequency-voltage relation by means of Feynman model. He was thus led to develop a rigorous semi-classical analysis to take into account the contribution due to the external circuit. We shall here give a simplified, but still rigorous, version of the more complete analysis given by Ohta. Starting from quantum mechanical considerations, Ohta first recovered Feynman's Hamiltonian. However, considering the classical nature of the problem (the system made of the external circuit and the Josephson junction), Ohta projected Feynman results onto the classical world. A way to do this, of course, is to consider the classically observable energy H_0 as derived from Feynman's model, by writing:

$$H_0 = \langle H_0 \rangle = \langle \Psi | H_0 | \Psi \rangle. \qquad (1.12)$$

We need to consider that the above energy term is only one portion of the classical Hamiltonian related to the whole system in Fig. 1.1. In fact, the remaining portion is given by the energy provided by the circuit, which can be written as $W = \int I V \, dt$. In this way, the classical complete Hamiltonian H is written as follows:

$$H = H_0 - W = E_1 N_1 + E_2 N_2 - 2K\sqrt{N_1 N_2} \cos \phi - W, \quad (1.13)$$

where we recall that $\phi = \theta_2 - \theta_1$. The transition to the classical world is now complete, so that a solution to the problem by classical mechanics can be found. First, let us notice that θ_k and $\hbar N_k$ are conjugate variables ($k = 1, 2$). Hamilton's equations thus give

$$\hbar \dot{N}_k = -\frac{\partial H}{\partial \theta_k}, \quad (1.14a)$$

$$\dot{\theta}_k = \frac{1}{\hbar}\frac{\partial H}{\partial N_k}, \quad (1.14b)$$

with $k = 1, 2$. By now defining the coupling energy as $E_C = -2K\sqrt{N_1 N_2} \cos \phi$, and by setting $E_R = E_C - W$, we may rewrite the above equations as follows:

$$\hbar \dot{N}_k = -\frac{\partial E_R}{\partial \theta_k}, \quad (1.15a)$$

$$\dot{\theta}_k = \frac{E_k}{\hbar} + \frac{1}{\hbar}\frac{\partial E_R}{\partial N_k}, \quad (1.15b)$$

with $k = 1, 2$. Let us now consider the time derivative of E_R:

$$\dot{E}_R = \sum_{k=1}^{2}\left(\frac{\partial E_R}{\partial \theta_k}\dot{\theta}_k + \frac{\partial E_R}{\partial N_k}\dot{N}_k\right). \quad (1.16)$$

By substituting for the partial derivatives of E_R as obtained by Eq. (1.15a) and (1.15b), we may write:

$$\dot{E}_R = \sum_{k=1}^{2} E_k \dot{N}_k. \quad (1.17)$$

As in Feynman's model, the Cooper pairs in the two condensates occupy distinct energy levels, so that $E_2 - E_1 = 2eV$. In this way, by setting $E_1 = -eV$ and $E_2 = +eV$, we have

$$E_R = eV \int (\dot{N}_2 - \dot{N}_1)\, dt. \quad (1.18)$$

We may now consider the two Josephson junctions immersed in a thermal bath, so that we can consider the number of Cooper pairs in each condensate to be constant. Therefore, we set $\dot{N}_1 = \dot{N}_2 = 0$. In this way, the charge conservation relation $\dot{N}_1 + \dot{N}_2 = 0$ becomes a mere identity. Furthermore, because of Eq. (1.18), the quantity E_R is seen to be exactly zero. In addition, for constant values of E_1 and E_2 the Hamiltonian H is a constant of the motion, so that the energy of the system is conserved. These results could appear trivial at first sight. However, these conditions allow us to write down the strict frequency-voltage relation (1.4b). In fact, by Eq. (1.15b) and by $E_R = 0$, we have:

$$\dot{\theta}_k = \frac{E_k}{\hbar}, \tag{1.19}$$

for $k = 1, 2$. The above relation leads exactly to the strict voltage-frequency relation.

We can also determine the CPR of the Josephson junction by noticing that conservation of energy ($\dot{H} = 0$) gives the following relation:

$$\dot{E}_C = \dot{W} \Rightarrow 2\dot{\phi}K\sqrt{N_1 N_2}\sin\phi = IV. \tag{1.20}$$

By use of Eq. (1.4b), already proven to be true by means of Eq. (1.19), we finally get the phase-current relation in Eq. (1.4a).

Therefore, we have seen that a model which considers a Josephson junction as an isolated quantum system does not lead to the strict voltage-frequency Josephson relation. However, by making use of Ohta's semi-classical analysis, we can write the correct Josephson equations for the superconducting phase difference across the electrodes of a single Josephson junction.

1.2. The Josephson effects and the RCSJ model

Going back to the Josephson equations (1.4a) and (1.4b), we may notice that the d. c. Josephson effect appears when a non-dissipative current flows at zero voltage, as it can be seen by setting $V = 0$ in (1.4b), so that

$$\phi = \sin^{-1} I/I_{J0} = \text{constant}. \tag{1.21}$$

In this way, I_{J0} represents the maximum value of I flowing in the junction in the zero-voltage state.

In the a. c. Josephson effect, on the other hand, the voltage across the Josephson junction is kept at a fixed non-zero value V_0. Integrating both sides of Eq. (1.4b) we obtain

$$\phi(t) = \frac{2\pi}{\Phi_0}V_0 t + \phi_0, \tag{1.22}$$

where ϕ_0 is the constant of integration. Therefore the current I is seen to oscillate at a frequency $\omega_J = 2\pi V_0/\Phi_0$. In fact, by substituting Eq. (1.22) into Eq. (1.4a), we may write:

$$I = I_{J0}\sin(\omega_J t + \phi_0). \tag{1.23}$$

In order to describe the dynamics of the superconducting phase difference ϕ in a Josephson junction, a Resistively and Capacitively Shunted Junction (RCSJ) model can be adopted (Barone & Paternò, 1982).

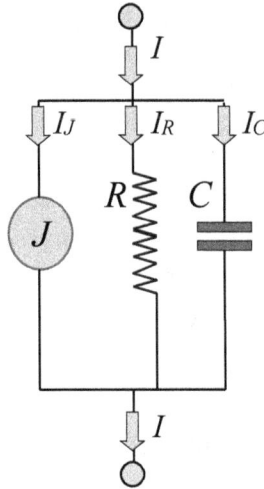

Fig. 1.2 Equivalent circuit describing the dynamics of a Josephson junction. The Resistively and Capacitively Shunted Junction (RCSJ) model is described by a parallel connection of a resistor with resistance R, a capacitor with capacitance C, and an ideal Josephson element J. In the latter circuit element a current obeying the Josephson current-phase relation can flow.

In this model, an ideal Josephson element J, carrying a current I expressed in terms of ϕ as in Eq. (1.4a), is placed in parallel with a resistor of resistance R and a capacitor of capacitance C, as shown in Fig. 1.2.

In what follows, we shall only consider equivalent circuits in which the capacitive element is absent; i. e., we shall only deal with overdamped Josephson junctions. However, in order to detect the conditions in which this limit is realized, let us proceed as follows.
By current conservation, we first write

$$I = I_C + I_R + I_J, \tag{1.24}$$

where the currents I_C, I_R, and I_J flow in the capacitor, in the resistor and in the Josephson element, respectively, and the current I is the current injected in the Josephson junction, as shown in Fig. 1.2. If V is the voltage across the junction, recalling that the Josephson element can carry a superconducting current obeying Eq. (1.4a), we may rewrite Eq. (1.24) as follows:

$$I = C\frac{dV}{dt} + \frac{V}{R} + I_{J0}\sin\phi. \tag{1.25}$$

Because of Eq. (1.4b), the voltage V across the junction can be expressed in terms of the derivative of ϕ with respect to time. In this way, Eq. (1.25) can be cast in the following form:

$$I = \frac{\Phi_0 C}{2\pi}\frac{d^2\phi}{dt^2} + \frac{\Phi_0}{2\pi R}\frac{d\phi}{dt} + I_{J0}\sin\phi. \tag{1.26}$$

Let us now divide both members of Eq. (1.26) by the maximum Josephson current I_{J0}, denoting with a small character the currents obtained by normalizing with respect to I_{J0}. In this way, we set $i = I/I_{J0}$. Furthermore, by rescaling the time variable, we write:

$$\tau = \frac{2\pi R I_{J0}}{\Phi_0}t. \tag{1.27}$$

Therefore, Eq. (1.26) can be rewritten as follows:

$$\beta_c \frac{d^2\phi}{d\tau^2} + \frac{d\phi}{d\tau} + \sin\phi = i. \tag{1.28}$$

where β_c is the Stewart-McCumber parameter (Barone & Paternò, 1982) defined as follows:

$$\beta_c = \frac{2\pi R^2 I_{J0} C}{\Phi_0}. \tag{1.29}$$

By looking at Eq. (1.28), one immediately sees that, if $\beta_c \ll 1$, the simplified Resistively Shunted Junction (RSJ) model can be adopted to describe the Josephson dynamics. In the following section and in the rest of the present book we shall only consider the over-damped case.

1.3.   Mechanical analog of an over-damped Josephson Junction and the RSJ Model

It is possible to exhibit a mechanical analog of Josephson junction by recognizing that Eq. (1.28) is the differential equation governing the dynamics of a pendulum in the presence of a viscous force.

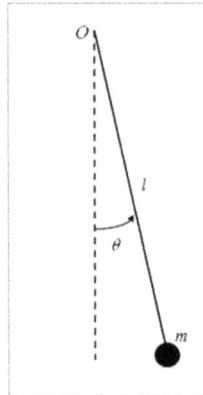

Fig. 1.3. Mechanical analog of a Josephson junction: the pendulum. Here the sphere is taken to have mass m and radius r. The massless rod of length l is hinged in O and attached to the surface of the sphere.

In fact, let us consider the simple pendulum in Fig. 1.3 hinged in O and consisting of a massless rod of length l and a spherical body of mass m and radius r. Assuming that the sphere is moving in a fluid of density ρ_f, it would be subject to the buoyancy force of intensity

$$F_B = \frac{4\pi r^3}{3} g \, \rho_f, \tag{1.30}$$

where g is the acceleration due to gravity. In addition, by assuming validity of Stokes' law (Sears, Zemansky, & Young, 1977), the sphere is taken to be subject to a viscous force, opposing its velocity and of intensity

$$F_R = 6\pi\eta r(r+l) \frac{d\theta}{dt}, \tag{1.31}$$

η being the coefficient of viscosity of the medium in which the sphere is moving and θ the angular displacement of the pendulum from its equilibrium position. The spherical body is subject, finally, to the tension in the massless rod of length l and to its weight. By taking moments with respect to point O, we may write:

$$I_0\ddot{\theta} + 6\pi\eta r(r+l)^2 \, \dot{\theta} + m^*g(r+l)\sin\theta = M_0(t), \tag{1.32}$$

where, as before, the dot stands for "derivative with respect to time", I_0 is the moment of inertia about the horizontal axis passing through point O, $m^* = m - 4\pi r^3 \rho_f/3$, and $M_0(t)$ is the applied torque. Here we take $m^* > 0$. By now proceeding as in the case of a Josephson junction, a normalized moment m_0 and a rescaled time variable may be defined as follows:

$$m_0(t) = \frac{M_0(t)}{m^*g(r+l)}, \tag{1.33a}$$

$$\tau = \frac{m^*g}{6\pi\eta r(r+l)} t. \tag{1.33b}$$

Therefore, Eq. (1.32) may be rewritten as follows:

$$\frac{m^*g \, I_0}{(6\pi\eta r)^2(r+l)^3} \frac{d^2\theta}{d\tau^2} + \frac{d\theta}{d\tau} + \sin\theta = m_0(t). \tag{1.34}$$

The previous expression is formally identical to Eq. (1.28). This close correspondence is sufficient to establish that the pendulum is a mechanical analog of a Josephson junction, whose dynamics is described by means of the RCSJ model.

In particular, it is possible to consider the over-damped limit of the RCSJ model, obtained for $\beta_c \ll 1$ in Eq. (1.28). This limit, corresponds to the following homologous condition on the mechanical model:

$$\frac{m^* g \, l_0}{(6\pi\eta r)^2 (r+l)^3} \ll 1. \qquad (1.35)$$

Therefore, while the limit $\beta_c \ll 1$ leads to the Resistively Shunted Junction (RSJ) model for a Josephson junction, the condition in Eq. (1.35) provides the analogous limit in the case of an over-damped pendulum. The dynamics of both systems may be represented by the following non-linear first-order ordinary differential equation, which we here explicitly write in terms of the angular displacement θ:

$$\frac{d\theta}{d\tau} + \sin\theta = m_0(t). \qquad (1.36)$$

In what follows, we shall adopt this model to qualitatively and quantitatively describe the behavior of Josephson junction devices. Successively, by utilizing the RSJ model for all Josephson junctions in superconducting networks, the physical properties of the latter systems will be analyzed. Finally, it should be remarked that the mechanical analog can be of great conceptual help, as we shall see in the following section.

1.4.   Current – Voltage Characteristics

In the present section we shall adopt the mechanical analog to derive the current – voltage $(I - V)$ characteristics of an over-damped Josephson junction. First, we need to define the following correspondences:

$$\theta \leftrightarrow \phi; \quad \frac{d\theta}{d\tau} \leftrightarrow \frac{d\phi}{d\tau} = \frac{V}{RI_{J0}}; \quad m_0 \leftrightarrow i. \qquad (1.37)$$

With the above prescription we will be able to go back, at the end of our analysis, to the electrodynamic properties of the Josephson junction.

Let us then consider a constant forcing term, writing $m_0(t) = m_0$, for the over-damped simple pendulum in Fig. 1.3. In this case, we can obtain analytic solutions for the differential equation (1.36) written in terms of m_0, namely:

$$\frac{d\theta}{d\tau} + \sin\theta = m_0. \tag{1.38}$$

We notice that, for $m_0 < 1$, two constant solutions exist in the interval $[0, \pi]$: one stable, one unstable. As it can be argued by means of the phase-plane analysis shown in Fig. 1.4, the stable solution is

$$\theta^* = \sin^{-1} m_0, \tag{1.39}$$

with $0 \leq \theta^* < \pi/2$. The unstable solution, on the other hand, is given by $\chi = \pi - \theta^*$. In fact, the stability regime of the system changes as the angle θ crosses the value $\pi/2$, as it can be noticed by analyzing the sign of the derivative $d\theta/d\tau$ about these fixed points in Fig. 1.4.

For $m_0 = 1$ we have an half-stable solution: The pendulum may swirl around O whenever an arbitrary small positive perturbation arises. We may finally notice that, for $m_0 > 1$, the function $\theta = \theta(\tau)$ is monotonically increasing, given that the curves in Fig. 1.4 lie above the θ-axis and the derivative $d\theta/d\tau$ is always positive.

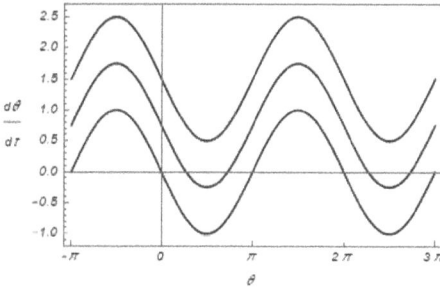

Fig. 1.4. Phase-plane analysis for the over-damped pendulum. The constant forcing term is $m_0 = 0.0$ (bottom curve), $m_0 = 0.75$ (middle curve), and $m_0 = 1.50$ (top curve).

In this "running state" we solve Eq. (1.38) by separation of variables (Barone & Paternò, 1982), by writing:

$$\int_{\theta_0}^{\theta(\tau)} \frac{d\theta}{m_0 - \sin\theta} = \tau, \tag{1.40}$$

where $\theta_0 = \theta(0)$.

By the substitution $x = \tan(\theta/2)$, we can write the integral in Eq. (1.40) as follows:

$$\frac{2}{m_0} \int_{\tan\left(\frac{\theta_0}{2}\right)}^{\tan\left(\frac{\theta(\tau)}{2}\right)} \frac{dx}{\left(x - \frac{1}{m_0}\right)^2 + 1 - \frac{1}{m_0^2}} = \tau, \tag{1.41}$$

where we completed the square in the denominator. The integral on the left hand side of Eq. (1.41) can now be solved. In fact, by defining

$$\alpha_0 = \tan^{-1}\left[\frac{m_0}{\sqrt{m_0^2 - 1}} \left(\tan\frac{\theta_0}{2} - \frac{1}{m_0}\right)\right], \tag{1.42}$$

after having solved the integral in Eq. (1.41), we may write:

$$\frac{2}{\sqrt{m_0^2 - 1}} \left\{\tan^{-1}\left[\frac{m_0}{\sqrt{m_0^2 - 1}} \left(\tan\frac{\theta(\tau)}{2} - \frac{1}{m_0}\right)\right] - \alpha_0\right\} = \tau. \tag{1.43}$$

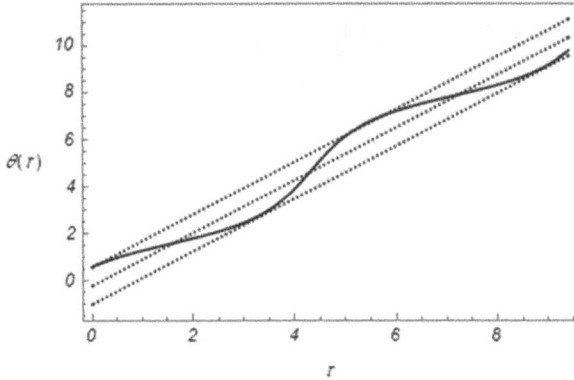

Fig. 1.5. Time–dependence of the angular variable θ of an over-damped pendulum (full-line curve) with constant driving term $m_0 = 1.5$ and with $\theta_0 = 0$. The top and bottom dashed lines enclose the undulatory behavior of θ, whose oscillations take place about the middle dashed line.

By now extracting the solution $\theta(\tau)$ from Eq. (1.43), we finally write:

$$\theta(\tau) = 2\tan^{-1}\left[\frac{1}{m_0} + \frac{\sqrt{m_0^2-1}}{m_0}\left(\tan\left(\frac{\sqrt{m_0^2-1}}{2}\tau\right) + \alpha_0\right)\right] + 2k\pi, \quad (1.44)$$

where k is an integer describing the number of 2π rotations needed to attain the correct increasing angular position $\theta(\tau)$, given that the value of the first addendum on the right hand side of the above expression is comprised in the interval $[-\pi, \pi]$.

The solution for the angular position $\theta(\tau)$ is represented in Fig. 1.5, as obtained from numerical integration of Eq. (1.38) for $m_0 = 1.5$ with initial condition $\theta_0 = 0$. We notice that this function oscillates within two dotted lines of equation $\theta_k(\tau) = \sqrt{m_0^2 - 1}\,\tau + \gamma_k$, for $k = 1, 2$, where $\gamma_k = \theta_k(\tau_k) - \sqrt{m_0^2 - 1}\,\tau_k$, τ_k being the two instants of time at which these lines are tangent to the oscillating curve $\theta(\tau)$ in the interval $[0, 2\pi]$. The quantities τ_k ($k = 1, 2$) can be found by a straightforward, but rather cumbersome, calculation. In this way, the curves of $\theta(\tau)$ are seen to oscillate about a central line $\theta_A(\tau) = \sqrt{m_0^2 - 1}\,\tau + \gamma_A$, whose intercept γ_A is the average value of γ_1 and γ_2. The solution of Eq. (1.38) is also represented in Fig. 1.6, for various values of the constant forcing term m_0, along with the central lines $\theta_A(\tau)$ obtained by the procedure described above.

Let us now study the time average $\langle d\theta/d\tau \rangle$ of the angular frequency $d\theta/d\tau$ as a function of the constant forcing term m_0. The m_0 versus $\langle d\theta/d\tau \rangle$ curves give, by analogy, the $I - V$ characteristics of an over-damped Josephson junction. In fact, according to Eq. (1.37), m_0 corresponds to the normalized current i and the normalized time averaged voltage $\langle v \rangle = \langle d\phi/d\tau \rangle$ corresponds to the quantity $\langle d\theta/d\tau \rangle$. We may start by considering the function $d\theta/d\tau$, represented in Fig. 1.7 for $m_0 = 1.5$ along with the value of the slope $\sqrt{m_0^2 - 1}$ of the central line running through the solution as seen in detail in Fig. 1.5. This slope corresponds to the time-average value of the curve in Fig. 1.7, as we shall see in the following page.

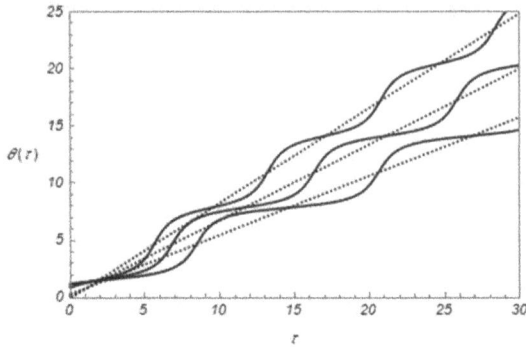

Fig. 1.6. Time–dependence of the angular variable θ of an over-damped pendulum (full-line curves) represented together with the central dashed line about which oscillations take place. The constant forcing terms are as follows: $m_0 = 1.125$ (lower curve), $m_0 = 1.200$ (middle curve), $m_0 = 1.300$ (upper curve).

We first notice that the function $d\theta/d\tau$ is periodic with period T equal to

$$T = \frac{2\pi}{\sqrt{m_0^2-1}}, \tag{1.45}$$

as it can be formally proven by calculating, from Eq. (1.44), the derivative of $\theta(T)$ with respect to τ. The slope $\sqrt{m_0^2 - 1}$ of the central line can thus be written as $2\pi/T$. On the other hand, the time-averaged value of $d\theta/d\tau$ can be calculated as follows:

$$\langle\frac{d\theta}{d\tau}\rangle = \frac{1}{T}\int_0^T \frac{d\theta}{d\tau}d\tau = \frac{\theta(T)-\theta(0)}{T} = \frac{2\pi}{T} = \sqrt{m_0^2 - 1} . \tag{1.46}$$

It is therefore proven that the average value of the angular frequency curves is $\sqrt{m_0^2 - 1}$. From Eq. (1.46) we can argue that

$$m_0 = \sqrt{\langle\frac{d\theta}{d\tau}\rangle^2 + 1} . \tag{1.47}$$

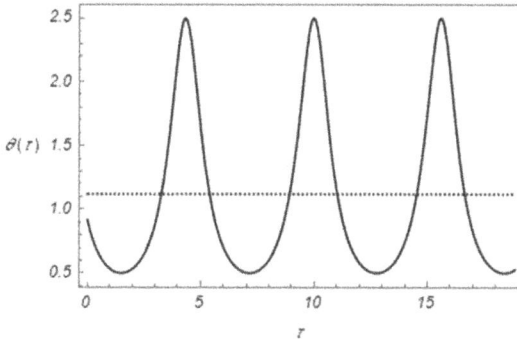

Fig. 1.7. Time–dependence of the angular frequency (full line) of an over-damped pendulum subject to a constant forcing equal to $m_0 = 1.50$. Notice that the curve is periodic, the period being $T = 2\pi/\sqrt{m_0^2 - 1}$. The dashed line represents the slope $\sqrt{m_0^2 - 1}$ of the central line of the primitive curve in Fig. 1.5 and, at the same time, the average value of the full-line curve represented in this figure.

The m_0 versus $\langle d\theta/d\tau \rangle$ curve is represented in Fig. 1.8. We soon notice the role played by the static solution in Eq. (1.39). In fact, for $0 \le m_0 < 1$, the pendulum is in static equilibrium, so that $\langle d\theta/d\tau \rangle = 0$.

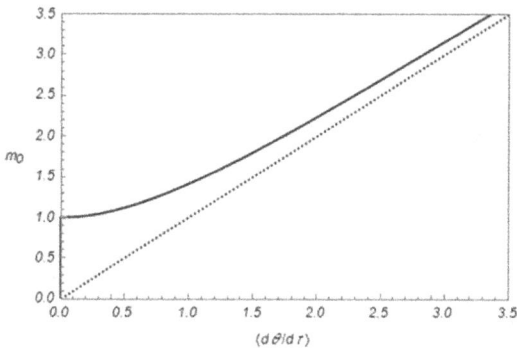

Fig. 1.8. Normalized forcing term versus the time average of the angular frequency (full line) of an over-damped pendulum. Notice that the curve is anchored at null angular frequency if $m_0 \le 1$. On the other hand, for $m_0 > 1$, the values of the curve tend toward the asymptote, given by the dashed line, as the abscissa increases. This universal curve is at all similar to the $I - V$ characteristics of over-damped Josephson junctions.

The same happens in a Josephson junction: when the value of the normalized bias current i is less than one, the junction is said to be in the superconducting or zero-voltage state. Therefore, no current flows in the resistive branch of the RSJ model in Fig. 1.2, so that the curve climbs vertically from 0 to 1 just as shown in Fig. 1.8. However, when $m_0 > 1$, the resistive branch is activated and a finite voltage appears across the junction in the way represented in Fig. 1.8. We also notice that the m_0 versus $\langle d\theta/d\tau\rangle$ curve presents the oblique asymptote $m_0 = \langle d\theta/d\tau\rangle$. In fact, for large enough values of m_0, this driving moment becomes predominant with respect to the nonlinear sine term in Eq. (1.38), thus justifying the observed asymptotic behavior.

For the analogy established in the previous section, we may affirm that the $I - V$ characteristics of over-damped Josephson junctions are given by the following expression

$$i = \sqrt{\langle v\rangle^2 + 1} \,, \tag{1.48}$$

to which the same arguments developed for an over-damped pendulum apply. In this rather fortunate situation, the expression given in Eq. (1.48) can be obtained analytically. Other features of an over-damped Josephson junction can be analyzed by means of the corresponding physical properties of an over-damped simple pendulum, as we shall see in the following section.

1.5. The Washboard Potential

Additional physical insight can be attained from the analogy between a Josephson junction and a pendulum in the over-damped regime by writing down the energy balance equation for the latter system. We start by noticing that energy is furnished to the pendulum from the externally applied torque at a rate $P_{ext} = M_0 \, d\theta/dt$. Here we have gone back to the laboratory time t. A power $P_d = -6\pi\eta r(r + l)^2(d\theta/dt)^2$ is dissipated because of the presence of the viscous force F_R, the minus sign meaning that energy is flowing from the system to the outer environment (the viscous fluid, for instance). Therefore, the mechanical energy E_M of the pendulum, defined as the sum of the kinetic energy $K = I_0(d\theta/dt)^2/2$ and of the potential energy $U = m^*g(r + l)(1 - \cos\theta)$, varies in time according to the following energy balance equation

$$\frac{dE_M}{dt} = P_{ext} + P_d. \tag{1.49}$$

By explicitly writing down some of the above terms, we have:

$$\frac{d}{dt}\left[\frac{1}{2}I_0\left(\frac{d\theta}{dt}\right)^2 + m^*g(r+l)(1 - \cos\theta) - M_0\theta\right] = P_d, \tag{1.50}$$

where we have taken M_0 to be constant and have included the external forcing term under the derivative operator on the left hand side. Of course, we can obtain the dynamic equation (1.34) from Eq. (1.50) by factoring out the angular frequency and by considering M_0 to be constant. However, we are here only interested in highlighting the role of the forcing term in the system.

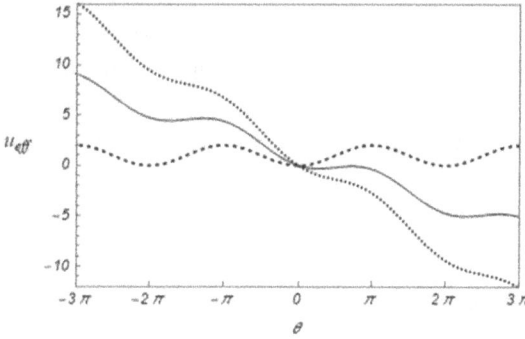

Fig. 1.9. Normalized effective potential as a function of the angle θ for the following three values of the parameter m_0: 0.0 (dashed line); 0.75 (full line); 1.5 (dotted line). Notice how the parameter m_0 determines the degree of tilting and stretching of the undulating curves.

Therefore, we may consider a normalized effective potential u_{eff} defined from Eq. (1.50) as follows:

$$u_{eff} = (1 - \cos\theta) - m_0\theta. \tag{1.51}$$

This normalized potential, called washboard potential because of its shape, is represented in Fig. 1.9 as a function of the variable θ and for various values of the parameter m_0. The above representation is useful, since it clarifies, once more, the crossover from static to dynamic solutions of the system. In fact, by looking at Fig. 1.9, we first notice that the parameter m_0 affects the degree of tilting and stretching of the washboard potential. This can be seen by starting from the dashed curve obtained for $m_0 = 0$ and by considering the remaining curves obtained for increasing values of this parameter. In the horizontal washboard all minima fall exactly at $\theta = 2n\pi$, with n integer. The number of minima fitting in the graph shown in Fig. 1.9 are three. The same number of minima, though their abscissa are slightly displaced with respect to the above specified positions, are still present in the stretched and tilted curve for $m_0 = 0.75$ (full line in Fig. 1.9). Therefore, a point-like body could still be in static equilibrium in the angular positions corresponding to the minima given by Eq. (1.39) in the interval $[0, 2\pi]$. Static equilibrium is not anymore possible for point-like particles on the washboard potential for $m_0 = 1.5$ (dotted curve in Fig. 1.9), because of excessive tilting and stretching of the curve. This feature can also be derived analytically from Eq. (1.51), by taking the derivative with respect to θ and by setting it to zero. Of course, this corresponds to finding the fixed points of the dynamic equation (1.38). Naturally, this type of approach leads to the same results found in the previous section.

1.6.　Shapiro Steps

In order to complete our discussion on the $I - V$ characteristics of over-damped Josephson junctions, we cannot miss mentioning the effects that an oscillating forcing term has on the current-voltage relation in these devices (Shapiro, Janus, & Holly, 1964). Let us then recall this analysis and turn our attention to the $I - V$ characteristics in the presence of r. f. radiation. We assume that the whole superconducting system is driven by an external oscillating voltage of the following form:

$$V(t) = V_0 + V_1 \cos \omega t, \tag{1.52}$$

where V_0 is the d. c. voltage component, and V_1 is the amplitude of the oscillating part, having angular frequency ω. Let us again adopt the RSJ model (Barone & Paternò, 1982) and normalize the voltage V in (1.52) by dividing it by $R\,I_{j0}$, where R is the resistance parameter of the

Josephson junction. Let us also rescale the time variable t, defining the dimensionless time τ as in Eq. (1.27). By integrating the Josephson voltage-frequency relation $v = V/RI_{J0} = d\phi/d\tau$, we have

$$\phi(\tau) = \phi_0 + v_0\tau + a \sin \tilde{\omega}\tau, \tag{1.53}$$

where: $v_0 = V_0/R\,I_{J0}$, $\tilde{\omega} = \Phi_0\omega/2\pi RI_{J0}$, $a = 2\pi V_1/\omega\Phi_0$. The $I - V$ curves are seen to show, at constant voltage, current steps of amplitude Δi.

The amplitude Δi can be calculated by the following procedure. First, we write the time-average of Eq. (1.28) in the over-damped limit:

$$\left\langle\frac{d\phi}{d\tau}\right\rangle + \langle\sin\phi\rangle = i. \tag{1.54}$$

Since the steps appear at constant voltage, the current steps have amplitude

$$\Delta i = \Delta\langle\sin\phi\rangle. \tag{1.55}$$

Let us then calculate $\langle\sin\phi\rangle$, by means of Eq. (1.53), by writing:

$$\langle\sin\phi\rangle = Im\left\{\frac{1}{T}\int_0^T e^{i\phi_0}e^{iv_0\tau}e^{ia\sin\tilde{\omega}\tau}d\tau\right\}, \tag{1.56}$$

where T is the period of the instantaneous voltage $v = d\phi/d\tau$, as deduced from Eq. (1.45). The integral in Eq. (1.56) can be calculated by setting (Abramowitz & Stegun, 1965):

$$e^{ia\sin\tilde{\omega}\tau} = \sum_{k=-\infty}^{+\infty} J_k(a)e^{ik\tilde{\omega}\tau}, \tag{1.57}$$

thus obtaining:

$$\langle\sin\phi\rangle = Im\left\{\frac{e^{i\phi_0}}{T}\sum_{k=-\infty}^{+\infty}J_k(a)\int_0^T e^{iv_0\tau}e^{ik\tilde{\omega}\tau}d\tau\right\}. \tag{1.58}$$

We therefore notice that the integral in (1.58), evaluated over the period T, gives null result except for $v_0 + k\tilde{\omega} = 0$, with k integer. In this way, we have:

$$\langle\sin\phi\rangle = J_k(a)\sin\phi_0, \tag{1.59}$$

with $k = -v_0/\tilde{\omega} = -2\pi V_0/\omega\Phi_0$.

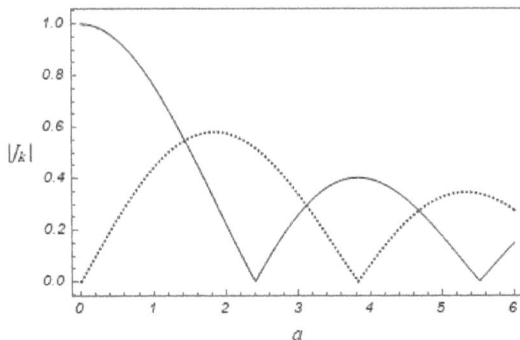

Fig. 1.10. Amplitude of the zero-voltage Shapiro step (full line) and of the successive step (dotted line) as a function of the quantity $a = 2\pi V_1/\omega\Phi_0$.

Therefore, only for voltages V_0 being integer multiples of the ratio $\nu\Phi_0$, where $\nu = \omega/2\pi$, Shapiro steps appear. In this case, we speak about "integer Shapiro steps". Finally, in order to get the amplitude Δi of the current steps, we maximize $\langle \sin\phi \rangle$ with respect to ϕ_0, obtaining:

$$\Delta i = |J_k(a)|, \tag{1.60}$$

with $k = -2\pi V_0/\omega\Phi_0$, as above. Therefore, the amplitude Δi of the Shapiro steps appearing in the $I - V$ characteristics depends both on V_0 and V_1. The d. c. voltage V_0 determines the position of the steps relative to the V −axis, on which the average voltage $\langle v \rangle = \langle d\phi/d\tau \rangle$ is reported. In fact, being the index k of the Bessel function $J_k(a)$ an integer, the voltages V_{0k} at which the steps are observed can be expressed as follows:

$$V_{0k} = \frac{|k|\Phi_0}{2\pi}\omega. \tag{1.61}$$

The amplitude V_1 of the oscillating part of the driving voltage $V(t)$ in Eq. (1.52) appears in the argument of the Bessel function $J_k(a)$ in Eq. (1.60). In fact, by the definition of the parameter $a = 2\pi V_1/\omega\Phi_0$, the term V_1 determines the amplitude of Shapiro steps, according to Eq. (1.60). For example, by taking $k = 0$, we see that the zero-voltage state in irradiated

JJs can be present up to the maximum Josephson current i_0, given, according to Eq. (1.60), by the following relation:

$$i_0 = |J_0(a)|. \tag{1.62}$$

Dependence on the parameter a of the zero-voltage step, representing the current i_0, is shown in Fig. 1.10; the dependence on a of the successive Shapiro step, of amplitude $i_1 = |J_1(a)|$, is also shown. Notice that we can use a positive index, because of Eq. (1.60) and of the following property of Bessel functions (Abramowitz & Stegun, 1965): $J_{-n}(a) = (-1)^n J_n(a)$, where n is a natural number. From Fig. 1.10 it can be noticed that, while for very low values of the quantity a the amplitude of the zero-voltage step is approximately equal to unity, it reduces to zero at $a_0 = 2.4048$, which corresponds to the first root of the Bessel function $J_0(a)$. In the vicinity of a_0, on the other hand, the amplitude i_1 attains its maximum value. By this simple example we may thus argue that the amplitude of Shapiro steps do not necessarily decrease for increasing indices k.

Chapter 2

Superconducting Systems in the presence of a Magnetic Field

2.1. Simply and Doubly Connected Superconductors

The phenomenon of expulsion of magnetic field from the inner region of simply connected type-I superconductors was first observed by Meissner and Ochsenfeld, 22 years after the discovery of superconductivity (Meissner & Ochsenfeld, 1933). The two German scientists showed that these superconducting systems, when cooled below the corresponding critical temperature T_c in the presence of low magnetic fields, behave like perfect diamagnetic materials. On the other hand, it was later realized that type-II superconductors show different features (Abrikosov, 1957): they lose the property of perfect diamagnetic behavior above a first critical field H_{c1}, while maintaining superconducting properties in a portion of the bulk. The formation of Abrikosov vortices above H_{c1} explains the magnetic properties of this class of materials (Schmidt, 1997).

With the discovery of high-T_c superconductivity in layered perovskites by Bednorz and Müller (Müller, Tagashike, & Bednorz, 1987) the magnetic properties of these novel superconducting systems were analyzed in detail, starting from the end of 1980's up to the beginning of 1990's. In particular, much attention was devoted to the so called "paramagnetic Meissner Effect" (PME), also known as "Wohlleben effect" (Braunish, et al., 1993) first reported by a group of German researchers in 1993. In this apparently contradictory definition of the observed phenomenon, the field-cooled susceptibility of high-T_c granular superconductors was observed to be positive for low measuring fields. It thus became evident that the polycrystalline structure in sintered high-T_c materials could play a role in explaining this experimental outcome (Li, 2003). In fact, considering sintered superconducting systems as a collection of weakly coupled micrometer sized granules, one may describe the magnetic properties of granular superconductors by means of equivalent networks of Josephson junctions (Müller, Tagashike, & Bednorz, 1987). We shall later dedicate the entire Chapter 7 to the topic of granular superconductors.

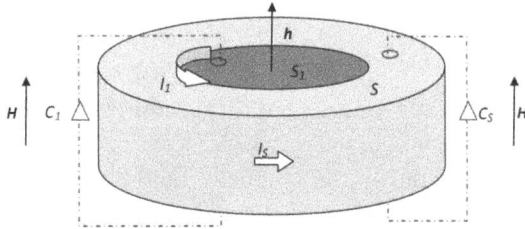

Fig. 2.1. Current distribution in a hollow superconducting cylinder of total cross section S (doubly connected system). An axial magnetic field H is applied to the whole system and a magnetic field induction h is present in the hole of section S_1. The current I_s shields the outer superconducting region from the external magnetic field H, while the current I_1 shields the inner superconducting region from the trapped field h. Two different paths, C_1 and C_S, are shown.

We here describe the fundamental mechanisms by which PME might arise. At the same time, we dwell in some depth on the magnetic behavior of simple superconducting systems. We therefore consider a multiply connected system: a type-I superconducting hollow cylinder (De Luca, 2000) or a type-II superconductor at very low temperatures and in the presence of low magnetic field. It can be shown that the magnetic response of these systems exhibits a diamagnetic character for most values of the measuring field magnitude H, while it may give a positive value of the field-cooled susceptibility for well-defined intervals of the applied magnetic flux, if the normal fraction, defined as the ratio between the volume of the hole and the total volume of the sample, is greater than $1/2$. In order to see how this property arises, let us consider the hollow superconducting cylinder shown in Fig. 2.1. If a uniform magnetic field of constant magnitude H is applied along the cylindrical axis, the current distribution can be simplified as follows: an external shielding current I_S flows in the outer surface of the cylinder; a second current I_1 shields the inner part of the superconductor from the field h inside the hole.

By applying Ampere's law, following path C_S, and by noticing that the magnetic induction in the superconducting region is zero, we can write:

$$Hd = -I_S. \tag{2.1}$$

where d is the height of the cylinder and C_S is a path going through the superconductor in a region sufficiently far from the outer surface. The latter hypothesis is necessary, in order to avoid considering the decaying magnetic field inside the superconductor, due to the existence of finite penetration lengths in these materials (Schmidt, 1997). We also neglect, for simplicity, demagnetization effects due to the finite size of the cylinder. Similarly, applying Ampere's law and following path C_1, we can write:

$$(H - h)d = -(I_S + I_1). \tag{2.2}$$

By now applying Eq. (2.1), we see that Eq. (2.2) reduces to the following:

$$hd = I_1. \tag{2.3}$$

Let us now write down the magnetic energy E_M due to the circulating currents as follows:

$$E_M = \frac{1}{2} L_1 I_1{}^2 + \frac{1}{2} L_S I_s{}^2 + M I_1 I_S, \tag{2.4}$$

where, denoting the permeability of vacuum as μ_0, the inductance coefficients pertaining to the two virtual loops followed by I_1 and I_S can be approximated, respectively, by:

$$L_1 = \frac{\mu_0 S_1}{d}; \ L_S = \frac{\mu_0 S}{d}. \tag{2.5}$$

Moreover, in Eq. (2.4) M is the mutual inductance coefficient between these same two loops. By taking $M = L_1$ and by substituting what found in Eq. (2.4), we have:

$$E_M = \frac{1}{2}L_1 d^2 (h - H)^2 + k_H, \qquad (2.6)$$

where $k_H = (L_S - L_1)d^2 H^2 / 2$ is a constant.

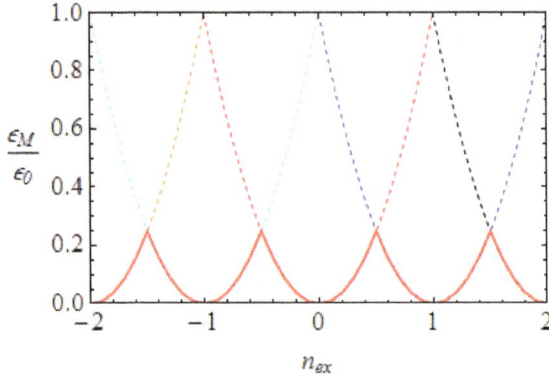

Fig. 2.2. Magnetic energy states in the hollow superconducting cylinder cooled in the presence of a constant measuring field of amplitude H. The low-lying states are shown as a collection of points on the red full-line curve. The parabolas $(n - n_{ex})^2$ are shown as dashed lines for the following values of n: $n = 0$ (red); $n = 1$ (blue); $n = -1$ (cyan); $n = 2$ (black), $n = -2$ (orange). The red full line curve is the result of a minimization procedure over all possible n-states, for fixed distribution in a hollow superconducting cylinder of total cross section S.

By now introducing the flux numbers $n = \mu_0 h S_1 / \Phi_0$ and $n_{ex} = \mu_0 H S_1 / \Phi_0$, we can rewrite Eq. (2.6) in the following final form

$$\epsilon_M = \epsilon_0 (n - n_{ex})^2, \qquad (2.7)$$

where $\epsilon_M = E_M - k_H$ and $\epsilon_0 = \Phi_0^2 / 2L_1$. The possible quantized values of the trapped magnetic flux in a field cooling experiment have been given by Goodman and Deaver in 1970. The experimental results reported by these researchers can be summarized by the following simple non-linear expression (Goodman & Deaver, 1970):

$$n = \Omega(n_{ex}), \qquad (2.8)$$

where the function Ω, when applied to a real number x, gives the closest integer to x. This function is a consequence of flux quantization, as we shall notice in the following sections, and can be interpreted by considering the minima of the energy ϵ_M. In fact, by fixing the value of the applied flux number n_{ex}, the system arranges itself in the quantized flux state with n trapped fluxons inside the hole of area S_1 in such a way to minimize the energy ϵ_M. In this way, only the lower parts of all parabolas in Eq. (2.7) are chosen as possible magnetic states in the system. The result of this procedure, by which one chooses the possible magnetic energy states as n_{ex} varies, is shown in Fig. 2.2.

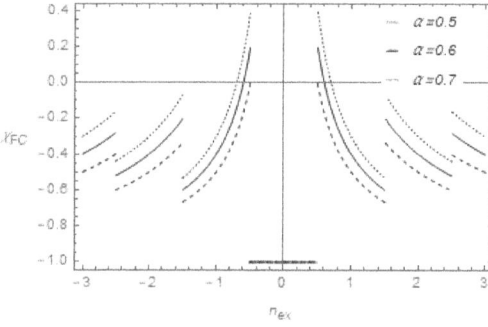

Fig. 2.3. Magnetic susceptibility of a hollow superconducting cylinder cooled in the presence of a constant measuring field H for various values of the normal fraction α, reported in the legend ($\alpha = 0.5, 0.6, 0.7$). Depending on the value of the measuring field, the susceptibility may attain positive values for $\alpha > 1/2$. Susceptibility values in the interval $(-1/2, 1/2)$ of the variable n_{ex} coincide for all three values of α.

Having specified the value of the flux number n in (2.8), the field distribution inside the cylinder can be summarized as follows: inside the hole, we may set

$$h = \frac{\Phi_0}{\mu_0 S_1} \Omega(n_{ex}), \qquad (2.9)$$

while $\boldsymbol{h}$ can be set equal to zero inside the superconducting region. In order to determine the field cooled susceptibility χ_{FC}, we need to find the average value $\langle h \rangle$ of the magnetic induction inside the hollow cylinder. By means of (2.9), we find $\langle h \rangle = \alpha \Phi_0 \Omega(n_{ex})/\mu_0 S_1$ where $\alpha = S_1/S$ is the normal fraction of the sample. Therefore, by setting the field-cooled magnetic susceptibility equal to

$$\chi_{FC} = \frac{\langle h \rangle}{H} - 1, \qquad (2.10)$$

we have:

$$\chi_{FC} = \frac{\alpha \Omega(n_{ex})}{n_{ex}} - 1. \tag{2.11}$$

In Fig. 2.3, the field-cooled magnetic susceptibility of a hollow cylinder as a function of the applied flux number n_{ex} is shown for various values of the normal fraction α. From Eq. (2.11), we notice that for $\alpha < 1/2$, the curves are always below the horizontal axis, so that the magnetic response is always diamagnetic, as it can be also argued from fig. 2.3. However, for $\alpha > 1/2$, positive values of χ_{FC} can appear in well-determined intervals of n_{ex} values. In fact, by considering the interval $1/2 < n_{ex} < 1$, for which $n = \Omega(n_{ex}) = 1$, we have $\chi_{FC} > 0$ if $1/2 < n_{ex} < \alpha$, provided $1/2 < \alpha < 1$. On the other hand, for $1 < n_{ex} < 3/2$, for which $n = \Omega(n_{ex}) = 2$, we cannot have positive values of the field-cooled susceptibility χ_{FC}. Therefore, we argue that, provided $1/2 < \alpha < 1$, the only field interval for which $\chi_{FC} > 0$ is given by the following simple relation

$$\frac{\Phi_0}{2S_1} < \mu_0 H < \alpha \frac{\Phi_0}{S_1}. \tag{2.12}$$

In order to detect the range in which this effect can be measured, we may notice that in a micro-cylinder (with a hole of inner radius of about 20 μm) the ratio Φ_0/S_1 is equal to 1.64 μT.

2.2. Triply Connected Superconductors

We may now apply the same analysis of the previous section to the triply-connected superconductor shown in Fig. 2.4. In this system, consisting of a cylinder of height d and total cross section S with two holes, one of area S_1, the second of area S_2, for a given value of H different flux numbers can be trapped in each hole (say, n_1 and n_2 in the holes of area S_1 and S_2, respectively). By applying Ampere's law following the three different paths in Fig. 2.4, we find the expressions for the currents I_S, I_1, and I_2 in terms of the various field values, so that we may write:

$$Hd = -I_S. \tag{2.13a}$$

$$h_1 d = I_1. \tag{2.13b}$$

$$h_2 d = I_2. \tag{2.13c}$$

The magnetic energy for this system can now be written as follows:

$$E_M = E_L + M_{1S} I_1 I_S + M_{2S} I_2 I_S + M_{12} I_1 I_2, \tag{2.14}$$

where $E_L = \left(L_S I_S{}^2 + L_1 I_1{}^2 + L_2 I_2{}^2 \right)/2.$

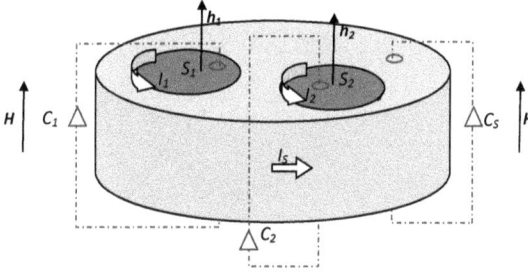

Fig. 2.4. Current distribution in a superconducting cylinder of total cross section S with two holes (triply-connected system). An axial magnetic field of amplitude H is applied to the whole system and magnetic field inductions h_1 and h_2 are present in the holes of section S_1 and S_2, respectively. The current I_S shields the outer superconducting region from the external magnetic field H, while the currents I_1 and I_2 shield the inner superconducting region from the trapped fields h_1 and h_2, respectively. Three different paths, C_1, C_2, and C_S, are shown.

As before, the inductance coefficients are approximated by the following expressions:

$$L_1 = \frac{\mu_0 S_1}{d}; L_2 = \frac{\mu_0 S_2}{d}; L_S = \frac{\mu_0 S}{d}. \tag{2.15}$$

The mutual inductance coefficients are M_{1S}, M_{12}, and M_{2S}. By proceeding as in the previous section, we define the following flux numbers

$$n_1 = \frac{\mu_0 h_1 S_1}{\Phi_0}; n_2 = \frac{\mu_0 h_2 S_2}{\Phi_0}; n_{ex}^{(1)} = \frac{\mu_0 H S_1}{\Phi_0}; n_{ex}^{(2)} = \frac{\mu_0 H S_2}{\Phi_0}. \tag{2.16}$$

By defining $\sigma = S_2/S_1$, we may further set $n_{ex}^{(1)} = n_{ex}$ and $n_{ex}^{(2)} = \sigma n_{ex}$. By now taking $M_{1S} = L_1$ and $M_{2S} = L_2$, we may write down the energy $\epsilon_M = E_M - \xi_H$, with $\xi_H = (L_S - L_1 - L_2)d^2 H^2/2$, as follows:

$$\epsilon_M = \epsilon_1 \left[(n_1 - n_{ex})^2 + \frac{1}{\sigma}(n_2 - \sigma n_{ex})^2 + 2\mu n_1 n_2 \right], \quad (2.17)$$

where $\epsilon_1 = \Phi_0^2/2L_1$ and $\mu = M_{12}/L_2$. From Eq. (2.17) it is then clear that, depending on the choice of n_1 and n_2, we obtain different parabolic dependence of ε_M as a function of n_{ex}.

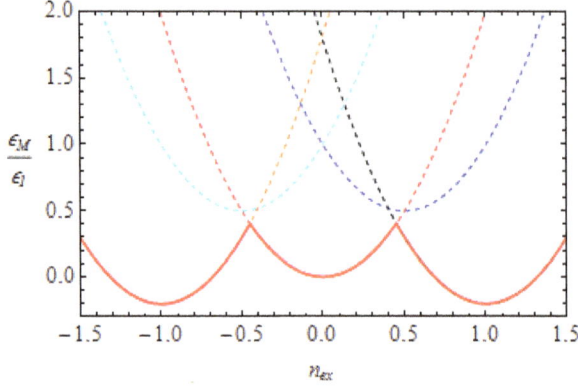

Fig. 2.5. Magnetic energy states in the hollow superconducting cylinder with two holes cooled in the presence of a constant measuring field H. These low-lying states are shown as a collection of points on the red full-line curve. The parabolas $(n_1 - n_{ex})^2 + (n_2 - \sigma n_{ex})^2/\sigma + 2\mu n_1 n_2$ are shown, for $\sigma = 1$ and $\mu = -0.1$, as dashed lines for the following couples (n_1, n_2): (0,0) (red); (0,1) (blue); (−1,0) (cyan); (1,1) (black), (−1,−1) (orange). The red full line curve is therefore the result of a minimization process over all possible couples (n_1, n_2), for fixed n_{ex} values.

As before, the system lies in the magnetic state that, for a given value of the forcing term n_{ex}, minimizes the energy ϵ_M. Therefore, by collecting the different parabolas, we shall choose only the low-lying states at a fixed value of n_{ex}. By applying a minimization procedure similar to the one explained in the previous section, these states are represented in Fig. 2.5 for $\sigma = 1$ and $\mu = -0.1$. For the choice of parameters in Fig. 2.5, not all parabola host low-lying energy states. However, different choices of the parameters σ and μ may alter the position of the minima of some parabolas and may also affect the shape of the resulting red curve.

By a similar algorithm, we can now choose to register, for a fixed value of n_{ex}, the couple (n_1, n_2) giving the parabola on which the minimum of the energy lies. In order to do this, for a fixed value of n_{ex} and for the particular values of the parameters σ and μ chosen, we look for the couple (n_1, n_2) giving the lowest energy ratio ϵ_M / ϵ_1, as expressed in Eq. (2.17).

In this way, by calculating the average value of the magnetic induction over the whole sample, we may define the field-cooled magnetic susceptibility as follows:

$$\chi_{FC} = \frac{\tilde{\alpha}\,(n_1 + n_2)}{(1 + \sigma)n_{ex}} - 1. \tag{2.18}$$

where the normal fraction is now defined as $\tilde{\alpha} = (S_1 + S_2)/S$. Therefore, by knowing the quantities n_1 and n_2, for a given value of n_{ex}, we can plot the χ_{FC} vs. n_{ex} curves as shown in Fig. 2.6a-c. For particularly simple choices of the parameters μ and σ, we could determine an explicit analytic expression for $n_k = n_k(n_{ex})$, with $k = 1, 2$. In general, however, a numerical algorithm has to be adopted. Comparison between the susceptibility curves in Fig. 2.6a and 2.6b shows that, for the same value of $\sigma = 1$, the various branches tend to rise toward positive values for the curves with a stronger magnetic coupling between the two holes. On the other hand, comparing Figs. 2.6a and 2.6c, for which the ratio σ is varied from 1.0 to 0.9, more branches in the χ_{FC} vs. n_{ex} curves are seen to appear in the latter, signifying a loss in the symmetry in the system. Moreover, in all Figs. 2.6a-c, we find intervals of n_{ex} for which χ_{FC} is positive even for $\tilde{\alpha} = 1/2$. This is in contrast with what found in the case of a hollow cylinder with a single hole, namely, that a single interval of the external forcing term n_{ex} was present only for $\alpha > 1/2$, as shown in Fig. 2.3.

A word of caution is to be spent about the above simple analysis, which is useful to shed some light on the intrinsic properties of multiply-connected superconductors. In this picture, the superconductor is seen as a perfectly diamagnetic entity and thus is either a type-I superconductor at very low temperatures or a type-II superconductor at very low temperatures and in the presence of very low magnetic fields. This assumption is made to exclude large penetration lengths inside the superconducting bulk and, in the case of type-II superconductors, the formation of vortices inside the superconducting matrix.

Generalization of the present analysis may be sought for more complex systems. However, the reader is now able to predict, following this approach, the low-field magnetic response of multiply connected superconductors.

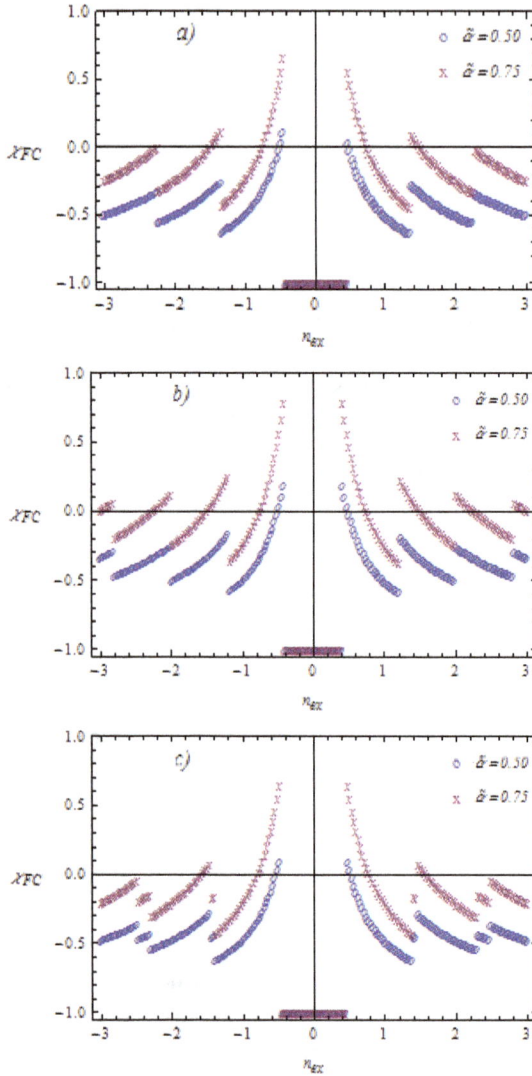

Fig. 2.6. Magnetic susceptibility of a hollow superconducting cylinder with two holes cooled in the presence of a constant measuring field H for two values of the normal fraction $\tilde{\alpha}$ $(0.50, 0.75)$ reported in the legend. Depending on the value of the measuring field, the susceptibility χ_{FC} may attain positive values. The parameters are chosen as follows: *a)* $\sigma = 1$ and $\mu = -0.1$; *b)* $\sigma = 1$ and $\mu = -0.2$; *c)* $\sigma = 0.9$ and $\mu = -0.1$. Susceptibility values in the interval $(-1/2, 1/2)$ coincide for the two values of $\tilde{\alpha}$.

2.3. A small Josephson junction in a magnetic field

We start this section by first giving an account of flux and fluxoid quantization in multiply-connected superconducting systems, in order to address the important topic of the behavior of a small Josephson junction in a magnetic field in a subsequent subsection.

2.3.1. Flux and fluxoid quantization

Let us define the current density $J_S = -2eJ$ flowing in a superconductor S in terms of the particle current density J and of the charge $-2e$ of a single Cooper pair, where e is the absolute value of the electron charge. As in any other quantum system described by a wave function Ψ, in a superconducting system J can be derived by considering the continuity equation

$$\vec{\nabla} \cdot J = -\frac{\partial}{\partial t}|\Psi|^2, \tag{2.19}$$

and the Schrödinger equation for a free particle of mass $m = 2m_e$:

$$i\frac{\partial}{\partial t}\Psi = -\frac{\hbar \nabla^2}{2m}\Psi. \tag{2.20}$$

By expanding the time derivative in Eq. (2.19) and by substituting Eq. (2.20) into (2.19), we have

$$\vec{\nabla} \cdot J = \frac{i\hbar}{2m}\left(\Psi \nabla^2 \Psi^* - \Psi^* \nabla^2 \Psi\right), \tag{2.21}$$

where Ψ^* is the complex conjugate of Ψ. In this way, it can be seen that $J = i\hbar\left(\Psi \vec{\nabla}\Psi^* - \Psi^* \vec{\nabla}\Psi\right)/2m$ and that:

$$J_S = \frac{ie\hbar}{2m_e}\left(\Psi^* \vec{\nabla}\Psi - \Psi \vec{\nabla}\Psi^*\right). \tag{2.22}$$

In the presence of a magnetic field we make the following minimal substitution in (2.22):

$$\mathbf{p} \rightarrow \mathbf{p} + 2e\mathbf{A}, \tag{2.23}$$

where $\mathbf{p}$ is the Cooper-pair momentum and $\mathbf{A}$ is the vector potential. Therefore, by setting $\mathbf{p} = -i\hbar\vec{\nabla}$ and by considering Eq. (2.23), we rewrite Eq. (2.22) in the following way:

$$J_S = \frac{ie\hbar}{2m_e}\left(\Psi^*\vec{\nabla}\Psi - \Psi\vec{\nabla}\Psi^*\right) - \frac{2e^2}{m_e}|\Psi|^2\mathbf{A}. \tag{2.24}$$

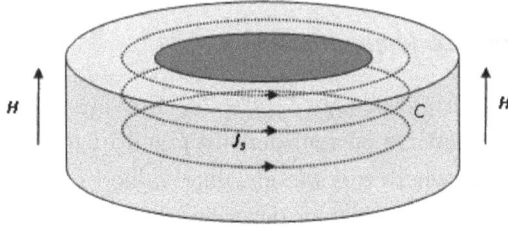

Fig. 2.7. A multiply-connected superconductor in the presence of a magnetic field $\mathbf{H}$. Well inside the superconductor, along any of the paths shown, the super-current J_S is zero. The middle path is labeled with the letter C.

The superconducting wave-function Ψ in a superconductor S can be expressed in terms of the number density of super-electrons n_S and of the superconducting phase θ (Schmidt, 1997), so that we may write:

$$\Psi = \sqrt{\frac{n_S}{2}}e^{-i\theta}, \tag{2.25}$$

where θ is the phase of the macroscopic quantum state. By substituting Eq. (2.25) into Eq. (2.24), the current density J_S can be written in the following form:

$$J_S = \frac{e\hbar n_S}{2m_e}\left(\vec{\nabla}\theta - \frac{2e}{\hbar}\mathbf{A}\right). \tag{2.26}$$

Consider now a multiply connected superconductor (at the absolute temperature T below the critical temperature T_c) in the presence of a magnetic field $\mathbf{H}$, as in Fig. 2.7. Well inside the superconductor, along the path C shown in Fig. 2.7, we can consider $J_S = 0$ so that, by Eq. (2.26) we may write:

$$\mathbf{A} = \frac{\Phi_0}{2\pi}\vec{\nabla}\theta. \tag{2.27}$$

where it is meant that inside the superconductor matrix the vector potential can be explicitly expressed in terms of the phase gradient $\vec{\nabla}\theta$. We recall that Φ_0 is the elementary flux quantum. We can now integrate Eq. (2.27) over the middle path C in Fig. 2.7 to obtain:

$$\Phi = \oint \mathbf{A} \cdot \mathbf{ds} = n\Phi_0, \tag{2.28}$$

where $\mathbf{ds}$ is an oriented element of the path C and n is an integer, recognizing that the line integral of the vector potential $\mathbf{A}$ over C is the magnetic flux Φ linked to this path and the homologous integral of the superconducting phase is equal to $2\pi n$. Eq. (2.28) expresses, in very simple terms, the quantization of trapped flux inside a multiply connected superconductor, as experimentally proven by Goodman and Deaver in 1970 (Goodman & Deaver, 1970).

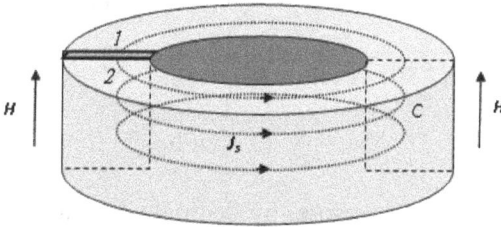

Fig. 2.8. A superconducting ring interrupted by a Josephson junction in the presence of a magnetic field. Well inside the superconductor, along any of the paths shown, the super-current is zero. The middle path is labeled with the letter C. This path crosses the insulating barrier (the gray cut within the ring) dividing, at one end, the two arms of the same ring. We label the two sides of the barrier as follows: side 1 (where the current enters the barrier) and side 2 (from which the current leaves the barrier).

Let us now consider a superconducting ring interrupted by a Josephson junction. As shown in Fig. 2.8, we may think the junction as a cut, consisting of a very thin insulating layer, between the two arms of the same superconducting ring. In this case, the line integral of $\mathbf{A}$ over the path C well inside the superconductor needs to be calculated into two parts: the first inside the superconducting region S, where Eq. (2.27) applies, the second across the thin insulating barrier B shown as the gray cut in Fig. 2.8. Therefore, we may write:

$$\oint \mathbf{A} \cdot d\mathbf{s} = \int_{2\,(S)}^{1} \mathbf{A} \cdot d\mathbf{s} + \int_{1\,(B)}^{2} \mathbf{A} \cdot d\mathbf{s}. \tag{2.29}$$

By again recognizing that the left-end side is the magnetic flux and by substituting Eq. (2.27) in the first line integral of the right-hand side, we have:

$$\Phi = \frac{\Phi_0}{2\pi} \int_{2\,(S)}^{1} \vec{\nabla}\theta \cdot d\mathbf{s} + \int_{1\,(B)}^{2} \mathbf{A} \cdot d\mathbf{s}. \tag{2.30}$$

where we follow the path C as prescribed by the right-hand screw rule. In this way, the first integral on the right-hand side of Eq. (2.30) is equal to the phase difference $\theta_1 - \theta_2$ plus an integer multiple of 2π, and the second is given by the presence of the magnetic field. By therefore defining a gauge-invariant phase difference ϕ across the junction as

$$\phi = \theta_2 - \theta_1 - \frac{2\pi}{\Phi_0} \int_{1\,(B)}^{2} \mathbf{A} \cdot d\mathbf{s}, \tag{2.31}$$

we may rewrite Eq. (2.30) as follows

$$\Phi + \frac{\phi}{2\pi}\Phi_0 = n\Phi_0, \tag{2.32}$$

where n is an integer. Therefore, by denoting as "fluxoid" the quantity $\Phi + \phi\Phi_0/2\pi$, by Eq. (2.32) we may state that in a superconducting ring interrupted by a Josephson junction the fluxoid is quantized.

2.3.2. *A Josephson junction in a magnetic field*

In the previous subsection, we have seen that the gauge-invariant superconducting phase difference ϕ across a Josephson junction interrupting a superconducting ring is related to the magnetic flux threading the ring itself. When considering an isolated extended Josephson junction in the presence of a magnetic field, we may notice that a similar relation exists between ϕ and the flux linked to the barrier. This property leads to a quantum interference phenomenon, similar to diffraction by a single slit in optics (Halliday, 2005), thus called the Fraunhofer-like pattern of the maximum Josephson current (Barone, 1982). We shall here see in details how this quantum interference phenomenon occurs.

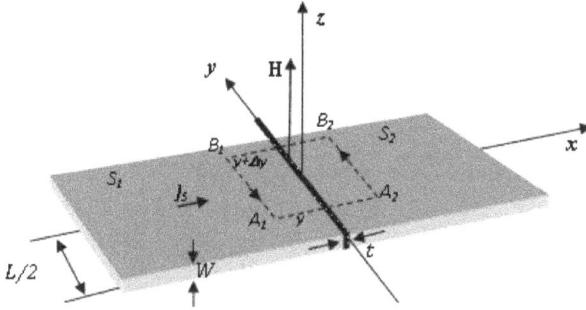

Fig. 2.9. A Josephson junction with a thin insulating barrier of thickness t, length L, and width W in the presence of a magnetic field **H**. Well inside the two superconducting electrodes S_1 and S_2, along the path $A_1A_2B_2B_1$, the induced super-current is zero. The magnetic flux linked to this path is $\mu_0 H d \Delta y$, d being the effective thickness of the barrier.

By referring to Fig. 2.9, we assume that the field **H** is uniform along the length L of the junction. This means that the junction is not able, by itself, to shield the externally applied field. In this respect, one can still talk of a small Josephson junction. We thus notice that the magnetic flux linked to the oriented rectangular path $A_1A_2B_2B_1$ is $\Delta\Phi = \mu_0 H d \Delta y$ where, considering the penetration lengths λ_1 and λ_2 inside S_1 and S_2, d is the effective thickness of the barrier, so that we set: $d = t + \lambda_1 + \lambda_2$. In this way, according to the definition of magnetic flux, we may write:

$$\Delta\Phi = \int_{A_1}^{A_2} \mathbf{A} \cdot \mathrm{d}\mathbf{s} + \int_{A_2}^{B_2} \mathbf{A} \cdot \mathrm{d}\mathbf{s} + \int_{B_2}^{B_1} \mathbf{A} \cdot \mathrm{d}\mathbf{s} + \int_{B_1}^{A_1} \mathbf{A} \cdot \mathrm{d}\mathbf{s}. \quad (2.33)$$

By now applying Eq. (2.27) only in the superconducting portion of the path (segments B_1A_1 and A_2B_2), rearranging terms and using definition (2.31) for the gauge-invariant superconducting phase difference ϕ, here evaluated at y and $y + \Delta y$, we have:

$$\frac{2\pi}{\Phi_0} \Delta\Phi = \phi(y + \Delta y) - \phi(y). \quad (2.34)$$

Recalling now the expression for $\Delta\Phi$, and setting $\Delta\phi = \phi(y + \Delta y) - \phi(y)$, we may rewrite (2.34) as follows:

$$\frac{\Delta\phi}{\Delta y} = 2\pi\frac{\mu_0 H d}{\Phi_0} = k, \tag{2.35}$$

where k is a constant.

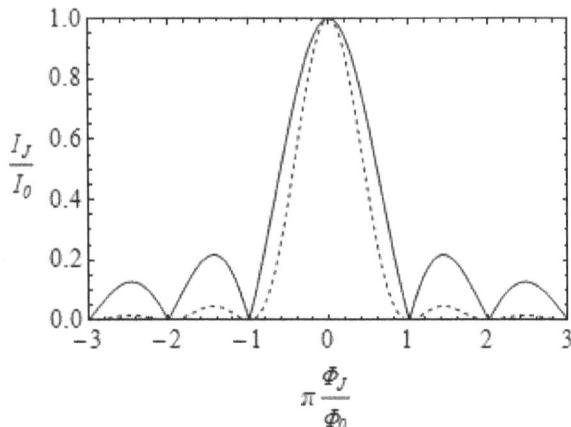

Fig. 2.10. Maximum Josephson current normalized to I_0 vs. the normalized applied flux Φ_J/Φ_0 (full line curve). For comparison the Fraunhofer pattern (dashed curve) describing the optical phenomenon of diffraction from a single slit is also shown.

By letting $\Delta y \rightarrow 0$, we may write the dependence on y of the gauge-invariant superconducting phase difference across the junction as follows:

$$\phi(y) = ky + \phi_0, \tag{2.36}$$

where ϕ_0 is an integration constant. By now taking, according to Eq. (1.4a), the current density J_S flowing in the element of junction area of amplitude $dS = W dy$ equal to $J_{S0} \sin\phi(y)$, the total current I flowing in the junction can be written as follows:

$$I = J_{S0}W \int_{-L/2}^{L/2} \sin\phi(y)\, dy, \tag{2.37}$$

where the choice of axes is shown in Fig. 2.9. By substituting Eq. (2.36) into Eq. (2.37), and by integrating, we have:

$$I = I_0 \frac{\sin\left(\pi\frac{\Phi_J}{\Phi_0}\right)}{\pi\frac{\Phi_J}{\Phi_0}} \sin \phi_0, \tag{2.38}$$

where $I_0 = J_{S0}WL$ and $\Phi_J = \mu_0 HdL$. The maximum value I_J of the current I is thus given by the following:

$$I_J = I_0 \left| \frac{\sin\left(\pi\frac{\Phi_J}{\Phi_0}\right)}{\pi\frac{\Phi_J}{\Phi_0}} \right|. \tag{2.39}$$

A graph of the above Fraunhofer-like function I_J/I_0 is reported in Fig. 2.10 (full line) along with the Fraunhofer pattern $F(x) = \sin^2(\pi x)/(\pi x)^2$ (dashed line) for comparison. The similarity of these curves is evident, especially because the minima are located at the same positions, i.e., at nonzero integer values of the independent variables Φ_J/Φ_0 and x.

2.4. Magnetic susceptibility of superconducting rings interrupted by a Josephson junction

It is well known that the electrodynamic properties of SQUIDs (Superconducting Quantum Interference Devices) are obtained by means of the dynamics of the Josephson junctions interrupting superconducting rings (Barone, 1982), (Likharev, 1986), (Clarke, 2004). Due to the intrinsic macroscopic coherence of superconductors, r. f. SQUIDs, whose basic features are obtained by considering the magnetic response of a superconducting ring interrupted by one Josepshon junction, have been proposed as basic units (qubits) in quantum computing (Bocko, 1997). In the realm of quantum computing non-dissipative quantum systems with small (or null) inductance parameter and finite capacitance of the Josephson junctions are usually considered (Crankshaw, 2001). The mesoscopic non-simply connected classical devices, on the other hand, are generally operated and studied in the over-damped limit with negligible capacitance of the junction and small (or null) values of the inductance parameter. Nowadays, r. f. SQUIDs find application in a large variety of fields, from biomedicine to aircraft maintenance (Clarke,

2004), thus justifying scientific interest in these systems. Nonetheless, as far as the aim of the present work is concerned, understanding the behavior of a superconducting ring interrupted by one Josephson junction is to be considered a first step toward the comprehension of more complex systems and, successively, of networks of junctions.

Let us start by giving a simple description of these systems. Imagine that we were asked to qualitatively describe the magnetic response of a superconducting ring interrupted by a Josephson junction, knowing the homologous properties of a doubly connected superconductor, as described in Section 2.1. We would therefore simply answer that, in principle, the magnetic response is similar. However, we need to consider what follows. Assume that we were to modify the flux state of the system. For a superconducting ring we need to raise the temperature above the critical temperature T_c of the material or, equivalently, we need to apply a sufficiently high energy to make a transition to the normal state. On the contrary, in superconducting rings interrupted by a Josephson junction, we can make fluxons enter the superconducting loop by simply varying the externally applied field. At relatively low magnetic field values the junction will suffer a phase slip (a 2π rotation of the superconducting phase difference across the electrodes), and the flux will rearrange to a new state, according to Eq. (2.32). How this can occur will be discussed in details in what follows.

We start by assuming that the dynamics of the Josephson junction is described by the RSJ model, as discussed in Section 1.2. For simplicity, we also assume that noise effects are negligible and that the maximum Josephson current of the junction is not affected by the very low fields applied to the system. In the case of a superconducting loop interrupted by one Josephson junction, the only forcing term is the externally applied flux $\Phi_{ex} = \mu_0 H S$, where S is the area of the hole, as we have noticed in Section 2.1. On semi-classical ground, we can describe a superconducting loop interrupted by one Josephson junction, also denoted as a one-junction quantum interferometer, by means of the following potential energy function:

$$E(\psi, \phi) = \epsilon_M (\psi - \psi_{ex})^2 + \epsilon_J (1 - \cos \phi). \qquad (2.40)$$

The first addendum on the right hand side of Eq. (2.40) pertains to the magnetic energy of the system, so that $\epsilon_M = \Phi_0^2/2L$, L being the inductance parameter of the superconducting loop, $\psi = \Phi/\Phi_0$ is the normalized flux inside the hole and $\psi_{ex} = \Phi_{ex}/\Phi_0$. This term can be derived, by an opportune change in notation, from Eq. (2.7). Here, in fact, the magnetic flux Φ is not quantized, so that we prefer to use, at least in the semi-classical regime, a different notation from the one adopted in Section 2.1. On the other side, the second addendum on the right hand side of Eq. (2.40) represents the Josephson coupling energy E_c of the junction interrupting the superconducting loop. Therefore, by recognizing that, from Eq. (1.20), the current I through the junction can be expressed as

$$I = \frac{2\pi}{\Phi_0}\frac{\partial E_C}{\partial \phi}, \qquad (2.41)$$

we may see that, in Eq. (2.40), $\epsilon_J = I_{J0}\,\Phi_0/2\pi$, I_{J0} being the maximum Josephson current of the junction. By now considering the fluxoid quantization condition (2.32), we can express the gauge-invariant superconducting phase difference ϕ in terms of the normalized flux ψ, so that Eq. (2.40) takes on the following final form

$$f_W(\psi) = (\psi - \psi_{ex})^2 + \frac{\beta}{\pi}[1 - \cos(2\pi\psi)], \qquad (2.42)$$

where f_W is the energy E normalized to ϵ_M, and where $\beta = \pi\,\epsilon_J/\epsilon_M$ is a relevant parameter of the quantum interferometer. The potential in Eq. (2.42) is commonly denoted as "parabolic washboard potential" and is represented, for $\beta = 0.3\,\pi$ and for three different values of ψ_{ex}, in Fig. 2.11. From this figure one can notice that the flux states, corresponding to the relative minima ψ_m of the potential f_W, are not quantized, since the ψ_m values may change continuously, at least in the interval of ψ_{ex} represented in Fig. 2.11. We shall see later how to derive a complete dependence of the minima ψ_m on the normalized external magnetic flux. We now derive the time evolution equation of the system, by means of the RSJ model. This derivation could be done by a power balance approach, which we shall also briefly mention. Start then by considering the normalized flux linked to the superconducting loop as the sum of the normalized applied and self-induced fluxes, so that:

$$\psi = \psi_{ex} + \beta\, i, \qquad (2.43)$$

where $i = I/I_{J0}$ is the normalized current.

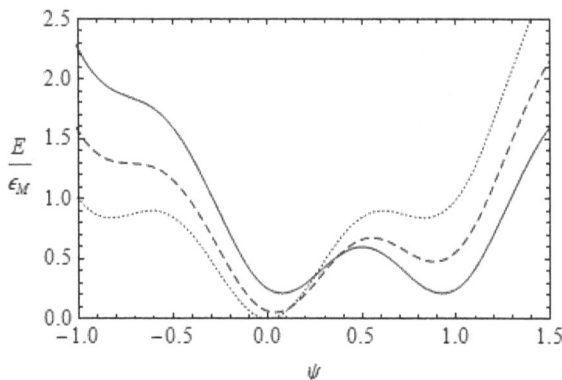

Fig. 2.11. Parabolic washboard potential for a superconducting loop interrupted by one Josephson junction for $\beta = 0.3\,\pi$. The values of the normalized applied flux are as follows: $\psi_{ex} = 0.00$ (dotted line), $\psi_{ex} = 0.25$ (dashed line), $\psi_{ex} = 0.50$ (full line).

On the other hand, by setting $\beta_c = 0$ in Eq. (1.28), by the RSJ model we can write:

$$\frac{d\phi}{d\tilde{\tau}} + \sin\phi = i, \qquad (2.44)$$

where $\tilde{\tau} = 2\pi R I_{J0} t/\Phi_0$, as defined in Eq. (1.27), R being the resistive parameter of the Josephson junction. By now taking into account the fluxoid quantization condition (2.32), we can rewrite Eq. (2.44) in terms of the variable ψ and of an appropriately normalized time $\tau = Rt/L$ as follows:

$$\frac{d\psi}{d\tau} + \beta\sin(2\pi\psi) + \psi = \psi_{ex}. \qquad (2.45)$$

The same dynamic equation (2.45) could be derived by a power balance-approach. In fact, by considering the power $P_d = -V^2/R$ dissipated in the resistor equal to the time derivative of the potential E, we may write:

$$\epsilon_M \frac{df_W}{dt} + \frac{V^2}{R} = 0. \tag{2.46}$$

Therefore, by the definition of ϵ_M and f_W, and by the Josephson voltage-frequency relation (1.4b), we have:

$$\frac{\Phi_0^2}{L}[\beta \sin(2\pi\psi) + (\psi - \psi_{ex})]\left(\frac{d\psi}{dt}\right) + \frac{\Phi_0^2}{R}\left(\frac{d\psi}{dt}\right)^2 = 0. \tag{2.47}$$

Aside from the stationary solution $d\psi/dt = 0$, which provides the minima ψ_m in the parabolic washboard potential, the time evolution equation, after factoring out the term $\Phi_0^2\, d\psi/dt$, is the same as in Eq. (2.45).

In order to derive an implicit dependence of the minima ψ_m, which from now on we shall denote simply as ψ, on the external magnetic flux ψ_{ex}, we proceed as follows. We first recognize that these minima may be obtained by setting $d\psi/d\tau = 0$. Therefore, by setting the time derivative of the quantity ψ equal to zero in Eq. (2.45), we have:

$$\beta \sin(2\pi\psi) + \psi = \psi_{ex}. \tag{2.48}$$

The above equation can also be obtained by setting $df_W/d\psi = 0$ and provides all fixed points (relative minima and maxima) in the parabolic washboard potential. Therefore, in order to have a minimum, the additional following relation, corresponding to $d^2 f_W/d\psi^2 > 0$, is to be satisfied:

$$2\pi\beta \cos(2\pi\psi) + 1 > 0. \tag{2.49}$$

For $2\pi\beta < 1$, the above inequality is always satisfied, meaning that the curve has always a positive curvature and that there can only exist one minimum. We then recognize that, although for $2\pi\beta < 1$, we can solve Eq. (2.48) unambiguously for the implicitly defined magnetic state ψ, the same cannot be done for $2\pi\beta \geq 1$. This can be understood by recalling that, when a system may occupy metastable magnetic states, its properties may become history-dependent.

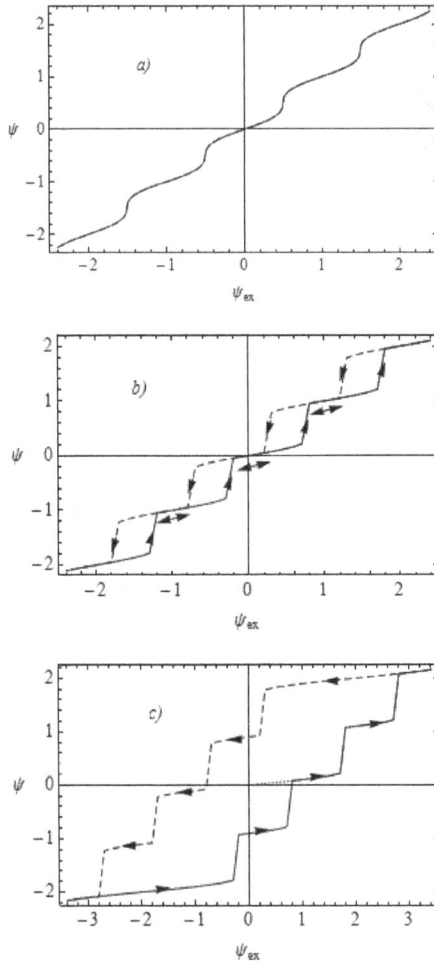

Fig. 2.12. Flux states in a superconducting ring interrupted by a Josephson junction as a function of the normalized externally applied flux ψ_{ex} for $\beta = 0.15$ (a), $\beta = 0.50$ (b), and $\beta = 1.50$ (c). In panels (b) and (c) the dotted curves (only visible in (c)) pertain to increasing values of ψ_{ex} in the interval $(0, \Psi_{ex})$, the dashed curves represent flux states obtained for decreasing values of ψ_{ex} in the interval $(\Psi_{ex}, -\Psi_{ex})$, and the full line curves are obtained for increasing values of ψ_{ex} in the interval $(-\Psi_{ex}, \Psi_{ex})$. Arrows indicate the way the field varies along the cycle. Double arrows in panel (b) indicate reversibility regions. All curves coincide in panel (a), since the system behaves reversibly for $\beta = 0.15 < 1/2\pi$.

Therefore, we need to numerically solve Eq. (2.48) in order to obtain, depending on the particular value of β, the magnetic states of the system. The numerical integration can thus be done as follows.

First, fix the value of the normalized externally applied flux ψ_{ex} to zero and raise it, step by step, to a maximum value, say Ψ_{ex}. After each step, record the result of the integration procedure performed for a sufficiently long time in order to let the system reach equilibrium after each small increase of ψ_{ex}. In this way, starting from zero trapped flux, a set of points $(\psi_{ex}, \psi(\psi_{ex}))$ can be collected in the interval $(0, \Psi_{ex})$. By next collecting the same couple of points from Ψ_{ex} to $-\Psi_{ex}$, and finally, starting from $-\Psi_{ex}$, the cycle can be completed going back to Ψ_{ex}. The results of numerical integration are shown in Fig. 2.12a–c for various values of the parameter β. In particular, in Fig. 2.12a the value of $\beta = 0.15 < 1/2\pi$ is chosen. In this case, no irreversible transition from one metastable state to another occurs and the system shows complete reversible magnetic behavior: the three curves coincide when the forcing term ψ_{ex} is varied in a cycle. When β increase above the crossover value $\beta_c = 1/2\pi$, however, the system may irreversibly go from one metastable magnetic state to another, so that hysteretic behavior in the ψ vs. ψ_{ex} curves appears. In fact, in Fig. 2.12b this hysteretic character appears in the ψ vs. ψ_{ex} curves obtained for $\beta = 0.50$ for well-defined field intervals. Notice that, in the same curve, reversibility regions – marked with a double arrow – are also present. Reversible behavior, on the other hand, completely disappears in the ψ vs. ψ_{ex} curves calculated for $\beta = 1.50$. In Fig. 2.12c the dotted line, corresponding to the first magnetization curve of the sample, is visible. Notice also that, by increasing the value of β, the ψ-ψ_{ex} cycles tend to enlarge. As a final comment on Fig. 2.12a-c, we may see how, by increasing the coupling between the two ends of the ring, i. e., by increasing the value of β, the ψ vs. ψ_{ex} curves tend to qualitatively resemble those for a superconducting ring observed by Goodman and Deaver (Goodman & Deaver, 1970). However, we need to remark an important difference once more: the n vs. ψ_{ex} curves obtained for a superconducting ring refer to flux trapping after cooling the sample below the transition temperature in the presence of an externally applied normalized flux ψ_{ex}. In the case of a superconducting ring interrupted by a Josephson junction, no temperature variation is assumed, while only the forcing term ψ_{ex} is varied. By letting the externally applied flux vary, in fact, the normalized flux threading the ring follows the curves in Fig. 2.12a-c, and the current i inside the ring changes, as it is arguable from Eq. (2.43). When the maximum Josephson current is reached, however, a flux transition occurs to reduce the flux gradient $(\psi - \psi_{ex})$. This mechanism will be clear when we shall see in details how the persistent currents in the system vary with ψ_{ex}.

Before devoting our attention to this aspect, let us calculate the approximate value $\psi_{ex}^{(0)}$ at which first irreversible flux penetration occurs in the one-junction interferometer. Let us then assume that the parameter β is greater than the crossover value $\beta_c = 1/2\pi$ and that no flux inside the ring is present at $\psi_{ex} = 0$. Successively, let us gradually apply an increasing field to our system, in such a way that the value of ψ slightly increases. We may argue from Fig. 2.12b-c, that when we further increase ψ_{ex}, the system reacts in such a way to allow flux penetrate inside the ring. This leads to a rather abrupt increase of ψ and, as a consequence, of the gauge invariant superconducting phase difference ϕ (which suffers a "phase slip"). Therefore, the first derivative of the curve $\psi = \psi(\psi_{ex})$ goes to relatively high values at $\psi_{ex} = \psi_{ex}^{(0)}$. The same type of behavior is observable for all transition points $\psi_{ex}^{(n)}$. In this way, by considering Eq. (2.48), we may equivalently set $d\psi_{ex}/d\psi \cong 0$ in such a way that we obtain all the transitions values ψ_n by the following relation:

$$\psi_n \cong \frac{1}{2\pi}\arccos\left(-\frac{1}{2\pi\beta}\right) + n. \tag{2.50}$$

Notice that, by the above relation, $\psi_n - \psi_{n-1} \cong 1$. This means that approximately one fluxon is allowed inside the ring when the normalized external flux varies between the values $\psi_{ex}^{(n-1)}$ and $\psi_{ex}^{(n)}$. These values can now be approximately found by substituting Eq. (2.50) back into Eq. (2.48), so that:

$$\psi_{ex}^{(n)} \cong \frac{1}{2\pi}\cos^{-1}\left(-\frac{1}{2\pi\beta}\right) + n + \sqrt{\beta^2 - \frac{1}{4\pi^2}}. \tag{2.51}$$

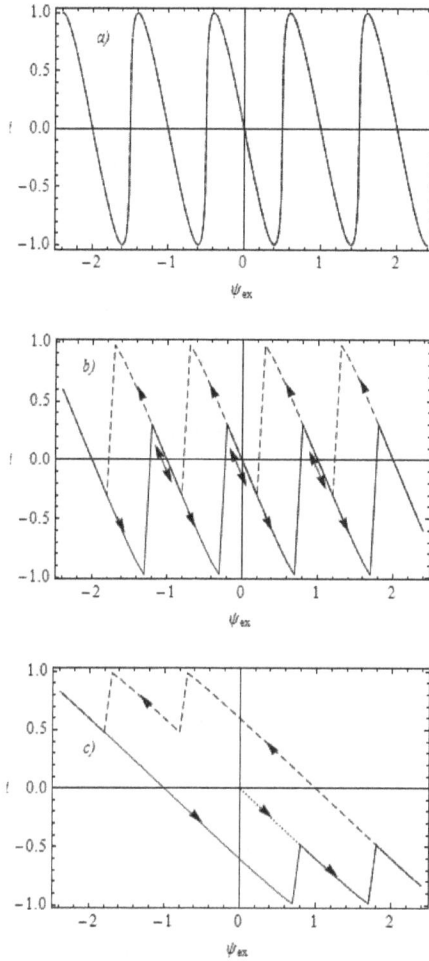

Fig. 2.13. Persistent currents in a superconducting ring interrupted by a Josephson junction as a function of the normalized externally applied flux ψ_{ex} for $\beta = 0.15$ (a), $\beta = 0.50$ (b), and $\beta = 1.50$ (c). In panels b and c the dotted curves (only visible in panel c) pertain to increasing values of ψ_{ex} in the interval $(0, \Psi_{ex})$, the dashed curves represent flux states obtained for decreasing values of ψ_{ex} in the interval $(\Psi_{ex}, -\Psi_{ex})$, and the full line curves are obtained for increasing values of ψ_{ex} in the interval $(-\Psi_{ex}, \Psi_{ex})$. Arrows indicate the way the field varies along the cycle. Double arrows in panel b indicate reversibility regions. All curves coincide in panel a, since the system behaves reversibly for $\beta = 0.15 < 1/2\pi$.

First irreversible flux penetration thus occurs, for $\beta > \beta_c$, at the following value of the normalized external field:

$$\psi_{ex}^{(0)} \cong \frac{1}{2\pi} \cos^{-1}\left(-\frac{1}{2\pi\beta}\right) + \sqrt{\beta^2 - \frac{1}{4\pi^2}}. \qquad (2.52)$$

This analysis, which may seem rather cumbersome, can be proven to be useful in understanding phenomena linked to flux penetration in granular superconductors.

Let us now turn our attention to the calculation of the persistent currents flowing inside the one-junction quantum interferometer represented in Fig. 2.8. This analysis, as already noted above, proves to be useful when one tries to interpret the shielding properties of superconducting granular system in the presence of very low magnetic fields. In particular, this type of approach clarifies the mechanism by which low-field flux penetration in granular superconductors occurs. From our initial perspective, however, we are only able to state, at the present stage, that a Josephson junction interrupting a superconducting ring acts like a bridge in allowing more and more flux quanta inside the superconducting loop, without any need of affecting the superconducting state, when the external magnetic field is increased.

Persistent currents flowing in the one-junction quantum interferometer of Fig. 2.8 are shown in Fig. 2.13a–c for the same values of the parameter β as in Fig. 2.12a–c. In Fig. 2.13a, all curves collapse within a single saw-tooth-like curve in which positive and negative current values alternate. Therefore, it appears clear that these types of systems may present magnetic behavior very much different from bulk superconductors.

From the persistent current curves, we may interpret the flux dynamics in the system. Let us then start by increasing the externally applied flux from zero. From Fig. 2.13a, in which $\beta = 0.15 < \beta_c$, we notice that the normalized current i decreases gradually from zero toward negative values until a conspicuous, but still reversible, flux penetration occurs inside the system when the inferior limiting value of $i = -1$ is approached. Looking at the increase of the normalized flux in Fig. 2.12a, we see that the initial slope of the ψ vs. ψ_{ex} curve is indeed replaced by a steeper gradient. Correspondingly, the normalized current grows to positive values up to the value of ψ_{ex} for which the superior limiting value $i = +1$ is approached. For still increasing values of ψ_{ex}, a smooth descent of the normalized current to -1 occurs, while ψ increases more slowly. When the externally applied flux is further increased and the current value is again close to $i = -1$, the same features are repeated. If the normalized applied magnetic flux ψ_{ex} is increased up to a maximum value Ψ_{ex} and then, starting from $\psi_{ex} = \Psi_{ex}$, is decreased to zero, the same path on the ψ vs. ψ_{ex} curve is traced. This means that reversible flux penetration is possible for $\beta < \beta_c$.

Irreversible flux penetration, on the other hand, occurs for $\beta = 0.50 > \beta_c$, as shown in Fig. 2.13b. In this figure, the values of ψ_{ex} where irreversible flux transitions inside the superconducting loop occur are well indicated by the lowest portion of the full-line curve. In particular, we notice that the normalized magnetic flux $\psi_{ex}^{(0)}$ at which first irreversible penetration occurs is close to the value of 0.8, as predicted by Eq. (2.52). The mechanism of flux penetration in this case is similar to what was described above. However, when the applied magnetic field is reversed, the system follows the dashed-line curve, now distinct from the full-line curve and thus visible in Fig. 2.13b. Therefore, when lowering the value of ψ_{ex} irreversible flux expulsion takes place when the normalized current approaches the uppermost value of $+1$. For the value of $\beta = 0.50$, ψ_{ex} intervals, in which reversible magnetic behavior is present, are indicated by the double arrow in Fig. 2.13b.

Reversibility intervals completely disappear, however, for higher values of the parameter β, as it is shown for $\beta = 1.50$ in Fig. 2.13c, where now the dotted curve is visible. This curve, we recall, represents initialization of the magnetic cycle, when the system, in the absence of applied field, is taken from a flux state $\psi = 0$ to different magnetic states, by gradually increasing ψ_{ex}. In Fig. 2.13c, we also notice that the normalized magnetic flux $\psi_{ex}^{(0)}$ at which first irreversible penetration occurs grows close to the value of 1.8, as also predicted by Eq. (2.52).

The magnetic susceptibility χ, calculated by considering Eq. (2.10), in the present case can be expressed as follows:

$$\chi = \frac{\alpha\psi - \psi_{ex}}{\psi_{ex}}, \qquad (2.53)$$

where α is the normal fraction of the system; i.e., α is given by the ratio between the cross section S of the hole and the total cross section area S_{TOT} of the whole system.

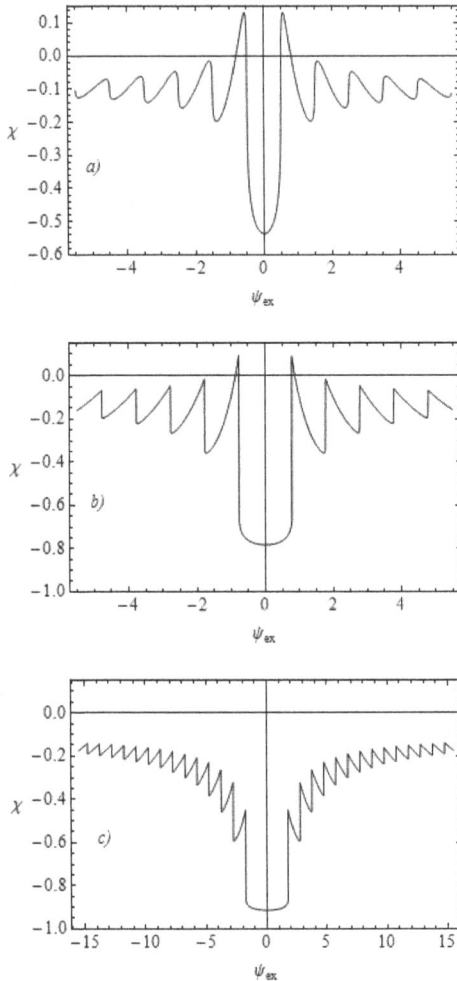

Fig. 2.14. Magnetic susceptibility curves in a superconducting ring interrupted by a Josephson junction for $\beta = 0.15$ (a), $\beta = 0.50$ (b), and $\beta = 1.50$ (c). In all curves $\alpha = 0.90$ and a zero-field cooled initial state is assumed. Therefore, in the interval $(0, \Psi_{ex})$ the externally applied normalized flux ψ_{ex} is gradually increased from zero. On the other hand, in the interval $(-\Psi_{ex}, 0)$ the forcing term ψ_{ex} is gradually decreased from zero.

The magnetic susceptibility curves for the one-junction quantum interferometer of Fig. 2.8 are shown in Fig. 2.14a–c for the same values of the parameters β as in Fig. 2.12a–c and Fig. 2.13a–c. The susceptibility is calculated, in order to avoid singularities, only starting from zero-field cooling conditions, i. e., by considering the system to be initially in a null magnetic state with $\psi = 0$ and $\psi_{ex} = 0$. Therefore, in the interval $(0, \Psi_{ex})$ the externally applied normalized flux ψ_{ex} is gradually increased from zero to $\psi_{ex} = \Psi_{ex}$. In the interval $(-\Psi_{ex}, 0)$ the forcing term ψ_{ex} is gradually decreased from zero to $\psi_{ex} = -\Psi_{ex}$. As a result, the susceptibility curve is symmetric with respect to the vertical axis. We take, for all curves, $\alpha = 0.90$. From these curves, we notice that a paramagnetic signal may be detected for $\beta = 0.15$ and $\beta = 0.50$, in very narrow intervals, as it also occurred in doubly connected superconductors. Another general feature in Fig. 2.14a-c is the smooth character of the χ vs. ψ_{ex} curve for $\beta = 0.15$, in the reversible magnetic regime, and the clear presence of first-derivative discontinuities in the irreversible magnetic regime in Fig. 2.14b and 2.14c. Notice also that the main lobes about $\psi_{ex} = 0$ tend to get larger as β increases, due to the increase of the magnetic field value at which first flux penetration occurs. These curves are not very much dissimilar to what was found for a doubly connected superconductor. However, we remark once more that the experimental conditions are now very much different. In fact, flux trapping inside a superconducting ring is obtained by cooling the sample below the superconducting transition temperature in the presence of a constant magnetic flux ψ_{ex}. In the case of a superconducting ring interrupted by a Josephson junction, the temperature is kept constant and the forcing term ψ_{ex} is varied, allowing the system to rearrange in different magnetic states, depending on the value of ψ_{ex}. The assumption of negligible noise effects made above imposes at least that the temperatures considered are very low. Noise effects in superconducting devices will be discussed in Chapter 4.

Chapter 3

Superconducting Quantum Interference Devices

3.1. The two-junction interferometer

We have seen that the magnetic properties of one-junction quantum interferometers can be analyzed by means of the notions acquired in studying the behavior of superconducting rings and of a single Josephson junction. On the basis of these principles, radio frequency Superconducting Quantum Interference Devices (r.f. SQUIDs) are operated (Barone, 1982) (Clarke, 2004). In this chapter, we shall see that the same fundamental mechanisms are used to describe the magnetic response of two-junction or multi-junction quantum interferometers.

Based on the same fundamental aspects reviewed in the previous sections, the characteristic properties of a d.c. Superconducting Quantum Interference Device (d. c. SQUID) can be understood. The static and dynamic properties of d. c. SQUIDs have been widely investigated in the literature (Barone, 1982) (Clarke, 2004) (Likharev, 1986). In the realm of quantum computing devices, superconducting interferometers containing three (Mooij, 1999) or more (Blatter, 2001) Josephson junctions have been proposed to obtain better defined minima in their potential energy, when compared to the two-junction interferometer states. The use of π-junctions (Ioffe, 1999) (Amin, 2005) (Yamashita, 2005), to obtain "quiet" devices operated around null magnetic field values, have been also proposed. The properties of the latter devices, which are more robust with respect to flux fluctuations, can be analyzed by adding only a minimal requirement in the fluxoid quantization condition (De Luca,

2001), so that the content of the present chapter will be useful also in grasping the properties of more exotic systems.

A two-junction quantum interferometer is schematically represented in Fig. 3.1. This system consists of a current-biased superconducting loop, of inductance L_1 and L_2, respectively, interrupted by two Josephson junctions, denoted as JJ1 and JJ2. The bias current I_B is seen to split in two branch currents, I_1 and I_2. A magnetic field H is applied perpendicularly to the plane of the quantum interferometer and assumed to be directed upward.

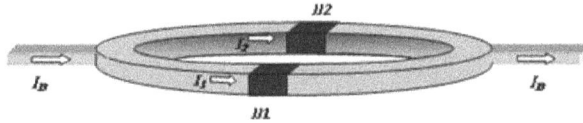

Fig. 3.1. Schematic representation of a two-junction quantum interferometer. The bias current I_B divides into two branch currents, I_1 and I_2. The applied magnetic field (not shown in the figure) is orthogonal to the plane in which the device lies and is directed upward.

We may start the analysis of the two-junction interferometer by writing the fluxoid quantization condition for the system, so that

$$\frac{2\pi}{\Phi_0}\Phi + \phi_1 - \phi_2 = 2\pi n, \qquad (3.1)$$

where Φ is the flux threading the superconducting loop, ϕ_1 and ϕ_2 are the gauge-invariant superconducting phase differences across JJ1 and JJ2, respectively, and n is an integer. The sign of the superconducting phase difference across JJ2 is negative, given that the oriented path around the superconducting loop, chosen according to the right-hand rule, opposes the assumed positive direction of the current I_2 in this junction. Let us also write the electrodynamic equation defining the flux Φ inside the loop as the sum of the induced flux and the applied flux Φ_{ex}, both defined as in the previous chapter:

$$\Phi = \Phi_{ex} + L_1 I_1 - L_2 I_2, \qquad (3.2)$$

where the minus sign in front of the last addendum on the right-hand side of Eq. (3.2) is now justified by the fact that the induced flux $L_2 I_2$ opposes the assumed positive flux Φ_{ex}. We may now write down the dynamic equation for each Josephson junction in the loop. By adopting the RSJ model (Barone, 1982) we may write one differential equation for each Josephson junction, so that, according to Eq. (2.44), we have:

$$\frac{\Phi_0}{2\pi R_1}\frac{d\phi_1}{dt} + I_{J1}\sin\phi_1 = I_1, \tag{3.3a}$$

$$\frac{\Phi_0}{2\pi R_2}\frac{d\phi_2}{dt} + I_{J2}\sin\phi_2 = I_2, \tag{3.3b}$$

where I_{J1} and R_1, I_{J2} and R_2 are the maximum Josephson currents and the resistive parameters of the junctions JJ1 and JJ2, respectively. The analysis of a non-symmetric ($L_1 \neq L_2$) and non-homogenous ($I_{J1} \neq I_{J2}$, $R_1 \neq R_2$) system is still possible (Clarke, 2004). However, for the sake of simplicity we here assume the system to be symmetric ($L_1 = L_2 = L$) and homogeneous in the junction parameters ($I_{J1} = I_{J2} = I_{J0}$, $R_1 = R_2 = R$). By this simplifying assumption, by defining the normalized time as

$$\tau = \frac{2\pi R I_{J0}}{\Phi_0}t, \tag{3.4}$$

and by setting, as usual, $i_1 = I_1/I_{J0}$, $i_2 = I_2/I_{J0}$, we may rewrite Eq. (3.3a) and (3.3b) as follows:

$$\frac{d\phi_k}{d\tau} + \sin\phi_k = i_k, \tag{3.5}$$

where $k = 1, 2$. By summing and subtracting homologous sides of the above equations, and by defining

$$\phi = \frac{\phi_1 + \phi_2}{2}, \quad \psi = \frac{\Phi}{\Phi_0} = \frac{\phi_2 - \phi_1}{2\pi} + n, \tag{3.6}$$

with n integer, we obtain the following two equations, equivalent to Eq. (3.3a) and (3.3b), for these new variables:

$$\frac{d\phi}{d\tau} + (-1)^n \cos\pi\psi \sin\phi = \frac{i_1 + i_2}{2}, \tag{3.7a}$$

$$\pi \frac{d\psi}{d\tau} + (-1)^n \sin \pi\psi \cos \phi = \frac{i_2 - i_1}{2}. \tag{3.7b}$$

By now taking $n = 0$ (zero flux when $\phi_1 = \phi_2 = 0$), by defining $i_B = I_B/I_{J0}$, by recognizing that $i_B = i_1 + i_2$, and by recalling Eq. (3.2), we may finally write:

$$\frac{d\phi}{d\tau} + \cos \pi\psi \sin \phi = \frac{i_B}{2}, \tag{3.8a}$$

$$\pi \frac{d\psi}{d\tau} + \sin \pi\psi \cos \phi + \frac{\psi}{2\beta} = \frac{\psi_{ex}}{2\beta}. \tag{3.8b}$$

where $\psi_{ex} = \Phi_{ex}/\Phi_0$ and $\beta = L I_{J0}/\Phi_0$. The simplest approach to the solution of the above dynamic equations is to assume that the normalized applied flux is equal to the flux number, so that $\Phi = \Phi_{ex}$ and only the first of the two above equations is needed in this approximation, namely:

$$\frac{d\phi}{d\tau} + \cos \pi\psi_{ex} \sin \phi = \frac{i_B}{2}, \tag{3.9}$$

which reduces to a special one-junction equation. In fact, by comparing Eq. (3.9) with Eq. (1.28) with null Stewart-McCumber parameter ($\beta_C = 0$), we may notice what follows. The effective one-junction model for $\beta_C = 0$, given in Eq. (3.9) can be obtained by considering a current-biased over-damped Josephson junction with a maximum Josephson current equal to $|\cos \pi\psi_{ex}|I_{J0}$. The constant normalized bias current i flowing in this effective junction needs to be taken half the value of the normalized current injected in the two-junction interferometer, i.e., $i_B/2$. In this way, the voltage-flux relation for the two-junction interferometer can be obtained by means of the notion of the single-junction current-voltage characteristics given in Eq. (1.48). In fact, going back to the derivation of the current-voltage characteristic of a an over-damped Josephson junction, we see that, for $0 < i_B/2 \leq |\cos \pi\psi_{ex}|$, Eq. (3.9) has a fixed point similar to the one represented in Eq. (1.39), namely:

$$\phi^* = \sin^{-1} \left(\frac{i_B}{2 \cos \pi\psi_{ex}} \right). \tag{3.10}$$

In this case, we have $\langle v \rangle = 0$. On the other hand, for $i_B/2 > |\cos \pi\psi_{ex}| > 0$, Eq. (3.9) can be solved by separation of variables, as done in Section 1.4. Here we need to consider that the normalized maximum Josephson current of the effective junction is equal to $|\cos \pi \psi_{ex}|$. Therefore, in order to obtain a solution of the ordinary differential equation (3.9) from the solution of Eq. (1.38), where m_0 is the forcing term, we may set

$$m_0 = \frac{i_B}{2|\cos \pi \psi_{ex}|} > 1, \tag{3.11}$$

by dividing both members of Eq. (3.9) by $|\cos \pi \psi_{ex}|$ and by defining the following time-variable:

$$\tau' = |\cos \pi \psi_{ex}| \, \tau. \tag{3.12}$$

In this way, we have

$$\frac{d\phi}{d\tau'} + \text{sgn}(\cos \pi \psi_{ex}) \sin \phi = \frac{i_B}{2|\cos \pi \psi_{ex}|}, \tag{3.13}$$

where $\text{sgn}(x)$ denotes the sign function, which extracts the sign of the real argument x. For $\cos \pi \psi_{ex} > 0$, the solution is immediately found and corresponds to the right hand side of Eq. (1.44) with τ replaced by τ', so that:

$$\phi(\tau') = 2\tan^{-1}\left[\frac{1}{m_0} + \frac{\sqrt{m_0^2-1}}{m_0}\left(\tan\left(\frac{\sqrt{m_0^2-1}}{2}\tau'\right)+\alpha_0\right)\right] + 2k\pi, \tag{3.14}$$

where k is an integer and where

$$\alpha_0 = \tan^{-1}\left[\frac{m_0}{\sqrt{m_0^2-1}}\left(\tan\frac{\phi_0}{2} - \frac{1}{m_0}\right)\right], \tag{3.15}$$

with $\phi_0 = \phi(0)$. From Eq. (3.14) we can argue that the period T' of the function $\phi(\tau')$ is equal to $2\pi/\sqrt{m_0^2 - 1}$. Calling now V the voltage across the device and defining the normalized voltage v as follows

$$v = \frac{V}{RI_{J0}} = \frac{d\phi}{d\tau}, \tag{3.16}$$

we have:

$$\langle v \rangle = \frac{2\pi}{T} = \frac{2\pi}{T'}\frac{T'}{T} = |\cos \pi \psi_{ex}|\sqrt{m_0^2 - 1}\,, \tag{3.17}$$

where we used Eq. (3.12) to calculate the ratio T'/T. Therefore, the portion of the voltage-flux characteristics relative to the normalized flux intervals for which $\cos \pi \psi_{ex} > 0$ is expressively given by:

$$\langle v \rangle = \sqrt{\left(\frac{i_B}{2}\right)^2 - \cos^2(\pi \psi_{ex})} \;. \qquad (3.18)$$

For $\cos \pi \psi_{ex} < 0$, on the other hand, one needs to find solution to the following equation

$$\frac{d\phi}{d\tau'} - \sin \phi = m_0, \qquad (3.19)$$

recalling that $m_0 > 1$. Proceeding as in Section 1.4, we define

$$\gamma_0 = \tan^{-1}\left[\frac{m_0}{\sqrt{m_0{}^2-1}} \left(\tan \frac{\phi_0}{2} + \frac{1}{m_0}\right)\right]. \qquad (3.20)$$

The solution to Eq. (3.19) can thus be implicitly expressed as follows:

$$\tan^{-1}\left[\frac{m_0}{\sqrt{m_0{}^2-1}} \left(\tan \frac{\phi(\tau')}{2} + \frac{1}{m_0}\right)\right] = \frac{\sqrt{m_0{}^2-1}}{2}\tau' + \gamma_0. \qquad (3.21)$$

Extracting $\phi(\tau')$ from Eq. (3.21), we have:

$$\phi(\tau') = 2\tan^{-1}\left[\frac{\sqrt{m_0{}^2-1}}{m_0} \left(\tan \frac{\sqrt{m_0{}^2-1}}{2}\tau' + \gamma_0\right) - \frac{1}{m_0}\right] + 2k\pi, \qquad (3.22)$$

where k is an integer. The above expression could also be found by noticing that Eq. (3.19) is equal to Eq. (1.44) when we make the following substitutions in the former:

$$\tau' \to \tau; \; \phi \to -\theta; \; m_0 \to -m_0. \qquad (3.23)$$

From Eq. (3.22) we can argue that the period T' of the function $\phi(\tau')$ is equal to $2\pi/\sqrt{m_0{}^2 - 1}$, as in the previous case. Therefore, we conclude that the same expression (3.18) also represents the voltage-flux characteristics for $\cos \pi \psi_{ex} < 0$. The case $\cos \pi \psi_{ex} = 0$ is immediately solved by noticing that Eq. (3.9) directly gives $\langle v \rangle = i_B/2$, or can be obtained by imposing continuity in the $\langle v \rangle$ vs. ψ_{ex} curve. The voltage-flux characteristics and the $d\langle v \rangle/d\psi_{ex}$ vs. ψ_{ex} curve are shown in Fig.

(3.2a) and (3.2b), respectively, for various values of the normalized bias current i_B. The latter curve gives account of the entity of the "transfer coefficient" $F(\psi_{ex})$ of a d. c. SQUID: the highest the value of this quantity, the highest the voltage signal $\Delta\langle v \rangle$ registered by the device in the presence of a small flux variation $\Delta\psi_{ex}$.

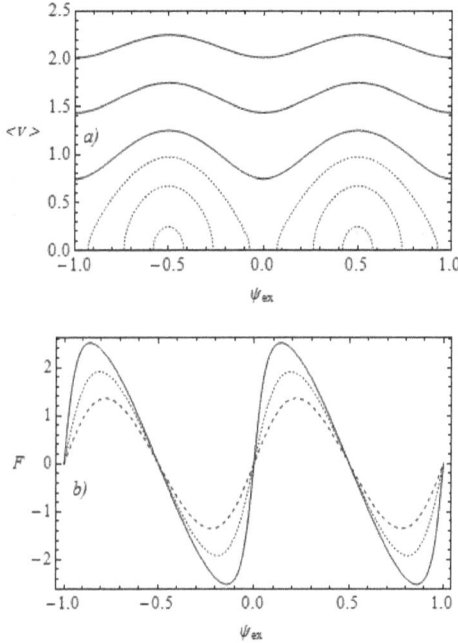

Fig. 3.2. (a) Voltage-flux characteristics and (b) transfer function $F(\psi_{ex})$ of a two-junction interferometer. In panel a the dotted curves are drawn (from bottom to top) for $i_B = 0.5, 1.35, 1.95$, while the full-line curves are drawn (from bottom to top) for $i_B = 2.5, 3.5, 4.5$. In panel b the values of i_B are chosen as follows: $i_B = 2.05$ (full line), $i_B = 2.25$ (dotted line), $i_B = 2.75$ (dashed line).

In Fig. (3.2a), the normalized bias current i_B is taken below (dotted curves) or above (full-line curves) the value of $l_B = 2.0$. In the first case ($i_B < 2$) the curves are non-zero only in those intervals in which $|\cos \pi \psi_{ex}| < i_B/2$. In the second case ($i_B > 2$), the inequality $|\cos \pi \psi_{ex}| < i_B/2$ cannot be satisfied for any value of ψ_{ex}, and the voltage curves all lie above the ψ_{ex}-axis. The periodicity of all curves in

Fig. (3.2a) and (3.2b) is exactly equal to $\Delta\psi_{ex} = 1.0$, as it can be argued from Eq. (3.18).

3.2. The effective one-junction model for $\beta > 0$

In the previous section, we have studied the analytic properties of the one-junction model for $\beta = 0$. While it was possible to derive the voltage-flux characteristics in this limit, there are other physical quantities, which cannot be immediately calculated by standard methods. For example, the superconducting circulating current i_S and the magnetic susceptibility χ in the superconducting state ($|\cos \pi\psi_{ex}| < i_B/2$) are two such quantities, which can be defined as follows:

$$i_S = i_1 - i_2 = \frac{\psi - \psi_{ex}}{\beta}, \tag{3.24a}$$

$$\chi = \frac{\alpha\psi - \psi_{ex}}{\psi_{ex}}, \tag{3.24b}$$

where, we recall, α is the normal fraction of the cylindrically symmetric two-junction interferometer schematically shown in Fig. 3.1. Therefore, while Eq. (3.24a) gives an ill-defined quantity for $\beta = 0$, the function χ in Eq. (3.24b) provides a strictly negative value for $\psi = \psi_{ex}$. This apparent rigidity can be overcome by introducing an effective one-junction model for $\beta > 0$, in the limit of small values of β.

In fact, we may consider Eq. (3.8a) and (3.8b) again. Let us also define two characteristic times

$$\tau_\psi = \frac{L}{R}, \quad \tau_\phi = \frac{\Phi_0}{2\pi R I_{J0}}, \tag{3.25}$$

representing the characteristic time scales of the variables ϕ and ψ, respectively. We may notice that the ratio between these two times is given by the following expression:

$$\tau_\psi/\tau_\phi = 2\pi\beta. \tag{3.26}$$

In this way, the flux dynamics for small values of β in Eq. (3.8b) can be considered very fast with respect to the equivalent junction dynamics given in Eq. (3.8a). Therefore, the superconducting phase ϕ can be assumed to be adiabatic, or slowly varying with respect to ψ in this limit. As a consequence, the equation of motion for ψ (3.8b) can be solved, by perturbation analysis, in terms of the quasi-static variable ϕ. This is the first step in obtaining the effective one-junction model for the two-junction interferometer with small values of β. The information for ψ obtained by solving Eq. (3.8b) can then be substituted back into Eq. (3.8a). The resulting effective dynamic equation for ϕ will thus represent the sought effective one-junction model. In what follows we shall construct this model step by step for $2\pi\beta \ll 1$.

We start by expressing the variable ψ as follows:

$$\psi = \psi_0 + \beta\psi_1 + \beta^2\psi_2 + O(\beta^3), \tag{3.27}$$

where the expression $O(\beta^3)$ means that we are neglecting terms of the order of β^3 and where ψ_0 is not necessarily time-independent. Let us then substitute Eq. (3.27) into Eq. (3.8b), obtaining to second order in β:

$$\pi\beta^2 \left(\frac{d\psi_1}{d\tau} + \psi_1\cos\pi\psi_0 \cos\phi \right) + \beta \left(\pi\frac{d\psi_0}{d\tau} + \sin\pi\psi_0 \cos\phi \right) + \frac{\psi_0}{2} +$$
$$+\beta\frac{\psi_1}{2} + \beta^2\frac{\psi_2}{2} \approx \frac{\psi_{ex}}{2}. \tag{3.28}$$

From the above expression we have the following three equations, one for each order in β:

$$\psi_0 = \psi_{ex}, \tag{3.29a}$$

$$\psi_1 = -2\pi\frac{d\psi_0}{d\tau} - 2\sin\pi\psi_0 \cos\phi, \tag{3.29b}$$

$$\psi_2 = -2\pi\frac{d\psi_1}{d\tau} - 2\pi\psi_1\cos\pi\psi_0 \cos\phi. \tag{3.29c}$$

Taking time-independent externally applied flux, and noticing that the derivative with respect to time of ψ_1 is proportional to $d\phi/d\tau$, which is zero under quasi-stationary conditions, equations (3.29b) and (3.29c) are rewritten as follows:

$$\psi_1 = -2\sin\pi\psi_{ex} \cos\phi, \tag{3.30a}$$

$$\psi_2 = 2\pi\sin 2\pi\psi_{ex} \cos^2\phi. \tag{3.30b}$$

In this way, we have:

$$\psi = \psi_{ex} - 2\beta \sin \pi\psi_{ex} \cos \phi + 2\pi\beta^2 \sin 2\pi\psi_{ex} \cos^2 \phi + O(\beta^3). \quad (3.31)$$

As specified above, we need to substitute this expression back into Eq. (3.8a), so that we need to expand to second order in β the following term: $\cos\pi\psi_{ex} - 2\pi\beta \sin \pi\psi_{ex} \cos \phi + 2\pi^2\beta^2 \sin 2\pi\psi_{ex} \cos^2 \phi$. By per-forming the expansion, Eq. (3.8a) can be rewritten as follows:

$$\frac{d\phi}{d\tau} + f(\phi, \psi_{ex}) \sin \phi = \frac{i_B}{2}, \quad (3.32)$$

where

$$f(\phi, \psi_{ex}) = f_0(\psi_{ex}) + \beta f_1(\phi, \psi_{ex}) + \beta^2 f_2(\phi, \psi_{ex}), \quad (3.33)$$

with

$$f_0(\psi_{ex}) = \cos \pi\psi_{ex}, \quad (3.34a)$$

$$f_1(\phi, \psi_{ex}) = 2\pi \sin^2 \pi\psi_{ex} \cos \phi, \quad (3.34b)$$

$$f_2(\phi, \psi_{ex}) = -6\pi^2 \sin^2 \pi\psi_{ex} \cos \pi\psi_{ex} \cos^2 \phi. \quad (3.34c)$$

Notice that higher-harmonic terms appearing in the current-phase relation of the effective one-junction model for a two-junction quantum interferometer is only due to the electromagnetic coupling of the two junctions. In fact, these terms arise when we do not neglect the magnetic energy term due to the circulating currents in the superconducting loop. This additional feature is thus given only by the first and higher order contributions of the inductive parameter and does not require any non-conventional current-phase relation in the Josephson junction.

By now turning back to the definitions of the circulating current i_S and of the magnetic susceptibility χ given in Eq. (3.24a) and in Eq. (3.24b), respectively, we may write:

$$i_S = -2 \sin \pi\psi_{ex} \cos \phi + 2\pi\beta \sin 2\pi\psi_{ex} \cos^2 \phi, \quad (3.35a)$$

$$\chi = -1 + \alpha \left[1 - 2\pi\beta \, \frac{\sin \pi\psi_{ex}}{\pi\psi_{ex}} \cos \phi + (2\pi\beta)^2 \, \frac{\sin 2\pi\psi_{ex}}{2\pi\psi_{ex}} \cos^2 \phi \right], \quad (3.35b)$$

where an expression for $\cos\phi$ needs to be found by the stationary solution of Eq. (3.32). Notice that, while in Eq. (3.35b) the first-order solution for $\cos\phi$ is sufficient to describe the function χ to second order in β, in Eq. (3.35a) we only get a first-order description of the circulating current i_S, which will be sufficient for our purposes. In fact, we shall see that the current maintains its qualitative and quantitative features for small variations of the parameter β.

In this way, we solve for the stationary phase states in Eq. (3.32) to first order in β, so that we write:

$$\cos\pi\psi_{ex}\sin\phi + \pi\beta\sin^2\pi\psi_{ex}\sin 2\phi = \frac{i_B}{2}. \qquad (3.36)$$

In order to obtain an approximated solution, we may set:

$$\phi = \phi_0 + \beta\phi_1 + O(\beta^2). \qquad (3.37)$$

By now substituting the above expression in Eq. (3.36) and by gathering terms of the same order in β, we have:

$$\sin\phi_0 = \frac{i_B}{2\cos\pi\psi_{ex}}, \qquad (3.38a)$$

$$\phi_1 = -2\pi\frac{\sin^2\pi\psi_{ex}}{\cos\pi\psi_{ex}}\sin\phi_0. \qquad (3.38b)$$

Therefore, by setting

$$\cos\phi = \cos\phi_0 - \beta\phi_1\sin\phi_0, \qquad (3.39)$$

we have:

$$\cos\phi^* = \pm\sqrt{1 - \left(\frac{i_B}{2\cos\pi\psi_{ex}}\right)^2} + 2\pi\beta\frac{\sin^2\pi\psi_{ex}}{\cos\pi\psi_{ex}}\left(\frac{i_B}{2\cos\pi\psi_{ex}}\right)^2, \qquad (3.40)$$

where we still have to decide upon the sign of the first addendum in Eq. (3.40). This can be done by considering Eq. (3.32) and by calculating the derivative $d\phi/d\tau$ in the vicinity of ϕ^*. Therefore, by setting $\phi = \phi^* + \delta\phi$ and by substituting in Eq. (3.32) we get, to leading order in β, the following expression:

$$\frac{d\phi}{d\tau} = -\cos \pi\psi_{ex} \cos \phi^* \, \delta\phi. \tag{3.41}$$

In order to detect a stable fixed point in $\phi = \phi^*$, we need to have negative sign of the derivative $d\phi/d\tau$ for positive variations $\delta\phi$ and, vice-versa, positive sign of $d\phi/d\tau$ for negative variations $\delta\phi$. Therefore, from (3.41) we can conclude that

$$\text{sign}(\cos \phi^*) = \text{sign}(\cos \pi\psi_{ex}). \tag{3.42}$$

In this way, the solution (3.40) takes then the following final form:

$$\cos \phi^* = \text{sign}(\cos \pi\psi_{ex})\sqrt{1 - \left(\frac{i_B}{2\cos \pi\psi_{ex}}\right)^2} + \frac{\pi\beta i_B^2 \tan^2 \pi\psi_{ex}}{2\cos \pi\psi_{ex}}. \tag{3.43}$$

By now substituting the above expression in Eq. (3.35a) we get

$$i_S = i_{S0} + 2\pi\beta \sin 2\pi\psi_{ex} - \frac{\pi\beta \, i_B^2 \sin \pi\psi_{ex}}{\cos^3 \pi\psi_{ex}}, \tag{3.44}$$

with

$$i_{S0} = -2 \sin \pi\psi_{ex} \, \text{sign}(\cos \pi\psi_{ex})\sqrt{1 - \left(\frac{i_B}{2\cos \pi\psi_{ex}}\right)^2}. \tag{3.45}$$

On the other hand, by substituting Eq. (3.43) in Eq. (3.35b), we get the following expression for the magnetic susceptibility:

$$\chi = \chi_0 + 2\pi\beta \, \chi_1 + (2\pi\beta)^2 \chi_2, \tag{3.46}$$

with

$$\chi_0 = \alpha - 1, \tag{3.47a}$$

$$\chi_1 = -\alpha \, \text{sign}(\cos \pi\psi_{ex})\frac{\sin \pi\psi_{ex}}{\pi\psi_{ex}}\sqrt{1 - \left(\frac{i_B}{2\cos \pi\psi_{ex}}\right)^2}. \tag{3.47b}$$

$$\chi_2 = -\alpha \left[\frac{\tan \pi\psi_{ex}}{\pi\psi_{ex}}\left(\frac{i_B}{2\cos \pi\psi_{ex}}\right)^2 - \frac{\sin 2 \pi\psi_{ex}}{2\pi\psi_{ex}}\right]. \tag{3.47c}$$

In Fig. 3.3a and 3.3b, the persistent current i_S is shown as a function of ψ_{ex}. In panel a the curves are shown for $\beta = 0.02$ and for the following values of the bias current: $i_B = 0.00, 0.50, 0.75, 1.00$; in panel b, on the other hand, the curves are shown for $i_B = 0.50$ and for the following values of the parameter β: 0.00, 0.02 , 0.04. All i_S vs. ψ_{ex} curves are seen to be piecewise continuous. From these figures it can be seen that the current i_S is equal to zero in the case the two junctions are in the resistive or running state ($i_B > 2|\cos \pi\psi_{ex}|$). In particular, in Fig. 3.3a, we notice that the full-line curve obtained for $i_B = 0$ does not allow any zero-current plateau, showing a discontinuity in correspondence with half-integer values of ψ_{ex}. As i_B increases, zero-current plateaus, centered around half-integer values of ψ_{ex} and having increasing amplitude, appear in the i_S vs. ψ_{ex} curves. The sharp discontinuity at $\psi_{ex} = (2k - 1)/2$, k being an integer, does not appear anymore for $i_B \neq 0$.

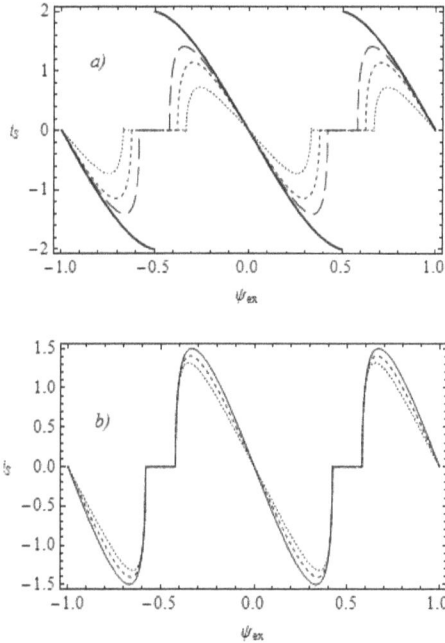

Fig. 3.3. Circulating current i_S as a function of the normalized applied flux ψ_{ex}. In panel a, the value of the parameter β is fixed to 0.02, while the four curves are obtained for the following values of the bias current: $i_B = 0.00$ (full line), $i_B = 0.50$ (coarsely dashed line), $i_B = 0.75$ (dashed line), $i_B = 1.00$ (dotted line). In panel b, the value of the bias current is fixed to $i_B = 0.50$, while the parameter β takes on the following values: $\beta = 0.00$ (full line), $\beta = 0.02$ (dashed line), $\beta = 0.04$ (dotted line).

In Fig. 3.4a and 3.4b, the magnetic susceptibility χ is shown as a function of ψ_{ex} for $\alpha = 0.95$. In panel a, the curves are shown for $\beta = 0.02$ and for the following values of the bias current: $i_B = 0.00, 0.50, 1.00$; in panel b, on the other hand, the curves are shown for $i_B = 0.50$ and for the following values of the parameter β: $0.00, 0.01, 0.02$.

All χ vs. ψ_{ex} curves are seen to be piecewise continuous. By taking $\alpha = 0.95$, we imagine that the superconducting loop is rather thin, so that a paramagnetic response may still be present.

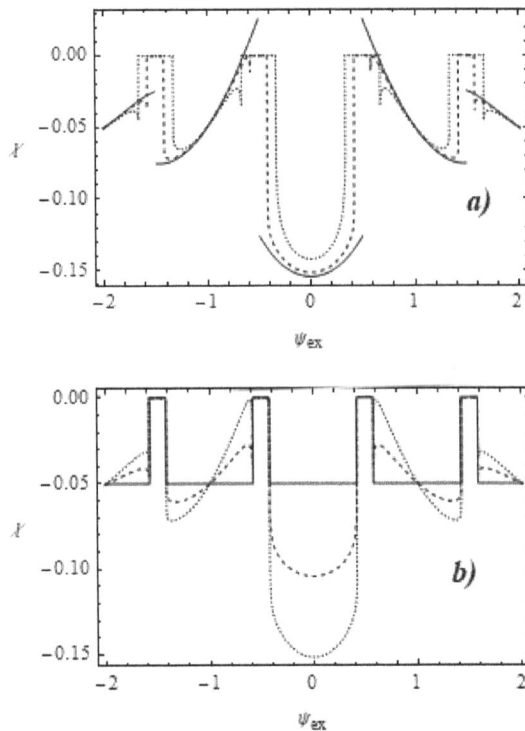

Fig. 3.4. Magnetic susceptibility χ as a function of the normalized applied flux ψ_{ex} for $\alpha = 0.95$. In panel a, the value of the parameter β is fixed to 0.02, while the three curves are obtained for the following values of the bias current: $i_B = 0.00$ (full line), $i_B = 0.50$ (dashed line), $i_B = 1.00$ (dotted line). In panel b, the value of the bias current is fixed to $i_B = 0.50$, while the parameter β takes on the following values: $\beta = 0.00$ (full line), $\beta = 0.01$ (dashed line), $\beta = 0.02$ (dotted line).

As in the case of the i_S vs. ψ_{ex} curves in Fig. 3.3a and 3.3b, we notice the appearance of null values of the magnetic susceptibility when the two junctions are in the resistive state $(i_B > 2|\cos \pi\psi_{ex}|)$. These zero-$\chi$ plateaus, centered about $\psi_{ex} = (2k - 1)/2$, k being an integer, increase in size when the bias current increases.

In Fig. 3.4a, the full-line curve obtained for $i_B = 0$ does not allow any plateau, so that the corresponding χ vs. ψ_{ex} curve shows a sharp discontinuity in correspondence to half-integer values of ψ_{ex}. Again, as noted for the circulating currents i_S, when the values of i_B increase, this sharp discontinuity at $\psi_{ex} = (2k - 1)/2$, k being an integer, does not anymore appear in the χ vs. ψ_{ex} curves. In Fig. 3.4a, we may notice that the piecewise continuous χ vs. ψ_{ex} curve obtained for $i_B = 0$ and $\beta = 0.02$ has portions of its branches in the symmetric ψ_{ex}-intervals $(1/2, 3/2)$ and $(-1/2, -3/2)$ above the ψ_{ex}-axis. This paramagnetic feature is not present in Fig. 3.4b, where we may detect the rectangular-wave pattern of the curve shown for $i_B = 0.50$ and $\beta = 0.00$. This feature could also be deduced from Eq. (3.46) and (3.47a) which give, for $\beta = 0$, $\chi = \alpha - 1$ if the system is in the superconducting state. On the other hand, $\chi = 0$ segments appear if the device is in the running state.

3.3. The washboard potential

As seen for one-junction interferometers, it is possible to describe the classical behavior of two-junction quantum interferometers by knowing the potential energy function W for the system. Let us then consider two identical Josephson junctions, both with maximum Josephson current I_{J0}, in a superconducting loop of total inductance $2L$ in which a bias current I_B is injected symmetrically with respect to the position of the junctions, as shown in Fig. 3.1. By considering the Josephson energy E_J of the two junctions, the magnetic energy E_M of the circulating current and the contribution of the bias current E_B, we may write:

$$W = E_J + E_M + E_B, \tag{3.48}$$

where:

$$E_J = -\frac{\Phi_0 I_{J0}}{2\pi}(\cos\phi_1 + \cos\phi_2), \tag{3.49a}$$

$$E_M = \frac{1}{4L}\left(\frac{\Phi_0}{2\pi}\right)^2 (\phi_2 - \phi_1 - 2\pi\psi_{ex})^2, \tag{3.49b}$$

$$E_B = -\frac{\Phi_0 I_{J0}}{4\pi} i_B(\phi_1 + \phi_2). \tag{3.49c}$$

Let us now prove that the potential $W = W(\phi_1, \phi_2)$ correctly provides the equations of the motion (3.3a) and (3.3b) for the system. By imposing power balance in the system, we may set:

$$\frac{dW}{dt} + \frac{V^2}{(R/2)} = 0, \tag{3.50}$$

where the effective resistance $R/2$ of the two Josephson junctions in the parallel configuration has been used. From the above equation we have:

$$\frac{\partial W}{\partial \phi_1}\dot{\phi}_1 + \frac{\partial W}{\partial \phi_2}\dot{\phi}_2 + \frac{1}{R}\left(\frac{\Phi_0}{2\pi}\right)^2 \left(\dot{\phi}_1^{\,2} + \dot{\phi}_2^{\,2}\right) = 0, \tag{3.51}$$

where the dots on ϕ_1 and ϕ_2 indicate the derivative with respect to time of these variables and where we have made use of the fact that $\dot{\phi}_1 = \dot{\phi}_1 = 2\pi V/\Phi_0$. By setting to zero each single-junction contribution in the power balance equation, we may write:

$$\frac{1}{R}\left(\frac{\Phi_0}{2\pi}\right)^2 \dot{\phi}_k + \frac{\partial W}{\partial \phi_k} = 0, \tag{3.52}$$

where $k = 1, 2$. By calculating the partial derivatives of W with respect to the phase variables ϕ_1 and ϕ_2, we have:

$$\frac{\Phi_0}{2\pi R I_{J0}}\dot{\phi}_1 + \sin\phi_1 = \frac{i_B}{2} + \frac{\Phi_0}{4\pi L I_{J0}}(\phi_2 - \phi_1 - 2\pi\psi_{ex}), \tag{3.53a}$$

$$\frac{\Phi_0}{2\pi R I_{J0}}\dot{\phi}_2 + \sin\phi_2 = \frac{i_B}{2} - \frac{\Phi_0}{4\pi L I_{J0}}(\phi_2 - \phi_1 - 2\pi\psi_{ex}). \tag{3.53b}$$

By now recalling the definition of ψ in Eq. (3.6) with $n = 0$, and by noticing that

$$i_1 + i_2 = i_B, \tag{3.54a}$$

$$i_1 - i_2 = \frac{\psi - \psi_{ex}}{\beta}, \qquad (3.54b)$$

we obtain:

$$i_1 = \frac{i_B}{2} + \frac{\Phi_0}{4\pi L I_{J0}}(\phi_2 - \phi_1 - 2\pi\psi_{ex}), \qquad (3.55a)$$

$$i_2 = \frac{i_B}{2} - \frac{\Phi_0}{4\pi L I_{J0}}(\phi_2 - \phi_1 - 2\pi\psi_{ex}). \qquad (3.55b)$$

The above expressions are equal to the right-hand sides of Eq. (3.53a) and (3.53b), so that, comparing the latter equations with Eq. (3.3a) and (3.3b) with homogeneous junction parameters ($I_{J1} = I_{J2} = I_{J0}$, $R_1 = R_2 = R$), we may consider our proof complete.

Let us now write the potential in terms of the variables ϕ and ψ. By recalling the definition of these variables given in Eq. (3.6), taking $n = 0$, we readily get:

$$W(\phi, \psi) = E_{J0}\left[-2\,\cos\phi\,\cos\pi\psi + \frac{\pi}{2\beta}(\psi - \psi_{ex})^2 - i_B\phi \right]. \qquad (3.56)$$

Having developed a second-order perturbation analysis with respect to the parameter β in the previous section, we are able to express the variable ψ in terms of the average superconducting phase ϕ. In this way, we may define, to second order in the parameter β, the effective potential W_P in terms of ϕ, by writing:

$$W_P(\phi, \psi_{ex}) = E_{J0}[w_{P0} + 2\pi\beta\, w_{P1} + (2\pi\beta)^2 w_{P2}], \qquad (3.57)$$

where:

$$w_{P0} = w_{P0}(\phi, \psi_{ex}) = -2\cos\pi\psi_{ex}\cos\phi - i_B\phi, \qquad (3.58a)$$

$$w_{P1} = w_{P1}(\phi, \psi_{ex}) = -\sin^2\pi\psi_{ex}\cos^2\phi, \qquad (3.58b)$$

$$w_{P2} = w_{P2}(\phi, \psi_{ex}) = \cos\pi\psi_{ex}\sin^2\pi\psi_{ex}\cos^3\phi. \qquad (3.58c)$$

The above expression is a washboard potential, similar to the one seen for a single Josephson junction in Eq. (1.51), with additional higher-order terms in β. The effective potential W_P is represented in Fig. 3.5a–c as a function of the variable ϕ. In particular, in Fig. 3.5a W_P is shown for

$\beta = 0.1$ and $i_B = 1.0$ and for various values of the normalized external applies flux ψ_{ex}. In this figure the overall tilting of the curves is fixed, being given by the value of i_B. Furthermore, the energy barrier separating two adjacent minima in the potential decreases as ψ_{ex} goes from zero to 0.25, while it disappears in the curve with $\psi_{ex} = 0.5$, meaning that the system is in the running state. In Fig. 3.5b, the effective potential is shown for $\beta = 0.1$ and $\psi_{ex} = 0.25$ and for various values of the normalized bias current i_B. In this figure, the overall tilting of the curves increases for increasing values of i_B. Furthermore, even though the external magnetic flux is fixed, the energy barrier separating two adjacent minima in the potential decreases as i_B increases. Notice, however, that the three curves represent a system in the superconducting state, since $i_B < 2|\cos \pi \psi_{ex}|$ for all three values of i_B (0.0, 0.5, 1.0). In Fig. 3.5c, the effective potential is shown for two different value of the parameter β and for $\psi_{ex} = 0.25$ and $i_B = 0.5$. In this figure, one may notice that the strictly defined washboard potential, obtained for $\beta = 0$, is only slightly modified when the value of this parameter increases to $\beta = 0.1$.

Let us now show that, by starting with the expression for the washboard potential in Eq. (3.57), we may derive the equation of the motion (3.32) for the superconducting average phase difference ϕ.

By following a procedure similar to the one described above, we might invoke validity of the following power balance relation:

$$\frac{dW_P}{dt} + \frac{2V^2}{R} = 0. \tag{3.59}$$

Notice now that:

$$\frac{dW_P}{dt} = E_{J0}\left[\frac{\partial w_{P0}}{\partial \phi} + 2\pi\beta \frac{\partial w_{P1}}{\partial \phi} + (2\pi\beta)^2 \frac{\partial w_{P2}}{\partial \phi}\right]\dot\phi, \tag{3.60}$$

and that

$$\frac{2V^2}{R} = \frac{\Phi_0^2}{2\pi^2 R}\dot\phi^2, \tag{3.61}$$

where the dot on ϕ stands for "derivative with respect to t".

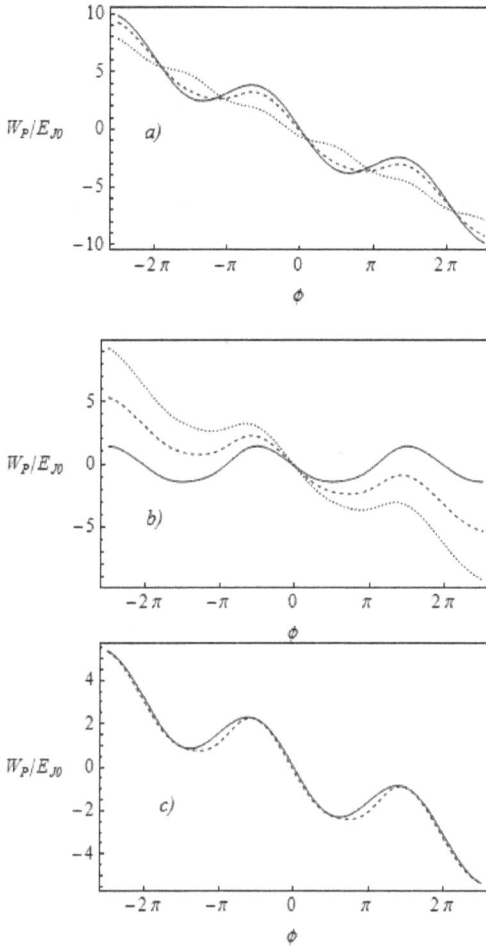

Fig. 3.5. Effective potential W_P as a function of the average superconducting phase difference ϕ. In panel a, the values of the fixed parameters are $\beta = 0.1$ and $i_B = 1.0$, while the three curves are obtained for the following values of the normalized external magnetic flux: $\psi_{ex} = 0.00$ (full line), $\psi_{ex} = 0.25$ (dashed line), $\psi_{ex} = 0.50$ (dotted line). In panel b, the values of the fixed parameters are $\beta = 0.1$ and $\psi_{ex} = 0.25$, while the three curves are obtained for the following values of the normalized bias current: $i_B = 0.0$ (full line), $i_B = 0.5$ (dashed line), $i_B = 1.0$ (dotted line). In panel c, the values of the fixed parameters are $i_B = 0.5$ and $\psi_{ex} = 0.25$, while the two curves are obtained for the following values of the parameter β: $\beta = 0.0$ (full line), $\beta = 0.1$ (dashed line).

By now using Eq. (3.60) and (3.61), Eq. (3.59) can be written, for $\dot{\phi} \neq 0$, as follows:

$$\frac{d\phi}{d\tau} = -\frac{1}{2}\frac{\partial w_{P0}}{\partial \phi} - \beta\left(\pi\frac{\partial w_{P1}}{\partial \phi}\right) - \beta^2\left(2\pi^2\frac{\partial w_{P2}}{\partial \phi}\right), \qquad (3.62)$$

where the normalized time τ has been introduced. By calculating the derivatives with respect to ψ in Eq. (3.62), we have:

$$\frac{1}{2}\frac{\partial w_{P0}}{\partial \phi} = \cos \pi\psi_{ex} \sin \phi - \frac{i_B}{2}, \qquad (3.63a)$$

$$\pi\frac{\partial w_{P1}}{\partial \phi} = 2\pi \sin^2 \pi\psi_{ex} \cos \phi \sin \phi, \qquad (3.63b)$$

$$2\pi^2\frac{\partial w_{P2}}{\partial \phi} = -6\pi^2 \cos \pi\psi_{ex} \sin^2 \pi\psi_{ex} \cos^2 \phi \sin \phi. \qquad (3.63c)$$

By now substituting in Eq. (3.62) the results obtained in Eq. (3.63a)-(3.63c) and by comparing the final outcome of this operation with Eq. (3.32), we may complete our proof.

It is now very instructive to utilize the potential W_P in Eq. (3.57) to derive the current $i_S = i_1 - i_2$, already obtained in the previous section. In fact, by defining:

$$i_S = -\frac{1}{\pi E_{J0}}\frac{\partial W_P}{\partial \psi_{ex}}, \qquad (3.64)$$

by means of Eq. (3.57), we may calculate the partial derivative of W_P with respect to ϕ (it suffices to take the expression of the effective potential to first order in β) obtaining:

$$i_S = -2 \sin \pi\psi_{ex} \cos \phi + 2\pi\beta \sin 2\pi\psi_{ex} \cos^2 \phi, \qquad (3.65)$$

which is identical to Eq. (3.35a).

3.4. Multi-junction superconducting systems

In the first section of the present Chapter, we have seen how to write the fluxoid quantization equation for a superconducting ring interrupted by two Josephson junctions. In this relation, the flux threading the superconducting loop can be expressed in terms of the gauge-invariant superconducting phase differences across the Josephson junctions, as specified in Eq. (3.1). An analogous relation can be derived by a multi-junction loop. On the same token, by introducing the RSJ model for all junctions in the ring, as done in the case of a two-junction interferometer, the dynamic properties of all superconducting phase differences can be defined. Therefore, the magnetic response of multi-junction quantum interferometers can be studied according to the same basic principles valid for superconducting ring interrupted by two Josephson junctions.

In this section, we shall first consider the magnetic properties of a superconducting ring interrupted by N Josephson junctions. We shall successively study the magnetic behavior of a current-biased superconducting loop with N_1 junctions in one branch and N_2 junctions in the second. Particular attention will be devoted to the special case of a superconducting quantum interferometer having N junctions in each branch.

3.4.1. *A multi-junction superconducting ring*

Consider a superconducting ring interrupted by N Josephson junctions, which we may label with the index $k = 1, 2, \ldots, N$. A schematic representation of this system is given in Fig. 3.6.

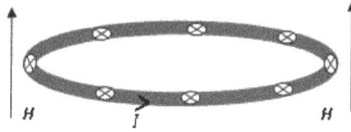

Fig. 3.6. Schematic representation of a superconducting ring interrupted by N Josephson junctions shown as crosses. The current I is shown to flow in the positive direction, according to the right hand rule. The externally applied field **H** is orthogonal to the plane in which the device lies.

Proceeding as in Section 3.1, we may start by writing the fluxoid quantization condition for the system, so that

$$\frac{2\pi}{\Phi_0} \Phi + \sum_{k=1}^{N} \phi_k = 2\pi n, \qquad (3.66)$$

where Φ is the flux threading the superconducting loop, ϕ_k is the gauge-invariant superconducting phase differences across the k-th junction in the loop, and n is an integer. Let us also write the electrodynamic equation defining the flux Φ threading the superconducting ring as the sum of the induced flux and the applied flux Φ_{ex} as follows:

$$\Phi = \Phi_{ex} + LI, \tag{3.67}$$

where the L is the inductance parameter of the superconducting loop and I is the persistent current flowing in the loop itself. The dynamic equation for each Josephson junction in the loop can be written by means of the RSJ model (Barone, 1982). We may write one differential equation for each Josephson junction, so that, according to Eq. (2.44), we have:

$$\frac{\Phi_0}{2\pi R_k}\frac{d\phi_k}{dt} + I_{Jk}\sin\phi_k = I, \tag{3.68}$$

where $k = 1, 2, \ldots, N$ and I_{Jk} and R_k are the maximum Josephson current and the resistive parameter of the k-th junction. For the sake of simplicity and clarity we here assume the system to be homogeneous in the junction parameters $(I_{Jk} = I_{J0},\ R_k = R)$. By this simplifying assumption, by relying on the ordinary definition of the normalized time

$$\tilde{\tau} = \frac{2\pi RI_{J0}}{\Phi_0}t, \tag{3.69}$$

we may rewrite Eq. (3.68) as follows:

$$\frac{d\phi_k}{d\tilde{\tau}} + \sin\phi_k = i, \tag{3.70}$$

where $k = 1, 2, \ldots, N$. From Eq. (3.70) we see that the same forcing term acts on each junction. Therefore, when the system is in its superconducting state $(i < 1)$, all junctions have the same static solution, namely:

$$\phi_k = \sin^{-1}i, \tag{3.71}$$

$k = 1, 2, \ldots, N$. On the other hand, for the running-state solution, we recall Eq. (1.44), from which we may argue that all superconducting phases evolve synchronously. In fact, being the forcing term common to all superconducting phases and starting the latter from a common static solution, if the current i is initially slowly varied, the Josephson junctions can be seen to be phase-locked. Therefore, it suffices to know the dynamics of only one representative junction to describe the dynamics of the entire system. As done in Chapter 2, where the flux dynamics was obtained by means of the fluxoid quantization relation, by taking $n = 0$ in Eq. (3.66) and by defining $\psi = \Phi/\Phi_0$, we have

$$\phi = -\frac{2\pi}{N}\psi, \tag{3.72}$$

where ϕ is the gauge-invariant superconducting phase difference across the representative junction. We can thus rewrite Eq. (3.72) in terms of the variable ψ and of an appropriately normalized time $\tau = NRt/L$ as follows:

$$\frac{d\psi}{d\tau} + \beta \sin\left(2\pi \frac{\psi}{N}\right) + \psi = \psi_{ex}, \tag{3.73}$$

where the parameter β is given as usual, by the ratio LI_{J0}/Φ_0. The above equation reduce to Eq. (2.45) if we divide by N, so that:

$$\frac{d}{d\tau}\left(\frac{\psi}{N}\right) + \frac{\beta}{N} \sin\left(2\pi \frac{\psi}{N}\right) + \frac{\psi}{N} = \frac{\psi_{ex}}{N}. \tag{3.74}$$

In this way, the multi-junction loop, when identical Josephson junctions are considered, behave in the same way as a single-junction loop, provided we rescale, in the equation of motion (2.45), the normalized flux variable ψ, the normalized applied flux ψ_{ex}, and the parameter β by $1/N$. The magnetic properties of this system can thus be obtained by the same type of analysis performed in Section 2.4.

3.4.2. *Current-biased superconducting loops*

Consider now a current-biased superconducting loop with N_1 over-damped junctions in one branch and N_2 over-damped junctions in the second, as shown in Fig. 3.7.

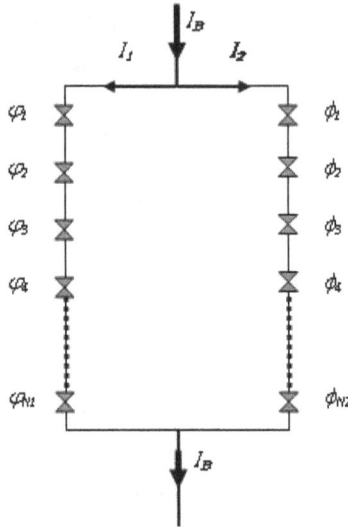

Fig. 3.7. Schematic representation of a multi-junction quantum interferometer. The bias current I_B divides into two branch currents, I_1 and I_2. The applied field (not shown in the figure) is orthogonal to the plane in which the device lies and is directed toward the reader.

The identical point-like over-damped Josephson junctions in the system are assumed to have resistive and coupling parameters R and I_{J0}, respectively. More general cases have been considered in the literature (Lewandowski, 1991). As shown in Fig. 3.7, the gauge-invariant superconducting phase differences across the Josephson junctions on the left branch and on the right branch of the device, as seen by the reader, are denoted as φ_k and ϕ_l, respectively, with $k = 1, 2, \ldots N_1$, and $l = 1, 2, \ldots N_2$. A bias current I_B is injected in the system, splitting into the two branch currents: I_1 in the left branch and I_2 in the right branch, so that: $I_B = I_1 + I_2$. Since the same current flows through all Josephson junctions on each branch, and given that all junctions are perfectly identical, we may consider all junctions on the left and all junctions on the right branch to be in two different phase-locked states. Therefore, for the same reasons set forth in the previous section, we may write:

$$\varphi_k = \varphi, \tag{3.75a}$$

$$\phi_l = \phi, \tag{3.75b}$$

for $k = 1, 2, \dots N_1$, and $l = 1, 2, \dots N_2$. Let now the inductance of the first and second branch to be $L_1 = N_1 L_0$ and $L_2 = N_2 L_0$, respectively. In this way, we may define $\beta_0 = L_0 I_{J0}/\Phi_0$. Let us also introduce the usual normalized quantities: $i_j = I_j/I_{J0}$, $j = 1, 2$, (normalized branch currents) and $\psi_{ex} = \Phi_{ex}/\Phi_0$ (normalized applied flux). By setting the total flux Φ threading the superconducting loop equal to the sum of the externally applied flux Φ_{ex} and of the induced flux $L_1 I_1 - L_2 I_2$, we have, in normalized form:

$$\psi = \psi_{ex} + \beta_0(N_1 i_1 - N_2 i_2), \tag{3.76}$$

where $\psi = \Phi/\Phi_0$. By fluxoid quantization, considering Eq. (3.75a) and (3.75b), we may write:

$$2\pi\psi + N_1\varphi - N_2\phi = 2\pi n, \tag{3.77}$$

where n is an integer, which we shall take equal to zero for simplicity. With the aid of the RSJ model (Barone, 1982) for the over-damped Josephson junctions in the device and in the absence of noise, the dynamic equations for the variables φ and ϕ can be written as follows:

$$\frac{d\varphi}{d\tau} + \sin\varphi = i_1, \tag{3.78a}$$

$$\frac{d\phi}{d\tau} + \sin\phi = i_2. \tag{3.78b}$$

Let us now define the following variables:

$$\phi_A = \frac{\varphi + \phi}{2}, \tag{3.79a}$$

$$\psi = \frac{N_2\phi - N_1\varphi}{2\pi}, \tag{3.79b}$$

where (3.79a) defines the average superconducting phase between the representative variables φ and ϕ, while (3.79b) is the definition of the normalized flux ψ threading the superconducting loop, as obtained from the fluxoid quantization relation. By rather obvious algebraic manipulations of Eq. (3.78a) and (3.78b), we may write

$$\frac{d\phi_A}{d\tau} + \cos\left(\frac{\varphi-\phi}{2}\right)\sin\phi_A = \frac{i_1+i_2}{2},$$ (3.80a)

$$2\pi\frac{d\psi}{d\tau} + N_2\sin\phi - N_1\sin\varphi = N_2 i_2 - N_1 i_1.$$ (3.80b)

By solving for φ and ϕ in Eq. (3.79a) and (3.79b) in terms of ϕ_A and ψ, by recalling Eq. (3.76) and the charge conservation relation $i_B = i_1 + i_2$, we may finally write Eq. (3.80a) and (3.80b) as follows:

$$\frac{d\phi_A}{d\tau} + \cos\left(\frac{(N_1-N_2)\phi_A+2\pi\psi}{N_1+N_2}\right)\sin\phi_A = \frac{i_B}{2},$$ (3.81a)

$$2\pi\frac{d\psi}{d\tau} + N_2\sin\left(\frac{2N_1\phi_A+2\pi\psi}{N_1+N_2}\right) - N_1\sin\left(\frac{2N_2\phi_A-2\pi\psi}{N_1+N_2}\right) = \frac{\psi_{ex}-\psi}{\beta_0}.$$ (3.81b)

It is possible to derive an effective model for the system by solving Eq. (3.81b) by means of a second-order expansion in the parameter β_0, as seen in Section 3.2. However, the calculations are rather cumbersome and lengthy. An immediate result, in terms of the analysis already carried out, can be obtained by considering $N_1 = N_2 = N$. In this case, the system becomes completely symmetric and the equation of the motion for of ϕ_A and ψ are written as follows:

$$\frac{d\phi_A}{d\tau} + \cos\left(\frac{\pi\psi}{N}\right)\sin\phi_A = \frac{i_B}{2},$$ (3.82a)

$$\pi\frac{d}{d\tau}\left(\frac{\psi}{N}\right) + \cos\phi_A\sin\left(\frac{\pi\psi}{N}\right) + \frac{\psi}{2N\beta_0} = \frac{\psi_{ex}}{2N\beta_0}.$$ (3.82b)

We may now realize that the above equations are similar to Eq. (3.8a) and (3.8b). In fact, by setting in the latter

$$\psi \to \frac{\psi}{N}, \psi_{ex} \to \frac{\psi_{ex}}{N},$$ (3.83)

we might immediately obtain the solution to Eq. (3.82b), by modifying Eq. (3.31) according to the substitutions suggested in Eq. (3.83), so that, to second-order in the parameter β_0, we have:

$$\psi = \psi_{ex} - 2N\beta_0\sin\frac{\pi\psi_{ex}}{N}\cos\phi_A + 2\pi N\beta_0^2\sin\frac{2\pi\psi_{ex}}{N}\cos^2\phi_A.$$ (3.84)

According to the analysis carried out in Section 3.2 on the effective one-junction model, taking into account Eq. (3.83), the adiabatic solution gives:

$$\cos\phi = \text{sign}\left(\cos\frac{\pi\psi_{ex}}{N}\right)\sqrt{1 - \left(\frac{i_B}{2\cos\frac{\pi\psi_{ex}}{N}}\right)^2} + \frac{\pi\beta i_B^2 \tan^2\frac{\pi\psi_{ex}}{N}}{2\cos\frac{\pi\psi_{ex}}{N}}. \qquad (3.85)$$

By again considering the substitutions indicated in Eq. (3.83), the persistent current can be expressed by means of (3.24a) as follows

$$i_s = i_{s0} + 2\pi\beta_0 \sin\frac{2\pi\psi_{ex}}{N} - \frac{\pi\beta_0 i_B^2 \sin\frac{\pi\psi_{ex}}{N}}{\cos^3\frac{\pi\psi_{ex}}{N}}, \qquad (3.86)$$

with

$$i_{s0} = -2\sin\frac{\pi\psi_{ex}}{N}\text{sign}\left(\cos\frac{\pi\psi_{ex}}{N}\right)\sqrt{1 - \left(\frac{i_B}{2\cos\frac{\pi\psi_{ex}}{N}}\right)^2}. \qquad (3.87)$$

On the other hand, by substituting Eq. (3.84) in Eq. (3.24b), we obtain the magnetic susceptibility to second order in β_0:

$$\chi = \chi_0 + 2\pi\beta_0\,\chi_1 + (2\pi\beta_0)^2\chi_2, \qquad (3.88)$$

with

$$\chi_0 = \alpha - 1, \qquad (3.89a)$$

$$\chi_1 = -\alpha\,\text{sign}\left(\cos\frac{\pi\psi_{ex}}{N}\right)\frac{\sin\frac{\pi\psi_{ex}}{N}}{\frac{\pi\psi_{ex}}{N}}\sqrt{1 - \left(\frac{i_B}{2\cos\frac{\pi\psi_{ex}}{N}}\right)^2}. \qquad (3.89b)$$

$$\chi_2 = -\alpha\left[\frac{\tan\frac{\pi\psi_{ex}}{N}}{\frac{\pi\psi_{ex}}{N}}\left(\frac{i_B}{2\cos\frac{\pi\psi_{ex}}{N}}\right)^2 - \frac{\sin\frac{2\pi\psi_{ex}}{N}}{\frac{2\pi\psi_{ex}}{N}}\right]. \qquad (3.89c)$$

After having obtained the above results, we may see that it suffices to make the substitution $\psi_{ex} \to \psi_{ex}/N$ in Eq. (3.44)-(3.45) and in Eq. (3.46)-(3.47a-c) in order to get the persistent current i_s and the magnetic susceptibility χ, respectively, for a multi-junction quantum inter-ferometer with exactly N junctions on each branch. In Fig. 3.8a and 3.8b, the persistent currents i_S for $N = 2$ and $N = 3$ are shown, respectively, as a function of ψ_{ex} for $\beta = 0.02$ and various values of the bias current.

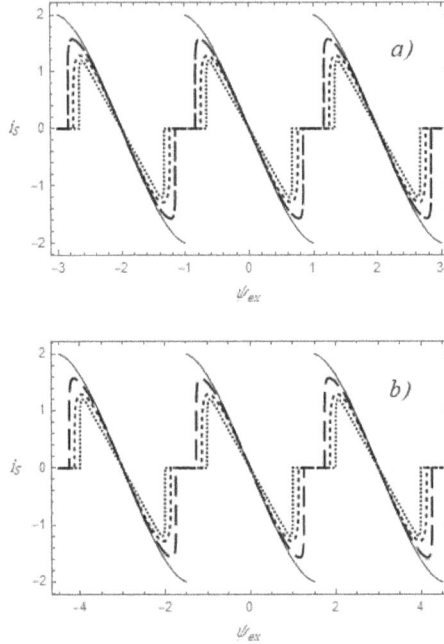

Fig. 3.8. Persistent current i_S in a current-biased loop containing N junction in each branch as a function of the normalized applied flux ψ_{ex}. In both panels the value of the parameter β is fixed to 0.02, while the four curves are obtained for the following values of the bias current: $i_B = 0.00$ (full line), $i_B = 0.50$ (coarsely dashed line), $i_B = 0.75$ (dashed line), $i_B = 1.00$ (dotted line). In panel a the number of junctions in each branch is $N = 2$, in panel b, on the other hand, $N = 3$. These curves are identical to the corresponding curves in Fig. 3.3a, apart the rescaling due to the substitution $\psi_{ex} \rightarrow \psi_{ex}/N$.

Comparing the curves in Fig. 3.3a, 3.8a, and 3.8b, we notice that, apart from the rescaling due to the substitution $\psi_{ex} \rightarrow \psi_{ex}/N$, these curves are identical, as it was to be expected. In this way, the discontinuities closer to the origin due to the term $\text{sign}(\cos \pi \psi_{ex})$, appearing at $\psi_{ex} = \pm 1/2$, in the curves in Fig. 3.3a are shifted to $\psi_{ex} = \pm 1$ in Fig. 3.8a, where $N = 2$, and to $\psi_{ex} = \pm 3/2$ in Fig. 3.8b, where $N = 3$.

In Fig. 3.9a and 3.9b, the magnetic susceptibility curves are shown as a function of ψ_{ex} for $N = 2$ and $N = 3$, respectively. In both panels, we have chosen $\alpha = 0.95$, $\beta = 0.02$, and $i_B = 0.00$, 0.50, 1.00. All the χ vs. ψ_{ex} curves are seen to be piecewise continuous.

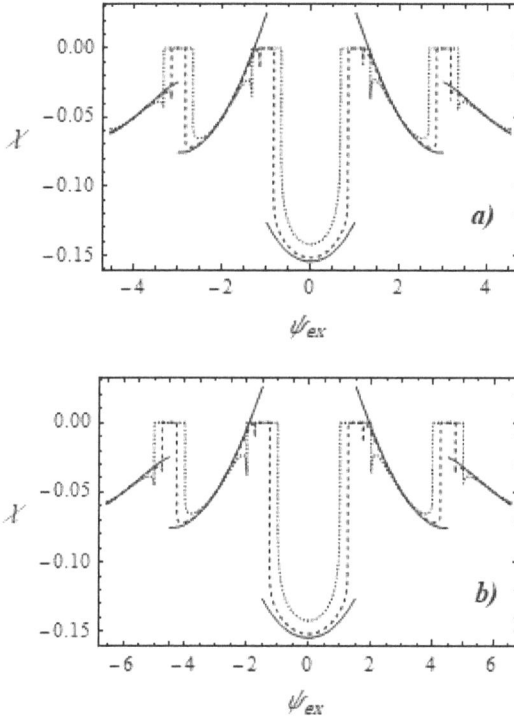

Fig. 3.9. Magnetic susceptibility χ in a current-biased loop containing N junction in each branch as a function of the normalized applied flux ψ_{ex} for $\alpha = 0.95$. In both panels, the value of the parameter β is fixed to 0.02, while the three curves are obtained for the following values of the bias current: $i_B = 0.00$ (full line), $i_B = 0.50$ (dashed line), $i_B = 1.00$ (dotted line). In panel a, the number of junctions in each branch is $N = 2$, in panel b, on the other hand, $N = 3$. These curves are identical to the corresponding curves in Fig. 3.4a, apart the rescaling due to the substitution $\psi_{ex} \rightarrow \psi_{ex}/N$.

As in the case of the i_S vs. ψ_{ex} curves in Fig. 3.8a-b, the susceptibility curves in Fig. 3.9a and 3.9b are identical, apart from the rescaling due to the substitution $\psi_{ex} \rightarrow \psi_{ex}/N$. Therefore, for $N = 2$, the discontinuities closer to the origin due to the term $\text{sign}(\cos \pi \psi_{ex})$ appear at $\psi_{ex} = \pm 1$ in Fig. 3.9a. On the other hand, for $N = 3$, the same discontinuities appear at $\psi_{ex} = \pm 3/2$ in Fig. 3.9b. Finally, we notice that, having taken $\alpha = 0.95$, a paramagnetic response appears in a small range of the normalized applied flux ψ_{ex} in the vicinity of $\psi_{ex} = N/2$.

Chapter 4

Thermal Fluctuations in Josephson Junctions and Superconducting Quantum Interference Devices

4.1. Thermal fluctuations in Josephson junctions

In the previous sections we have studied the magnetic behavior of single Josephson junctions and of two- and multi-junction superconducting quantum interference devices in the absence of noise. We have seen that, for small values of the parameters β and β_0, opportunely defined in the previous section for the two- and multi-junction superconducting quantum interference devices, respectively, the dynamics of both systems can be studied by means of a single-junction model. Therefore, owing to the possibility of using a single-junction description for these types of devices, we can, in principle, make use of the classical Ambegaokar and Halperin (Ambegaokar, 1969) analysis, when these systems are in the presence of non-negligible noise effects. We therefore start by considering the Ambegaokar and Halperin (AH) method in details and we successively specialize this approach to two-junction and multi-junction superconducting quantum interference devices.

In order to analyze the effect of thermal fluctuations in a Josephson junction, we need to first consider the dynamical mechanisms induced by these types of perturbations on the physical system under exam. Therefore, we may try to give a qualitative picture of these effects, before applying the AH method for a quantitative solution of the problem. For this purpose, consider the single Josephson junction washboard potential given in Eq. (1.51). Let us express this potential as a function of the variable ϕ and of the current i_B flowing in the device and normalized to the maximum Josephson current I_{J0} as follows:

$$W(\phi) = (1 - \cos\phi) - i_B\phi. \qquad (4.1)$$

By what is seen in section 1.5, the Josephson junction presents metastable superconducting states for $i_B < 1$. In the absence of noise, the system, depending on its history, lies exactly in one of the many minima of the potential $W(\phi)$. Therefore, under these conditions, the superconducting phase ϕ is a constant and the voltage V across the junction is null. Moreover, we have noticed that, as the normalized current i increases, the energy barrier separating two adjacent minima in the potential $W(\phi)$ decreases. The value of the energy barrier related to the potential $W(\phi)$ is the following: $W_B = 2\sqrt{1 - i_B^2} - 2i_B \cos^{-1} i_B$. The resistive (or running) state can thus be attained for a value of the bias current i_B greater than one. In this state, the junction dynamics evolves over the flattened $W(\phi)$ curve with an oscillating values of V, whose behavior can be argued, for instance, from Fig. 1.7. As said before, the measuring instruments are able to detect the time average value $\langle V \rangle$ of the voltage V. Being $\langle V \rangle \neq 0$, a finite voltage state is realized in the junction. The i_B vs. $\langle V \rangle$ curves are given by Eq. (1.48).

We can now ask what type of analysis needs to be set forth in the presence of thermal noise. For $i_B < 1$, when thermal fluctuations are not negligible, the junction dynamics cannot be described only by defining a single metastable state corresponding to the minimum of a single well. In addition to this, for normalized currents i_B approaching the value $i_B = 1$, a phase jump from one well to another could take place. This effect can be mimicked by introducing a white-noise normalized current i_f in addition to the bias current i_B, so that the deterministic equation governing the dynamics of the Josephson junction at a finite temperature T becomes:

$$\frac{d\phi}{d\tau} + \sin\phi = i_B + i_f(\tau), \qquad (4.2)$$

where $\tau = 2\pi R I_{J0} t / \Phi_0$, R being the resistive parameter of the junction, as in Eq. (3.4) and $\langle i_f(\tau) \rangle = 0$ and $\langle i_f(\tau) i_f(\tau') \rangle = 2k_B T \delta(\tau - \tau')/E_J$, with $E_J = I_{J0}\Phi_0/2\pi$, where the triangular brackets can be considered as statistical averages. Equation (4.2) can be written in the form of the Langevin equation (Kramers, 1940), so that:

$$\frac{d\phi}{d\tau} = -\frac{\partial W}{\partial \phi} + i_f(\tau), \tag{4.3}$$

where W is given in Eq. (4.1) and ϕ can be considered a stochastic process. In the realm of the general theory of Brownian motions (Kramers, 1940), Eq. (4.3) gives rise to the following Fokker-Planck equation for the probability density $p(\phi, \tau)$:

$$\frac{\partial p}{\partial \tau} = \frac{\partial}{\partial \phi}\left(p(\phi, \tau)\frac{\partial W}{\partial \phi}\right) + \Gamma\frac{\partial^2 p}{\partial \phi^2}, \tag{4.4}$$

where $\Gamma = k_B T/E_J$ and the probability density function $p(\phi, \tau)$ is normalized according to the following relation:

$$\int_0^{2\pi} p(\phi, \tau)d\phi = 1. \tag{4.5}$$

One can therefore look for a stationary solution of Eq. (4.4), by setting

$$\frac{\partial}{\partial \phi}\left(p(\phi)\frac{\partial W}{\partial \phi} + \Gamma\frac{\partial p}{\partial \phi}\right) = 0, \tag{4.6}$$

or, equivalently,

$$p(\phi)\frac{\partial W}{\partial \phi} + \Gamma\frac{\partial p}{\partial \phi} = \lambda, \tag{4.7}$$

where λ is a constant. A general solution to the above equation can be seen to be the following:

$$p(\phi) = \frac{1}{2\Gamma}e^{-\mu(\phi)}\left[2\lambda\int_0^\phi d\xi\, e^{\mu(\xi)} + B\right], \tag{4.8}$$

where B is a constant and:

$$\mu(\phi) = \frac{1}{\Gamma}\int_0^\phi d\xi\, \frac{\partial W}{\partial \xi} = \frac{W(\phi)}{\Gamma}. \tag{4.9}$$

In this way, we may obtain a formal solution to the stationary equation (4.7) by writing:

$$p(\phi) = \frac{\lambda}{\Gamma} e^{-\frac{W(\phi)}{\Gamma}} \left[\int_0^{\phi} d\xi \, e^{\frac{W(\xi)}{\Gamma}} - C \right].$$ (4.10)

The constant C can be found by imposing periodicity of the function $p(\phi)$, namely, by setting $p(0) = p(2\pi)$, so that:

$$C = \frac{F(2\pi)}{1 - e^{-2\pi i_B/\Gamma}},$$ (4.11)

where

$$F(\phi) = \int_0^{\phi} d\xi \, e^{\frac{W(\xi)}{\Gamma}}.$$ (4.12)

By using the normalization condition (4.5), we may now determine the constant λ, so that:

$$\lambda = \frac{\Gamma}{\int_0^{2\pi} d\phi \, e^{-\frac{W(\phi)}{\Gamma}} [F(\phi) - C]}.$$ (4.13)

In this way, the probability density function will take the following final form:

$$p(\phi) = \frac{e^{-\frac{W(\phi)}{\Gamma}} [F(\phi) - C]}{\int_0^{2\pi} d\xi \, e^{-\frac{W(\xi)}{\Gamma}} [F(\xi) - C]}.$$ (4.14)

We have now acquired the necessary analytic instruments to determine the response of a Josephson junction in the presence of thermal noise. As an example, in the following section we shall consider how the i_B vs. $\langle V \rangle$ curves, given by Eq. (1.48), modify in the presence of thermal fluctuations.

4.2. Current-voltage curves in the presence of thermal noise

In the previous section we have studied the analytic properties of the one-junction problem in the presence of thermal noise. We now give an example of how this analytic effort can be translated into practical results, by studying the current-voltage curves of these devices. Let us then start by considering the statistical average $\langle v \rangle$ of the normalized voltage $v = d\phi/d\tau$ across the junction. Then, because of Eq. (4.3), we can write:

$$\langle v \rangle = \int_0^{2\pi} p(\phi) \left(-\frac{\partial W}{\partial \phi} \right) d\phi. \tag{4.15}$$

Using Eq. (4.7), we may now rewrite (4.15) as follows:

$$\langle v \rangle = \int_0^{2\pi} \left(\Gamma \frac{\partial p}{\partial \phi} - \lambda \right) d\phi. \tag{4.16}$$

The integral in (4.15) can be broken into two addenda, the first addendum being null because of the periodicity conditions on the stationary probability density function $p(\phi)$. In this way, one writes:

$$\langle v \rangle = -2\pi\lambda. \tag{4.17}$$

By Eq. (4.13) we may thus write:

$$\langle v \rangle = \frac{-2\pi\Gamma}{\int_0^{2\pi} d\phi \, e^{-\frac{W(\phi)}{\Gamma}[F(\phi)-C]}}. \tag{4.18}$$

We now proceed to calculate the integral at the denominator by means of a modified Bessel function expansion of the exponential of the cosine term in $W(\phi)$, by setting:

$$e^{\frac{\cos \phi}{\Gamma}} = \sum_{k=-\infty}^{+\infty} I_k \left(\frac{1}{\Gamma} \right) e^{ik\phi}. \tag{4.19}$$

Let us proceed in order, so that we first evaluate the function $F(\phi)$. By Eq. (4.1), (4.12), and (4.19), and by recalling the basic properties of the modified Bessel's functions I_k, we may write:

$$F(\phi) = e^{\frac{1}{\Gamma}} e^{-\frac{i_B\phi}{\Gamma}} [S_0(\phi) + S_1(\phi) + S_2(\phi)], \tag{4.20}$$

where

$$S_0(\phi) = \frac{\Gamma}{i_B} I_0 \left(\frac{1}{\Gamma} \right) \left(e^{\frac{i_B\phi}{\Gamma}} - 1 \right), \tag{4.21a}$$

$$S_1(\phi) = \frac{2i_B}{\Gamma} e^{\frac{i_B\phi}{\Gamma}} \sum_{k=1}^{+\infty} \frac{(-1)^k I_k\left(\frac{1}{\Gamma}\right)}{k^2 + \left(\frac{i_B}{\Gamma}\right)^2}, \tag{4.21b}$$

$$S_2(\phi) = \sum_{k=1}^{+\infty} \frac{(-1)^k I_k\left(\frac{1}{\Gamma}\right)}{k^2 + \left(\frac{i_B}{\Gamma}\right)^2} \left[\left(ik - \frac{i_B}{\Gamma}\right) e^{-ik\phi} - \left(ik + \frac{i_B}{\Gamma}\right) e^{ik\phi}\right]. \tag{4.21c}$$

Having calculated the function $F(\phi)$, we can evaluate the constant C in (4.11), obtaining:

$$C = \frac{\Gamma}{i_B} e^{1/\Gamma} \left[I_0\left(\frac{1}{\Gamma}\right) + 2\left(\frac{i_B}{\Gamma}\right)^2 \sum_{k=1}^{+\infty} \frac{(-1)^k I_k\left(\frac{1}{\Gamma}\right)}{k^2 + \left(\frac{i_B}{\Gamma}\right)^2}\right]. \tag{4.22}$$

The expression in parenthesis in Eq. (4.22) is a recurrent term, as we shall see, so that it is convenient to reserve the special symbol K_B for it, where the index B recalls its dependence on the bias current i_B. In this way, we set:

$$K_B = I_0\left(\frac{1}{\Gamma}\right) + 2\left(\frac{i_B}{\Gamma}\right)^2 \sum_{k=1}^{+\infty} \frac{(-1)^k I_k\left(\frac{1}{\Gamma}\right)}{k^2 + \left(\frac{i_B}{\Gamma}\right)^2}. \tag{4.23}$$

We now need to go a step further to calculate $\langle v \rangle$ as Eq. (4.18) prescribes. In fact, we need to evaluate the integral in the denominator of Eq. (4.18). Let us calculate this term a bit at a time, by first performing the easiest integral:

$$\mu_0 = F(2\pi) = \int_0^{2\pi} d\phi \, e^{-\frac{W(\phi)}{\Gamma}} = \frac{\Gamma}{i_B} e^{1/\Gamma} \left(e^{\frac{2\pi i_B}{\Gamma}} - 1\right) K_B. \tag{4.24}$$

The above result has been obtained by recalling the definition of C in Eq. (4.11) and by Eq. (4.24). The second integral in the denominator of Eq. (4.18) can also be calculated following the procedure we briefly outline here. Let us first call this integral as follows:

$$\mu_1 = \int_0^{2\pi} d\phi \, F(\phi) e^{-\frac{W(\phi)}{\Gamma}}. \tag{4.25}$$

Then recall that $F(\phi)$ has been broken into the three pieces as in Eq. (4.20), so that the same can be done with μ_1, by writing:

$$\mu_1 = e^{\frac{1}{\Gamma}} \int_0^{2\pi} d\phi \, [S_0(\phi) + S_1(\phi) + S_2(\phi)] e^{\frac{\cos\phi}{\Gamma}}. \qquad (4.26)$$

Therefore, we are left with three more tractable integrals to deal with. By recalling Eq. (4.21a-c) and by adopting the same strategy as in previous calculations, we find:

$$e^{\frac{1}{\Gamma}} \int_0^{2\pi} d\phi \, S_0(\phi) e^{\frac{\cos\phi}{\Gamma}} = \frac{\Gamma}{i_B} I_0\left(\frac{1}{\Gamma}\right) \left[e^{\frac{1}{\Gamma}} \mu_0 - 2\pi I_0\left(\frac{1}{\Gamma}\right) \right]. \quad (4.27\text{a})$$

$$e^{\frac{1}{\Gamma}} \int_0^{2\pi} d\phi \, S_1(\phi) e^{\frac{\cos\phi}{\Gamma}} = 2\frac{i_B}{\Gamma} e^{\frac{1}{\Gamma}} \mu_0 \sum_{k=1}^{+\infty} \frac{(-1)^k I_k\left(\frac{1}{\Gamma}\right)}{k^2 + \left(\frac{i_B}{\Gamma}\right)^2}. \quad (4.27\text{b})$$

$$e^{\frac{1}{\Gamma}} \int_0^{2\pi} d\phi \, S_2(\phi) e^{\frac{\cos\phi}{\Gamma}} = -4\pi \frac{i_B}{\Gamma} \sum_{k=1}^{+\infty} \frac{(-1)^k I_k^2\left(\frac{1}{\Gamma}\right)}{k^2 + \left(\frac{i_B}{\Gamma}\right)^2}. \quad (4.27\text{c})$$

In this way, by summing up the three pieces, we have:

$$\mu_1 = \frac{\Gamma}{i_B} e^{\frac{1}{\Gamma}} \mu_0 K_B - 2\pi \frac{\Gamma}{i_B} \left[I_0^2\left(\frac{1}{\Gamma}\right) + 2\left(\frac{i_B}{\Gamma}\right)^2 \sum_{k=1}^{+\infty} \frac{(-1)^k I_k^2\left(\frac{1}{\Gamma}\right)}{k^2 + \left(\frac{i_B}{\Gamma}\right)^2} \right]. \quad (4.28)$$

Notice that, in Eq. (4.28) the term K_B appears again, while a different type of summation is also present in parenthesis. Let us then determine the denominator of (4.18) by writing:

$$\mu_1 - C\mu_0 = -2\pi \frac{\Gamma}{i_B} \left[I_0^2\left(\frac{1}{\Gamma}\right) + 2\left(\frac{i_B}{\Gamma}\right)^2 \sum_{k=1}^{+\infty} \frac{(-1)^k I_k^2\left(\frac{1}{\Gamma}\right)}{k^2 + \left(\frac{i_B}{\Gamma}\right)^2} \right]. \quad (4.29)$$

The average voltage is thus

$$\langle v \rangle = \frac{i_B}{I_0^2\left(\frac{1}{\Gamma}\right) + 2\left(\frac{i_B}{\Gamma}\right)^2 \sum_{k=1}^{+\infty} \frac{(-1)^k I_k^2\left(\frac{1}{\Gamma}\right)}{k^2 + \left(\frac{i_B}{\Gamma}\right)^2}}. \qquad (4.30)$$

which gives the $\langle v \rangle$ vs. i_B curves solely in terms of the parameter Γ. A representation of these curves for various values of Γ is given in Fig. 4.1. We notice that for increasing values of the parameter Γ, i.e., for increasing value of the temperature T, the curves tend to move away from the $T = 0$ K behavior, in such a way that the normalized maximum

Josephson current i_{max} of the device tends to decrease for increasing values of Γ. In fact, while we can estimate, by simple inspection of Fig. 4.1, values of $i_{max} = 0.7$ for $\Gamma = 0.1$, and of $i_{max} = 0.5$ for $\Gamma = 0.2$, the maximum Josephson current of the junction can be seen to go down to about 0.3 for $\Gamma = 0.3$.

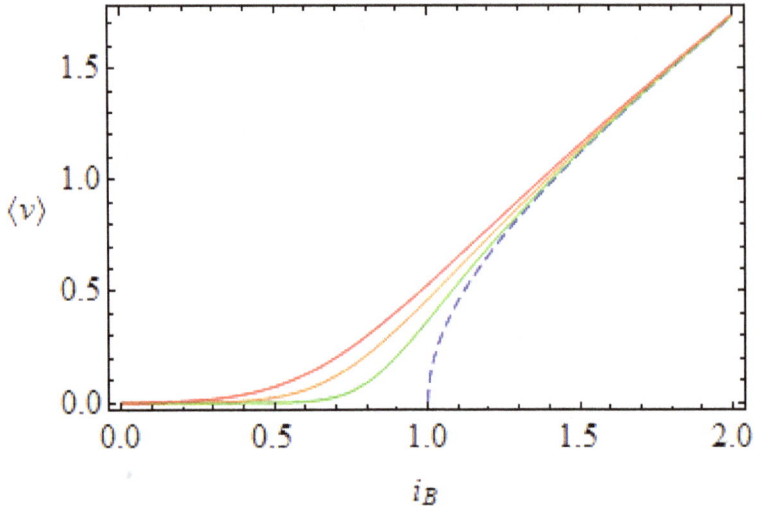

Fig. 4.1. Voltage-current characteristics of a single Josephson junction in the presence of thermal noise. The parameter Γ has been chosen as follows: $\Gamma = 0$ (dashed line), $\Gamma = 0.10$ (green line), $\Gamma = 0.20$ (orange line), $\Gamma = 0.30$ (red line). In the numerical evaluation of the curves, the series expansion in terms of modified Bessel function has been truncated at the integer $k_{max} = 200$.

4.3. The d.c. SQUID model in the presence of thermal fluctuations

In the previous section we have studied the analytic properties of the one-junction model in the presence of thermal noise. Here we develop a similar analysis for a superconducting quantum interference device (d. c. SQUID), based on essentially the same type of approach. Therefore, we shall first write down the equations for the two superconducting phases, ϕ_1 and ϕ_2, for the two junction of a d. c. SQUID with addition of a stochastic current term added to each of the two equations. By considering again Eq. (3.3a) and (3.3b), we may thus write:

$$\frac{\Phi_0}{2\pi R_1}\frac{d\phi_1}{dt} + I_{J1}\sin\phi_1 = I_1 + I_{f1}(t), \tag{4.31a}$$

$$\frac{\Phi_0}{2\pi R_2}\frac{d\phi_2}{dt} + I_{J2}\sin\phi_2 = I_2 + I_{f2}(t), \tag{4.31b}$$

where I_{J1} and R_1, I_{J2} and R_2 are the maximum Josepshon currents and the resistive parameters of the first and the second junction, respectively. As done in the previous chapter, we consider the system to be symmetric, so that the branch inductances are equal, or $L_1 = L_2 = L$, and homogeneous in the junction parameters, so that $I_{J1} = I_{J2} = I_{J0}$, and $R_1 = R_2 = R$. We may now consider, as in the single-junction case, ϕ_1 and ϕ_2 to be regarded as stochastic processes. The currents $I_{f1}(t)$ and $I_{f2}(t)$ represent two independent white noise terms such that

$$\langle I_{fk} \rangle = 0 \tag{4.32a}$$

$$\langle I_{fk}(t)I_{fm}(t') \rangle = 2\frac{k_B T}{R}\delta_{km}\,\delta(t-t') \tag{4.32b}$$

for $k, m = 1, 2$, where δ_{km} is the Kronecker delta. Here the triangular brackets can be again considered as statistical averages.

By considering the sets of equations (4.31a-b), assuming properties (4.32a-b) for the currents $I_{f1}(t)$ and $I_{f2}(t)$, Greenberg (Greenberg, 2001) and Chesca (Chesca, 1998) have developed a detailed analytic description of the electrodynamic response of d. c. SQUID by means of a two-dimensional Fokker-Plank equation (FPE). Here we point out that, in the limit of small values of the parameter $\beta = L I_{J0}/\Phi_0$, besides the possibility of adopting a perturbation approach to the two-dimensional FPE, one could distinguish between two types of dynamic behaviors: one fast, the other slow, which would allow to treat the two equations independently. Indeed, by considering the slowly varying process, which appears as frozen in the dynamic equation for the fast process, we are able to apply a similar approach as in the Ambegaokar and Halperin (Ambegaokar, 1969) analysis of a single Josephson junction in the presence of noise. In this way, a one-dimensional FPE can be adopted in solving for the probability density function of the process. Therefore, by restricting the system studied by Grønbech-Jensen et al. (Grønbech-Jensen, 2003) to overdamped junctions in the interferometer loop, we may extend the analysis carried out in absence of noise.

By now defining the normalized time as usual

$$\tau = \frac{2\pi R I_{J0}}{\Phi_0} t, \tag{4.33}$$

and by setting, $i_1 = I_1/I_{J0}$, $i_2 = I_2/I_{J0}$, $i_{f1} = I_{f1}/I_{J0}$, $i_{f2} = I_{f2}/I_{J0}$, we may rewrite Eq. (4.31a) and (4.31b) as follows:

$$\frac{d\phi_k}{d\tau} + \sin\phi_k = i_k + i_{fk}(\tau), \tag{4.34}$$

with $k = 1, 2$. By Dirac's delta properties, because of time rescaling, the correlation in Eq. (4.32b) needs to be rewritten as follows:

$$\langle i_{fk}(\tau) i_{fm}(\tau') \rangle = 2\Gamma \delta_{km} \, \delta(\tau - \tau'), \tag{4.35}$$

where $\Gamma = k_B T / E_J$ as before. Notice now that the definition of the flux Φ linked to the superconducting loop in terms of the normalized applied flux $\psi_{ex} = \Phi_{ex}/\Phi_0$ and of the normalized currents i_k, give, in normalized form:

$$\psi = \psi_{ex} + \beta(i_1 - i_2), \tag{4.36}$$

where $\psi = \Phi/\Phi_0$. By fluxoid quantization, we may also write:

$$2\pi\psi + \phi_1 - \phi_2 = \pi N, \tag{4.37}$$

where N is an even integer for 0-SQUID and an odd integer for π-SQUID. Here we shall take $N = 0$ for simplicity, assuming to deal exclusively with 0-SQUIDs. Let us now introduce the following stochastic processes:

$$x_1 = \frac{\phi_1 + \phi_2}{2}, \tag{4.38a}$$

$$x_2 = \frac{\phi_1 - \phi_2}{2} = -\pi\psi. \tag{4.38b}$$

By summing and subtracting homologous sides of the indexed equation (4.34) written explicitly for $k = 1$ and $k = 2$, the deterministic dynamic equations for a two-junction interferometer can be rewritten as follows:

$$\frac{dx_1}{d\tau} + \cos x_2 \sin x_1 = \frac{i_B}{2} + \frac{i_{f1}(\tau) + i_{f2}(\tau)}{2}, \qquad (4.39a)$$

$$\frac{dx_2}{d\tau} + \cos x_1 \sin x_2 + \frac{x_2}{2\pi\beta} = -\frac{\psi_{ex}}{2\beta} + \frac{i_{f1}(\tau) - i_{f2}(\tau)}{2}. \qquad (4.39b)$$

We now notice that the quantities

$$i_{f+}(\tau) = \frac{i_{f1}(\tau) + i_{f2}(\tau)}{2}, \qquad (4.40a)$$

$$i_{f-}(\tau) = \frac{i_{f1}(\tau) - i_{f2}(\tau)}{2}, \qquad (4.40b)$$

still represent two independent white noise terms. In fact, by Eq. (4.32a-b) and (4.35), one can prove that the following relations are true:

$$\langle i_{f+}(\tau) \rangle = \langle i_{f-}(\tau) \rangle = 0, \qquad (4.41a)$$

$$\langle i_{f+}(\tau) i_{f+}(\tau) \rangle = \langle i_{f-}(\tau) i_{f-}(\tau) \rangle = \Gamma \delta(\tau - \tau'), \qquad (4.41b)$$

$$\langle i_{f+}(\tau) i_{f-}(\tau) \rangle = \langle i_{f-}(\tau) i_{f+}(\tau) \rangle = 0. \qquad (4.41c)$$

The above equations can be written in a more compact form, by introducing the following effective potential $W(x_1, x_2)$:

$$W(x_1, x_2) = (1 - \cos x_1 \cos x_2) - \frac{i_B}{2} x_1 + \frac{(x_2 + \pi\psi_{ex})^2}{4\pi\beta}. \qquad (4.42)$$

In this way, we can rewrite Eq. (4.39a-b) as follows:

$$\frac{dx_1}{d\tau} = -\frac{\partial W(x_1, x_2)}{\partial x_1} + i_{f+}(\tau), \qquad (4.43a)$$

$$\frac{dx_2}{d\tau} = -\frac{\partial W(x_1, x_2)}{\partial x_2} + i_{f-}(\tau), \qquad (4.43b)$$

Eq. (4.43a-b) represent the starting point for constructing, as we shall do in the following section, the reduced model for a d. c. SQUID in the presence of thermal fluctuations.

4.4. The reduced model for a d. c. SQUID in the presence of thermal fluctuations

In the previous chapter, Section 3.2, we have argued that, in the absence of thermal fluctuations, the d. c. SQUID dynamics could be separated in "slow" and "fast". In fact, in Eq. (3.25) we defined two characteristic times, which we may redefine here as follows:

$$\tau_1 = \frac{\Phi_0}{2\pi R I_{J0}}, \quad \tau_2 = \frac{L}{R}. \tag{4.44}$$

The above quantities represent the characteristic time scales of the processes x_1 and x_2, respectively. We had also noticed that the ratio between these two times is equal to $2\pi\beta$. In this way, for small values of the parameter β, such that $2\pi\beta \ll 1$, the dynamics of x_2 can be considered very fast with respect to the characteristic evolution of the processes x_1. Therefore, the quantity x_1 can be assumed to be adiabatic, or slowly varying with respect to x_2. This will allow us to write a FPE corresponding to Eq. (4.43b) in which the quantity x_1 is seen as frozen. Attention must be paid in this respect, so that we proceed gradually.

First of all, let us write the FPE corresponding to the stochastic dynamics described in Eq. (4.43a-b):

$$\frac{\partial p(x_1, x_2, \tau)}{\partial \tau} = \nabla \cdot (p(x_1, x_2, \tau)\nabla W) + \frac{\Gamma}{2}\nabla^2 p(x_1, x_2, \tau), \tag{4.45}$$

where we notice that the factor of ½ in front of the parameter Γ appears because of the different choice of processes, so that the original correlation in Eq. (4.35) needs to be substituted, in this new picture, by Eq. (4.41a–c). By considering again the stationary solution $p(x_1, x_2)$ to Eq. (4.45), we may set:

$$\frac{\partial}{\partial x_1}\left[\left(p\frac{\partial W}{\partial x_1}\right) + \frac{\Gamma}{2}\frac{\partial p}{\partial x_1}\right] + \frac{\partial}{\partial x_2}\left[\left(p\frac{\partial W}{\partial x_2}\right) + \frac{\Gamma}{2}\frac{\partial p}{\partial x_2}\right] = 0. \tag{4.46}$$

The above relation can be satisfied only if

$$\frac{\partial}{\partial x_1}\left[\left(p\frac{\partial W}{\partial x_1}\right) + \frac{\Gamma}{2}\frac{\partial p}{\partial x_1}\right] = 0, \tag{4.47a}$$

$$\frac{\partial}{\partial x_2}\left[\left(p\frac{\partial W}{\partial x_2}\right)+\frac{\Gamma}{2}\frac{\partial p}{\partial x_2}\right]=0. \tag{4.47b}$$

In order to factorize the dynamics of the fast process, we set:

$$p(x_1,x_2)=q(x_1)p_2(x_1,x_2). \tag{4.48}$$

4.4.1. Fokker-Planck equation for the fast process

Having factorized the probability density function $p(x_1,x_2)$ as in Eq. (4.48), Eq. (4.47a-b) gives:

$$q(x_1)p_2(x_1,x_2)\frac{\partial W}{\partial x_1}+\frac{\Gamma}{2}\frac{\partial}{\partial x_1}[q(x_1)p_2(x_1,x_2)]=\lambda\ , \tag{4.49a}$$

$$p_2(x_1,x_2)\frac{\partial W}{\partial x_2}+\frac{\Gamma}{2}\frac{\partial p_2(x_1,x_2)}{\partial x_2}=0,\ \text{where}\ \lambda\ \text{is a constant.} \tag{4.49b}$$

It is then evident that Eq. (4.49b) corresponds to the FPE for the fast process x_2, when x_1 is considered to be a constant. Notice also that the right hand side of Eq. (4.49b) is null because of the factorization made in Eq. (4.48). According to what was already seen in Section 4.2, the probability density function $p_2(x_1,x_2)$ must fulfill the following normalization condition (Kramers, 1940):

$$\int_{-\infty}^{+\infty}p_2(x_1,x_2)dx_2=1. \tag{4.50}$$

We can now solve Eq. (4.49b), utilizing the same approach as in Section 4.1, obtaining:

$$p_2(x_1,x_2)=\frac{e^{-\frac{2}{\Gamma}W(x_1,x_2)}}{A(x_1)}, \tag{4.51}$$

where $A(x_1)$ is determined through Eq. (4.50), so that:

$$A(x_1)=\int_{-\infty}^{+\infty}dx_2\, e^{-\frac{2}{\Gamma}W(x_1,x_2)}. \tag{4.52}$$

The integral could be evaluated, at least in the lowest order of the parameter β. We shall do this later; for the time being, we leave the result for $p_2(x_1,x_2)$ as it appears in Eq. (4.51).

4.4.2. *Fokker-Planck equation for the slow process*

We now turn our attention to finding an effective potential $W_{eff}(x_1)$ for the slow process, by which we can write down the FPE for $q(x_1)$. Let us preliminary notice that, being $p(x_1, x_2) = q(x_1)p_2(x_1, x_2)$, the normalization requirement for $p_2(x_1, x_2)$ in Eq. (4.50) implies the following similar relation for $q(x_1)$:

$$\int_0^{2\pi} q(x_1)dx_1 = 1, \tag{4.53}$$

where we already recognize the 2π periodicity of $q(x_1)$. The above relation is necessarily true, since we need to require normalization of the overall probability density function $p(x_1, x_2)$. In order to write down the FPE for $q(x_1)$, we proceed as follows. First consider Eq. (4.47a) and Eq. (4.48), where all functions are assumed to be smooth. By integrating over x_2, we get:

$$\frac{\partial}{\partial x_1} \int_{-\infty}^{+\infty} \left\{ q(x_1)p_2(x_1, x_2)\frac{\partial W}{\partial x_1} + \frac{\Gamma}{2}\frac{\partial}{\partial x_1}[q(x_1)p_2(x_1, x_2)] \right\} dx_2 = 0. \tag{4.54}$$

By making use of Eq. (4.50) we have:

$$q(x_1) \int_{-\infty}^{+\infty} p_2(x_1, x_2)\frac{\partial W}{\partial x_1}dx_2 + \frac{\Gamma}{2}\frac{\partial q(x_1)}{\partial x_1} = \frac{\lambda_1}{2}, \tag{4.55}$$

where λ_1 is a constant to be determined by means of Eq. (4.53). By now recalling Eq. (4.51), we rewrite Eq. (4.55) as follows:

$$\frac{q(x_1)}{A(x_1)} \int_{-\infty}^{+\infty} e^{-\frac{2}{\Gamma}W(x_1, x_2)}\frac{\partial W}{\partial x_1}dx_2 + \frac{\Gamma}{2}\frac{\partial q(x_1)}{\partial x_1} = \frac{\lambda_1}{2}. \tag{4.56}$$

Let us then define:

$$W_{eff}(x_1) - D = -\Gamma \ln\left[\int_{-\infty}^{+\infty} e^{-\frac{2}{\Gamma}W(x_1, x_2)}dx_2\right] = -\Gamma \ln[A(x_1)], \tag{4.57}$$

where D is a constant. Eq. (4.56) can thus be cast in the following final form:

$$q(x_1)\frac{\partial W_{eff}(x_1)}{\partial x_1} + \Gamma\frac{\partial q(x_1)}{\partial x_1} = \lambda_1. \tag{4.58}$$

One can now soon recognize the above expression to be the FPE for the slow process, formally identical to Eq. (4.7).

We may now proceed to find the effective potential $W_{eff}(x_1)$, at least in the leading order in the parameter β, in order to successively solve Eq. (4.58) as in the AH approach.

We start by evaluating $A(x_1)$, substituting into Eq. (4.52) the expression for $W(x_1, x_2)$ given by Eq. (4.42), obtaining:

$$A(x_1) = e^{-\frac{2}{\Gamma}+\frac{i_B}{\Gamma}x_1}\Lambda(x_1). \qquad (4.59)$$

where

$$\Lambda(x_1) = \int_{-\infty}^{+\infty} e^{\frac{2}{\Gamma}\cos x_1 \cos x_2} e^{-\frac{(x_2+\pi\psi_{ex})^2}{2\pi\beta\Gamma}} dx_2. \qquad (4.60)$$

By now making the following change of variables

$$z = \frac{x_2 + \pi\psi_{ex}}{\sqrt{2\pi\beta\Gamma}}, \qquad (4.61)$$

the integral $\Lambda(x_1)$ in Eq. (4.60) can be written as follows:

$$\Lambda(x_1) = \sqrt{2\pi\beta\Gamma} \int_{-\infty}^{+\infty} e^{\frac{2}{\Gamma}\cos x_1 \cos(\sqrt{2\pi\beta\Gamma}z - \pi\psi_{ex})} e^{-z^2} dz. \qquad (4.62)$$

By simple trigonometry, we also have:

$$\frac{\Lambda(x_1)}{\sqrt{2\pi\beta\Gamma}} = \int_{-\infty}^{+\infty} e^{\frac{2}{\Gamma}\cos x_1 [\cos(\sqrt{2\pi\beta\Gamma}z)\cos(\pi\psi_{ex}) + \sin(\pi\psi_{ex})\sin(\sqrt{2\pi\beta\Gamma}z)]} e^{-z^2} dz. \qquad (4.63)$$

Up to this point, we have only dealt with exact results. However, if we wish to extract a useful (and significant) expression for the quantity $W_{eff}(x_1)$ we need to approximate the integral in Eq. (4.63) to the lowest order in β, for small values of the latter parameter. The reader might have already guessed the answer to this tedious calculation, which will be left as an exercise: the effective potential $W_{eff}(x_1)$ is equal, to lowest order in β and apart from an additive constant, to the parabolic washboard potential W_P defined in Eq. (3.57). We thus find:

$$W_{eff}(x_1) = w_0(x_1) + 2\pi\beta\, w_1(x_1) + D, \qquad (4.64)$$

where D is an additive constant and where:

$$w_0(x_1) = -2\cos \pi\psi_{ex} \cos x_1 - i_B x_1, \qquad (4.65a)$$

$$w_1(x_1) = -\sin^2 \pi\psi_{ex} \cos^2 x_1 - \frac{\Gamma}{2}\cos \pi\psi_{ex} \cos x_1. \qquad (4.65b)$$

Comparing Eq. (4.64) and (4.65a-b) with Eq. (3.57) and (3.58a-b), we notice the perfect agreement of these results with the simplest scheme of the reduced model in the absence of thermal fluctuations. Naturally, the second addendum in (4.65b) is not present at $T = 0$ K. This term can be considered as a rescaling parameter of the energy barrier height separating two metastable magnetic states of the system. Furthermore, we notice that the additive constant D does not affect the FPE (4.58) for the slow process, since only the derivative of the effective potential $W_{eff}(x_1)$ is present in Eq. (4.58).

4.4.3. *The probability density function for the slow process*

We now turn our attention to finding the probability density function $q(x_1)$. Let us then recall Eq. (4.7) and its general solution Eq. (4.10), so that, by comparison with Eq. (4.58), we may write:

$$q(x_1) = \frac{\lambda_1}{\Gamma} e^{-\frac{W_{eff}(x_1)}{\Gamma}} \left[\int_0^\phi d\xi\, e^{\frac{W_{eff}(\xi)}{\Gamma}} - C' \right]. \qquad (4.66)$$

The constant C' can be found by imposing periodicity of the function $q(x_1)$, namely, by setting $q(0) = q(2\pi)$, so that:

$$C' = \frac{F(2\pi)}{1 - e^{-2\pi i_B/\Gamma}}, \qquad (4.67)$$

where

$$F(x_1) = \int_0^{x_1} d\xi\, e^{\frac{W_{eff}(\xi)}{\Gamma}}. \qquad (4.68)$$

By now using the normalization condition (4.53) we may determine the constant λ_1, so that:

$$\lambda_1 = \frac{\Gamma}{\int_0^{2\pi} dx_1\, e^{-\frac{W_{eff}(x_1)}{\Gamma}[F(x_1)-C']}}. \tag{4.69}$$

In this way, the probability density function will take the following final form:

$$q(x_1) = \frac{e^{-\frac{W_{eff}(x_1)}{\Gamma}[F(x_1)-C']}}{\int_0^{2\pi} d\xi\, e^{-\frac{W_{eff}(\xi)}{\Gamma}[F(\xi)-C']}}. \tag{4.70}$$

4.5. Current-voltage curves of a d. c. SQUID in the presence of thermal fluctuations

By recalling the results in Section 1.4, we can express the statistical average of the normalized voltage across the two junctions of a d. c. SQUID as follows:

$$\langle v \rangle = \langle \frac{dx_1}{d\tau} \rangle = -\pi\lambda_1, \tag{4.71}$$

where a factor ½ on the right hand side must be included in order to correctly take account of the dynamic equation (4.39a). By Eq. (4.69) we may thus write:

$$\langle v \rangle = \frac{-\pi\Gamma}{\int_0^{2\pi} dx_1\, e^{-\frac{W_{eff}(x_1)}{\Gamma}[F(x_1)-C']}}. \tag{4.72}$$

Let us first consider the zero-order approximation in the parameter β ($\beta = 0$) of the effective potential, by setting $W_{eff}(x_1) = w_0(x_1)$. After having mastered the technique, we shall consider the first-order approximation in β, giving a calculation scheme to the reader. As in Section 1.3, we need to calculate the integral at the denominator by means of a modified Bessel function expansion of the exponential of the cosine terms in $w_0(x_1)$, by setting:

$$e^{\frac{2\cos \pi\psi_{ex} \cos x_1}{\Gamma}} = \Sigma_{k=-\infty}^{+\infty} I_k\left(\frac{2\cos \pi\psi_{ex}}{\Gamma}\right) e^{ikx_1}. \tag{4.73}$$

Comparing now Eq. (4.72) with Eq. (4.18) and Eq. (4.73) with Eq. (4.19), we conclude that we may get the final result by replacing the quantity $1/\Gamma$ with $2\cos \pi \psi_{ex}/\Gamma$ in the argument of the modified Bessel functions. Therefore, the average voltage for $\beta = 0$ is given by

$$\langle v_0 \rangle = \frac{i_B/2}{I_0^2\left(\frac{2\cos \pi \psi_{ex}}{\Gamma}\right) + 2\left(\frac{i_B}{\Gamma}\right)^2 \sum_{k=1}^{+\infty} \frac{(-1)^k I_k^2\left(\frac{2\cos \pi \psi_{ex}}{\Gamma}\right)}{k^2 + \left(\frac{i_B}{\Gamma}\right)^2}}. \qquad (4.74)$$

Representations of these curves for various values of Γ and for two values of ψ_{ex}, namely $\psi_{ex} = 0.0$ and $\psi_{ex} = 0.25$, are given in Fig. 4.2a-b. In particular, in Fig. 4.2a the $\langle v_0 \rangle$ vs. i_B curves are shown for $\psi_{ex} = 0.0$ and for $\Gamma = 0.15, 0.30, 0.45$. We notice that for increasing values of the parameter Γ, i.e., for increasing value of the temperature T, the curves tend to move away from the $T = 0$ K behavior, represented by the dashed line. In this way, the normalized maximum Josephson current i_{max} of the device tends to decrease for increasing values of Γ. In Fig. 4.2b, on the other hand, the $\langle v_0 \rangle$ vs. i_B curves are shown for $\psi_{ex} = 0.25$ and again for $\Gamma = 0.15, 0.30, 0.45$. We notice the normalized maximum Josephson current i_{max} of the device is less than 2.0 even at $T = 0$ K (dashed curve) because of the normalized magnetic flux value. As in the previous figure, the value of i_{max} tends to decrease for increasing values of Γ.

The $\langle v_0 \rangle$ vs. i_B curves for various values of ψ_{ex} and for two values of Γ, namely $\Gamma = 0.20$ and $\Gamma = 0.30$, are given in Fig. 4.3a-b. In particular, in Fig. 4.3a the $\langle v_0 \rangle$ vs. i_B curves are shown for $\Gamma = 0.20$ and for $\psi_{ex} = 0.05, 0.15, 0.25, 0.35$. We notice that for all values of ψ_{ex} the curves tend asymptotically toward the $\psi_{ex} = 0$ curve (dashed line) some from above, some from below. The normalized maximum Josephson current i_{max} of the device tends to decrease, in this interval of ψ_{ex}, for increasing values of this same quantity. In Fig. 4.3b, on the other hand, the $\langle v_0 \rangle$ vs. i_B curves are shown for $\Gamma = 0.30$ and for the same four values of ψ_{ex}.

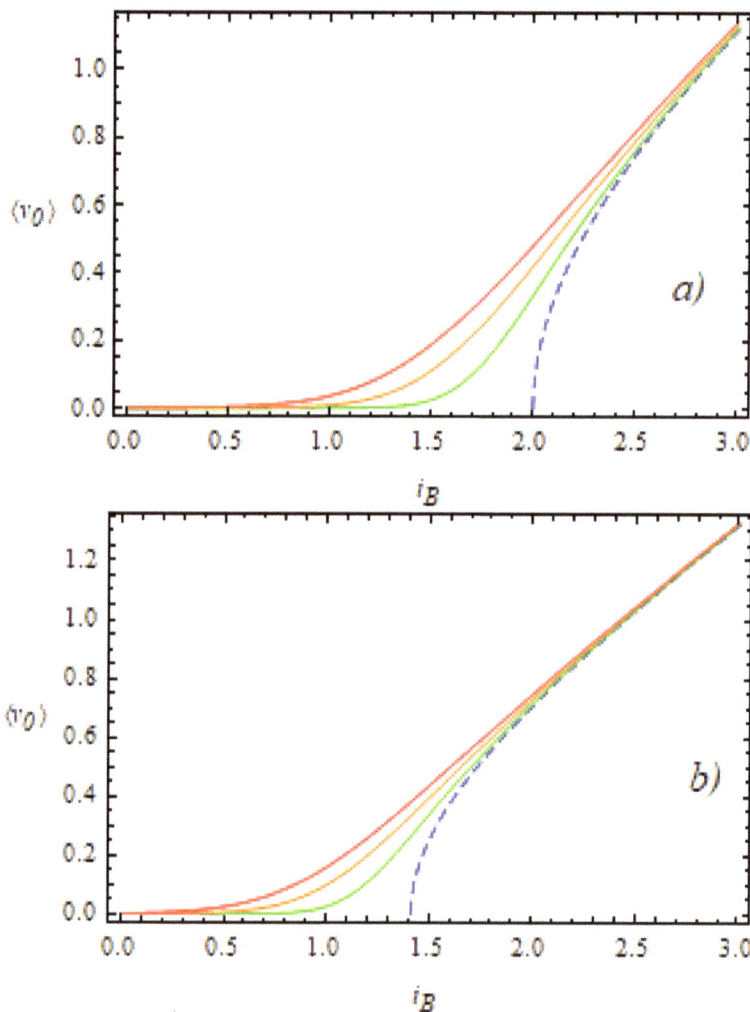

Fig. 4.2. Voltage-current characteristics of a d. c. SQUID with parameter $\beta = 0$ in the presence of thermal fluctuations for $\psi_{ex} = 0.0$ in panel a) and $\psi_{ex} = 0.25$ in panel b). The parameter Γ has been chosen as follows for both panels a) and b): $\Gamma = 0$ (dashed line), $\Gamma = 0.15$ (green line), $\Gamma = 0.30$ (orange line), $\Gamma = 0.45$ (red line). In the numerical evaluation of the curves, the series expansion in terms of modified Bessel function has been truncated at the integer $k_{max} = 100$.

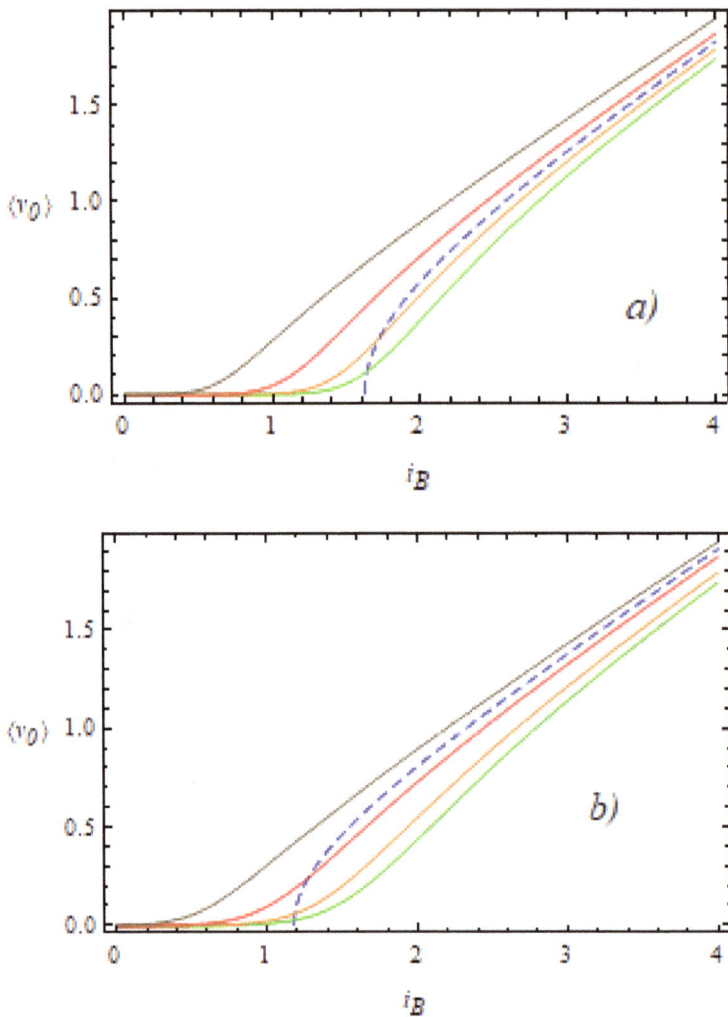

Fig. 4.3. Voltage-current characteristics of a d. c. SQUID with parameter $\beta = 0$ in the presence of thermal fluctuations for $\Gamma = 0.20$ in panel a) and $\Gamma = 0.30$ in panel b). The normalized external flux ψ_{ex} has been chosen as follows for both panels a) and b): $\psi_{ex} = 0.0$ (dashed line), $\psi_{ex} = 0.05$ (green line), $\psi_{ex} = 0.15$, (orange line), $\psi_{ex} = 0.25$ (red line), $\psi_{ex} = 0.35$ (red line). In the numerical evaluation of the curves, the series expansion in terms of modified Bessel function has been truncated at the integer $k_{max} = 100$.

When comparing these curves to the ones shown in Fig.4.3a, a similar qualitative behavior is detected. However, in Fig.4.3b a higher number of curves lie below the $\psi_{ex} = 0$ curve (dashed line).

We are now ready to tackle the calculation of the voltage-current curves to first order in the parameter β. We shall make use of the results obtained in Section 1.2 when possible. We start by giving a way to evaluate the integral in Eq. (4.69)

$$\frac{\Gamma}{\lambda_1} = \int_0^{2\pi} dx_1 \, e^{-\frac{w_0(x_1)}{\Gamma}} e^{-\frac{2\pi\beta \, w_1(x_1)}{\Gamma}} [F(x_1) - C']. \qquad (4.75)$$

By now writing

$$F(x_1) = F_0(x_1) + \frac{2\pi\beta}{\Gamma} F_1(x_1), \qquad (4.76)$$

where for the exponential containing $w_1(x_1)$ has been given a first-order approximation in β, and where

$$F_0(x_1) = \int_0^{x_1} dx \, e^{\frac{w_0(x)}{\Gamma}}, \qquad (4.77a)$$

$$F_1(x_1) = \int_0^{x_1} dx \, w_1(x) \, e^{\frac{w_0(x)}{\Gamma}}. \qquad (4.77b)$$

By imposing periodicity for $q(x_1)$ and performing calculations as in Eq. (4.20), (4.21a-c) and (4.22), we may write the expression for the constant C' in Eq. (4.75):

$$C' = -K_B + \frac{2\pi\beta}{\Gamma}\left(\sin^2 \pi\psi_{ex} K_B'' - \frac{\Gamma}{2}\cos \pi\psi_{ex} K_B'\right) = C_0' + \frac{2\pi\beta}{\Gamma} C_1', \, (4.78)$$

with the implicit definition of $C_0' = -K_B$ and $C_1' = \sin^2 \pi\psi_{ex} K_B'' - \Gamma \cos \pi\psi_{ex} K_B'/2$, where:

$$K_B = \sum_{k=-\infty}^{+\infty} \frac{(-1)^k I_k(z)}{ik - \frac{i_B}{\Gamma}}, \qquad (4.79a)$$

$$K_B' = \sum_{k=-\infty}^{+\infty} \frac{(-1)^k I_k'(z)}{ik - \frac{i_B}{\Gamma}}, \tag{4.79b}$$

$$K_B'' = \sum_{k=-\infty}^{+\infty} \frac{(-1)^k I_k''(z)}{ik - \frac{i_B}{\Gamma}}, \tag{4.79c}$$

with

$$z = \frac{2\cos\pi\psi_{ex}}{\Gamma}. \tag{4.80}$$

The prime on the modified Bessel functions stands for the derivative of the same function with respect to the argument z. Here we are adopting a different definition of K_B compared to the corresponding quantity defined for a single junction. In this way, by Eq. (4.67), we also have:

$$F(2\pi) = \left(1 - e^{-2\pi i_B/\Gamma}\right) C'. \tag{4.81}$$

Recalling Eq. (4.68), the expression (4.75) can be written, to first-order in β, as follows:

$$\frac{\Gamma}{\lambda_1} = s_0 - b_0 C_0' + \frac{2\pi\beta}{\Gamma} [s_1 - \tilde{s}_1 + b_1 C_0' - b_0 C_1'], \tag{4.82}$$

where:

$$b_0 = \int_0^{2\pi} dx_1 \, e^{-\frac{w_0(x_1)}{\Gamma}}, \tag{4.83a}$$

$$b_1 = \int_0^{2\pi} dx_1 \, w_1(x_1) e^{-\frac{w_0(x_1)}{\Gamma}}, \tag{4.83b}$$

$$s_{0,1} = \int_0^{2\pi} dx_1 \, e^{-\frac{w_0(x_1)}{\Gamma}} F_{0,1}(x_1), \tag{4.83c}$$

$$\tilde{s}_1 = \int_0^{2\pi} dx_1 \, e^{-\frac{w_0(x_1)}{\Gamma}} F_0(x_1) w_1(x_1). \tag{4.83d}$$

All the integrals should be patiently calculated, also adopting some mathematical tricks. We thus proceed in finding $b_{0,1}$, $s_{0,1}$ and $\tilde{s}_1$ appearing in Eq. (4.78). By being very careful in our calculations, we find:

$$b_0 = \left(e^{2\pi i_B/\Gamma} - 1\right) \sum_{k'=-\infty}^{+\infty} \frac{I_{k'}(z)}{\left(ik' + \frac{i_B}{\Gamma}\right)}, \tag{4.84a}$$

$$b_1 = -\sin^2 \pi\psi_{ex} \, b_0'' - \frac{\Gamma}{2} \cos \pi\psi_{ex} \, b_0', \tag{4.84b}$$

$$s_0 = 2\pi N_B - b_0 K_B, \tag{4.84c}$$

$$s_1 = -\sin^2 \pi\psi_{ex} \left(2\pi N_B'' - b_0 K_B''\right) + \frac{\Gamma}{2} \cos \pi\psi_{ex} \left(2\pi N_B' - b_0 K_B'\right) \tag{4.84d}$$

$$\tilde{s}_1 = -\sin^2 \pi\psi_{ex} \left(2\pi N_B'' - b_0'' K_B\right) - \frac{\Gamma}{2} \cos \pi\psi_{ex} \left(2\pi N_B' - b_0' K_B\right), \tag{4.84e}$$

where

$$N_B = \sum_{k=-\infty}^{+\infty} \frac{(-1)^k I_k^2(z)}{ik - \frac{i_B}{\Gamma}}, \tag{4.85a}$$

$$N_B' = \sum_{k=-\infty}^{+\infty} \frac{(-1)^k I_k'(z) I_k(z)}{ik - \frac{i_B}{\Gamma}}. \tag{4.85b}$$

$$N_B'' = \sum_{k=-\infty}^{+\infty} \frac{(-1)^k I_k''(z) I_k(z)}{ik - \frac{i_B}{\Gamma}}. \tag{4.85c}$$

The primes on b_0 in (4.84b) are directly positioned on the modified Bessel functions in (4.84a), as for K_B in Eq. (4.79a-c). However, the same is not true for N_B. In fact, in Eq. (4.85b-c) the prime is only used for definition sake.

The tricks used in the calculations are mostly two. The first is to write the multiplicative terms $\cos x_1$ and $\cos^2 x_1$ in the expression for $w_1(x_1)$ as a first and second derivative with respect to z of the exponential of the cosine in $w_0(x_1)$, respectively, taking care of the sign in the case we treat $\cos x_1$. The second trick is to calculate b_0 and b_1 by means of the results obtained for $\Gamma_0(2\pi)$ and $\Gamma_1(2\pi)$, respectively, by the following substitutions: $z \to -z$ and $i_B \to -i_B$.

It is now useful to notice that the zero-order term $s_0 - b_0 C_0' = 2\pi N_B$ is the same as the one previously calculated, giving $\langle v_0 \rangle$ in Eq. (4.74). As for the first-order term, we find:

$$s_1 - \tilde{s}_1 + b_1 C_0' - b_0 C_1' = 2\pi\Gamma \cos \pi\psi_{ex} N_B' = \pi\Gamma \cos \pi\psi_{ex} \frac{\partial}{\partial z} N_B. \quad (4.86)$$

The calculation is rather long and is left as a useful exercise for the reader. In this way, from Eq. (4.82) we have:

$$\frac{\Gamma}{\lambda_1} = 2\pi N_B \left[1 + \pi\beta \cos \pi\psi_{ex} \frac{\partial}{\partial z} \ln N_B\right]. \quad (4.87)$$

Recognizing now that $\langle v_0 \rangle = -\Gamma/(2N_B)$, we may finally write the average voltage defined in Eq. (4.71) as follows:

$$\langle v \rangle = \frac{\langle v_0 \rangle}{1 + 2\pi\beta \cos \pi\psi_{ex}\frac{N_B'}{N_B}} \approx \langle v_0 \rangle \left(1 - \pi\beta \cos \pi\psi_{ex} \frac{\partial}{\partial z} \ln N_B\right). \quad (4.88)$$

By finally noticing that

$$\frac{\partial}{\partial z} \ln N_B = -\frac{\partial}{\partial z} \ln\langle v_0 \rangle, \quad (4.89)$$

we can cast Eq. (4.88) in the following final form only in terms of the known voltage $\langle v_0 \rangle$:

$$\langle v \rangle \approx \langle v_0 \rangle \left(1 + \pi\beta \cos \pi\psi_{ex} \frac{\partial}{\partial z} \ln\langle v_0 \rangle\right). \quad (4.90)$$

The above expression agrees with the first-order approximation of $\langle v \rangle$ as derived by De Luca *et al.* (De Luca 2010) with a different analytic approach.

Let us now explicitly write the quantities N_B and N_B' in Eq. (4.88) in terms of summations on the index k as :

$$N_B = -\frac{\Gamma}{i_B} \left[I_0^2(z) + 2\left(\frac{i_B}{\Gamma}\right)^2 \sum_{k=1}^{+\infty} \frac{(-1)^k I_k^2(z)}{k^2 + \left(\frac{i_B}{\Gamma}\right)^2}\right]. \quad (4.91a)$$

$$N_B' = -\frac{\Gamma}{i_B} \left[I_0(z)I_0'(z) + 2\left(\frac{i_B}{\Gamma}\right)^2 \sum_{k=1}^{+\infty} \frac{(-1)^k I_k(z)I_k'(z)}{k^2 + \left(\frac{i_B}{\Gamma}\right)^2}\right]. \quad (4.91b)$$

Having got rid of the complex quantities in the definition of N_B and N_B', we can obtain a numerically more tractable expression for $\langle v \rangle$.

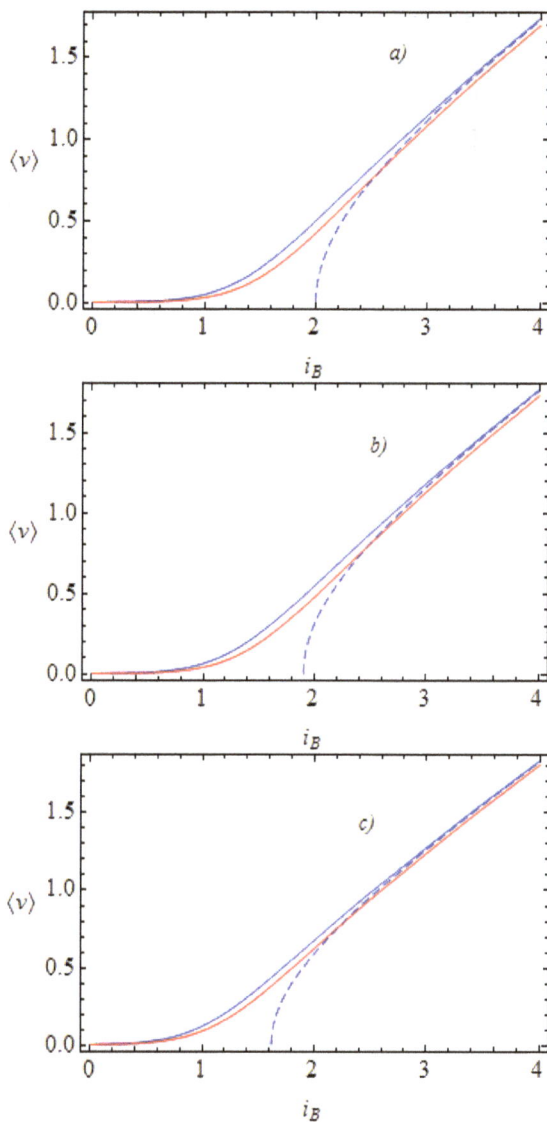

Fig. 4.4. Voltage-current characteristics of a d. c. SQUID for $\psi_{ex} = 0.0, 0.1, 0.2$ in panels *a*), *b*), and *c*), respectively In all panels: $\Gamma = 0.50$ (full-line curves); the dashed curves are calculated at $T = 0$ K; the blue and red curves are for $\beta = 0.0$ and $\beta = 0.1$, respectively. In the numerical evaluation of the curves, the series expansion in terms of modified Bessel function has been truncated at the integer $k_{max} = 100$.

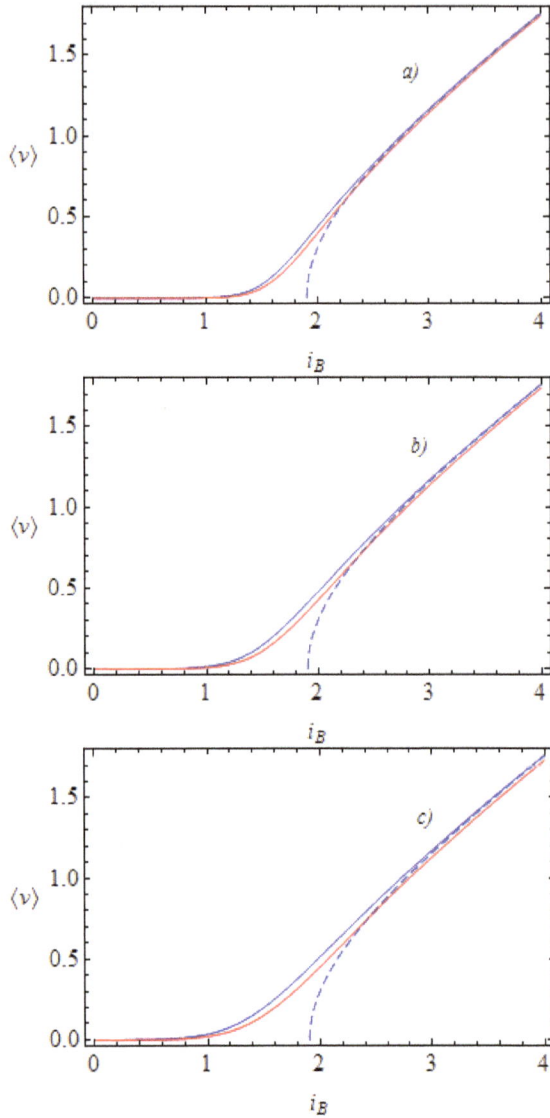

Fig. 4.5. Voltage-current characteristics of a d. c. SQUID for $\Gamma = 0.10, 0.20, 0.30$ in panels a), b), and c), respectively In all panels: $\psi_{ex} = 0.10$ (full-line curves); the dashed curves are calculated at $T = 0$ K; the blue and red curves are for $\beta = 0.0$ and $\beta = 0.1$, respectively. In the numerical evaluation of the curves, the series expansion in terms of modified Bessel function has been truncated at the integer $k_{max} = 100$.

Therefore, by implementing a calculation routine for these two quantities, we can obtain the graphical representation of the $\langle v \rangle$ vs. i_B curves in Fig. 4.4a-c and Fig. 4.5a-c. In particular, in Fig. 4.4a-c we give a comparison between the voltage-current curves for $\beta = 0.0$ and $\beta = 0.1$ when calculated for $\Gamma = 0.50$ and various values of the normalized externally applied flux, namely, $\psi_{ex} = 0.00$, $\psi_{ex} = 0.10$, and $\psi_{ex} = 0.10$ in panels a), b), and c), respectively. In Fig. 4.5a-c, on the other hand, a comparison between the $\langle v \rangle$ vs. i_B curves for $\beta = 0.0$ and $\beta = 0.1$ is given when the voltage-current curves are calculated for $\psi_{ex} = 0.10$ and various values of the parameter Γ, namely, $\Gamma = 0.20$, $\Gamma = 0.30$, and $\Gamma = 0.40$ in panels a), b), and c), respectively. Along with all the curves represented in Fig. 4.4a-c and Fig. 4.5a-c, dashed curves representing the zero-temperature behavior of the system are reported. From these graphs we may detect some overall features. The main feature is that the qualitative behavior of a d. c. SQUID in the presence of thermal fluctuations does not vary, with respect to the simplifying assumption $\beta \approx 0$, when the parameter β is small ($\beta = 0.10$ in our case). However, a systematic slight depression of the $\langle v \rangle$ vs. i_B curves with respect to the $\beta = 0.0$ case is observed for $\beta = 0.10$.

4.6. Circulating currents and magnetic susceptibility of d. c. SQUIDs in the presence of thermal fluctuations

By recalling the definition of the circulating current i_S and of the magnetic susceptibility χ of d. c. SQUIDs at $T = 0$ K given in Eq. (3.24a) and (3.24b), respectively, we may redefine these quantities in terms of the new variable , by writing:

$$i_S = -\frac{x_2 + \pi\psi_{ex}}{\pi\beta}, \tag{4.92a}$$

$$\chi = -\frac{\alpha x_2 + \pi\psi_{ex}}{\pi\psi_{ex}}, \tag{4.92b}$$

where α is the normal fraction of the cylindrically symmetric two-junction interferometer. We might notice that the above quantities do not depend on the variable x_1. Therefore, for a generic function $f(x_2)$, we have, by Eq. (4.48):

$$\langle f \rangle = \int_0^{2\pi} dx_1 q(x_1) \int_{-\infty}^{+\infty} dx_2 f(x_2) p_2(x_1, x_2). \quad (4.93)$$

Recall now that in Eq. (4.51) we have set:

$$p_2(x_1, x_2) = \frac{1}{A(x_1)} e^{-\frac{2}{\Gamma} W(x_1, x_2)},$$

where $A(x_1)$ is defined by Eq. (4.52). Therefore, we find, to first order in β:

$$A(x_1) = \sqrt{2\pi^2 \beta \Gamma} \, (1 - \pi\beta \cos(\pi\psi_{ex}) \cos x_1) e^{-W_0(x_1)}. \quad (4.94)$$

where for the sake of simple notation $z = 2 \cos(\pi\psi_{ex}) / \Gamma$. In this way, Eq. (4.93) can be written as follows:

$$\langle f \rangle = \int_0^{2\pi} dx_1 \frac{q(x_1)}{A(x_1)} \int_{-\infty}^{+\infty} dx_2 f(x_2) e^{-\frac{2}{\Gamma} W(x_1, x_2)}, \quad (4.95)$$

By now defining:

$$H_f(x_1) = \int_{-\infty}^{+\infty} dx_2 f(x_2) e^{-\frac{2}{\Gamma} W(x_1, x_2)}, \quad (4.96)$$

we have:

$$\langle f \rangle = \frac{1}{\sqrt{2\pi^2 \beta \Gamma}} \int_0^{2\pi} dx_1 q(x_1) e^{W_0(x_1)} \frac{H_f(x_1)}{1 - \frac{\pi\beta\Gamma}{2} z \cos x_1}. \quad (4.97)$$

Therefore, in order to find the statistical average $\langle f \rangle$ one can proceed by first finding the function $H_f(x_1)$ and by then calculating the integral in Eq. (4.97).

We give a general procedure to find $H_f(x_1)$. This calculation is rather similar to what was done for finding the expression for the effective potential $W_{eff}(x_1)$ in Eq. (4.64). First substitute the expression for $W(x_1, x_2)$ as given by Eq. (4.42), neglecting the constant term, in Eq. (4.96):

$$H_f(x_1) = e^{\frac{i_B}{\Gamma} x_1} \int_{-\infty}^{+\infty} dx_2 f(x_2) e^{\frac{\cos x_1 \cos x_2}{\Gamma}} e^{-\frac{(x_2 + \pi\psi_{ex})^2}{2\pi\beta\Gamma}}. \quad (4.98)$$

By now making the change of variables

$$y = \frac{x_2 + \pi\psi_{ex}}{\sqrt{2\pi\beta\Gamma}}, \tag{4.99}$$

the integral $H_f(x_1)$ in Eq. (4.98) can be written as follows:

$$\frac{H_f(x_1)}{\sqrt{2\pi\beta\Gamma}e^{\frac{i\beta}{\Gamma}x_1}} = \int_{-\infty}^{+\infty} f\left(\sqrt{2\pi\beta\Gamma}y - \pi\psi_{ex}\right) \, e^{\frac{2}{\Gamma}\cos x_1 \cos(\sqrt{2\pi\beta\Gamma}y - \pi\psi_{ex})} e^{-y^2} dy. \tag{4.100}$$

By elementary trigonometry, we may write:

$$e^{\frac{2}{\Gamma}\cos x_1 \cos(\sqrt{2\pi\beta\Gamma}y - \pi\psi_{ex})} =$$

$$e^{\frac{2}{\Gamma}\cos x_1 [\cos(\sqrt{2\pi\beta\Gamma}y)\cos(\pi\psi_{ex}) + \sin(\pi\psi_{ex})\sin(\sqrt{2\pi\beta\Gamma}y)]}. \tag{4.101}$$

To first order in β, we may set the above expression equal to:

$$e^{z \cos x_1}(1 - 2\pi\beta \cos(\pi\psi_{ex}) \cos x_1 y^2)$$

$$\left(1 + 2\sqrt{\frac{2\pi\beta}{\Gamma}} \sin(\pi\psi_{ex}) \cos x_1 y\right). \tag{4.102}$$

In this way, the exponential in Eq. (4.101) can be approximated, to first order in β, by the following expression:

$$e^{z \cos x_1}\left(1 + 2\sqrt{\frac{2\pi\beta}{\Gamma}} \sin(\pi\psi_{ex}) \cos x_1 y - 2\pi\beta \cos(\pi\psi_{ex}) \cos x_1 y^2\right). \tag{4.103}$$

Furthermore, the term $f\left(\sqrt{2\pi\beta\Gamma}y - \pi\psi_{ex}\right)$, for small values of the parameter β, can be written as follows:

$$f\left(\sqrt{2\pi\beta\Gamma}y - \pi\psi_{ex}\right) \approx f_0 + f_0'\sqrt{2\pi\beta\Gamma}y + f_0''\pi\beta\Gamma y^2. \tag{4.104}$$

where $f_0 = f(-\pi\psi_{ex})$, $f_0' = f'(-\pi\psi_{ex})$, and $f_0'' = f''(-\pi\psi_{ex})$. In order to find $H_f(x_1)$, we now need to multiply the terms in Eq. (4.103) and (4.104) and to integrate over all real numbers using the measure

$e^{-y^2} dy$. When we do this, all odd terms in y give null contribution, while other terms are of the following two types:

$$\int_{-\infty}^{+\infty} e^{-y^2} dy = \sqrt{\pi}, \qquad \int_{-\infty}^{+\infty} y^2 e^{-y^2} dy = \frac{\sqrt{\pi}}{2}. \qquad (4.105)$$

Carrying out these steps to first order in β, we thus find:

$$\frac{H_f(x_1)}{1 - \frac{\pi\beta\Gamma}{2} z \cos x_1} = \sqrt{2\pi^2\beta\Gamma} e^{-w_0(x_1)} S_f(x_1). \qquad (4.106)$$

where

$$S_f(x_1) = f_0 + 2\pi\beta \left[\sin(\pi\psi_{ex}) \cos x_1 f_0' + \frac{\Gamma}{4} f_0'' \right]. \qquad (4.107)$$

The reader might have got a bit worried by the length of the calculation. However, we need to remark that this procedure is carried out once and for all functions $f(x_2)$. In fact, by this same procedure we are able to obtain the general expression for the statistical average $\langle f \rangle$ to lowest order in β:

$$\langle f \rangle = \int_0^{2\pi} dx_1 q(x_1) S_f(x_1). \qquad (4.108)$$

Therefore, the effective quantity, whose statistical average in the slow process is to be calculated, is $S_f(x_1)$ given by the simple expression (4.107). Therefore:

$$\langle f \rangle = f_0 + 2\pi\beta \left[\sin(\pi\psi_{ex}) \langle \cos x_1 \rangle f_0' + \frac{\Gamma}{4} f_0'' \right]. \qquad (4.109)$$

4.6.1. *The circulating currents*

Being the circulating current a function only of the variable x_2, we may proceed to finding its statistical average by means of Eq. (4.109), setting, according to Eq. (4.92a):

$$f_0 = f_0'' = 0, \qquad f_0' = -\frac{1}{\pi\beta}. \qquad (4.110)$$

Therefore, from Eq. (4.109) we have:

$$\langle i_S \rangle = -2 \sin(\pi\psi_{ex}) \langle \cos x_1 \rangle. \qquad (4.111)$$

The above expression closely recalls the leading order term in β for the circulating current i_S given in Eq. (3.35a) where the cosine of the variable ϕ is now replaced by the statistical average $\langle \cos x_1 \rangle$. In order to calculate the latter quantity, let us recall Eq. (4.78) and (4.70) and set:

$$\langle \cos x_1 \rangle = \frac{\lambda_1}{\Gamma} \int_0^{2\pi} dx_1 \cos x_1 \, e^{-\frac{w_0(x_1)}{\Gamma}} e^{-2\pi\beta \frac{w_1(x_1)}{\Gamma}} [F(x_1) - C']. \quad (4.112)$$

The leading order term is thus

$$\langle \cos x_1 \rangle = \frac{\lambda_1}{\Gamma} \int_0^{2\pi} dx_1 \cos x_1 \, e^{\frac{i_B}{\Gamma} x_1} e^{z \cos x_1} [F_0(x_1) - C_0']. \quad (4.113)$$

Recognizing that the above expression can be written as follows:

$$\langle \cos x_1 \rangle = \frac{\lambda_1}{\Gamma} \int_0^{2\pi} dx_1 e^{\frac{i_B}{\Gamma} x_1} \frac{\partial}{\partial z} (e^{z \cos x_1}) [F_0(x_1) - C_0'], \quad (4.114)$$

and using the results obtained in the previous section, we write

$$\langle \cos x_1 \rangle = \frac{\lambda_1}{\Gamma} (2\pi N_B' - b_0' K_B - b_0' C_0') = 2\pi N_B' \frac{\lambda_1}{\Gamma}. \quad (4.115)$$

By taking the leading order term in Eq. (4.87), we have:

$$\langle \cos x_1 \rangle = \frac{N_B'}{N_B}, \quad (4.116)$$

so that:

$$\langle i_S \rangle = -2 \sin(\pi\psi_{ex}) \frac{N_B'}{N_B}. \quad (4.117)$$

Representation of the circulating currents as a function of the externally applied flux ψ_{ex} are given in Fig.4.6a and Fig.4.6b for various values of the parameter Γ and i_B, respectively. In particular, in Fig.4.6a, the value of the bias current is fixed to $i_B = 0.20$ while the parameter Γ is varied as follows: $\Gamma = 0.10, 0.50, 0.90$. In Fig.4.6b the value of the parameter Γ is kept fixed to 0.10 and the bias current is varied as follows: $i_B = 0.20$, 0.80, 1.40. We notice that the curves are similar to the ones obtained in the absence of thermal fluctuations in Chapter 3. In fact, if we look at the circulating currents in Fig.3.3a we notice that the sharp discontinuities present at $T = 0$ K now disappear because of thermal fluctuations. In fact, the function defined in Eq. (4.117) does not need any range specification, as in its $T = 0$ K counterpart in Eq. (3.45), where we need to restrict the domain of i_S to those values of ψ_{ex} for which the square root does exist. Once more we notice the role played by thermal fluctuations on smoothing out the sharp point of the physical quantities obtained for $T = 0$ K.

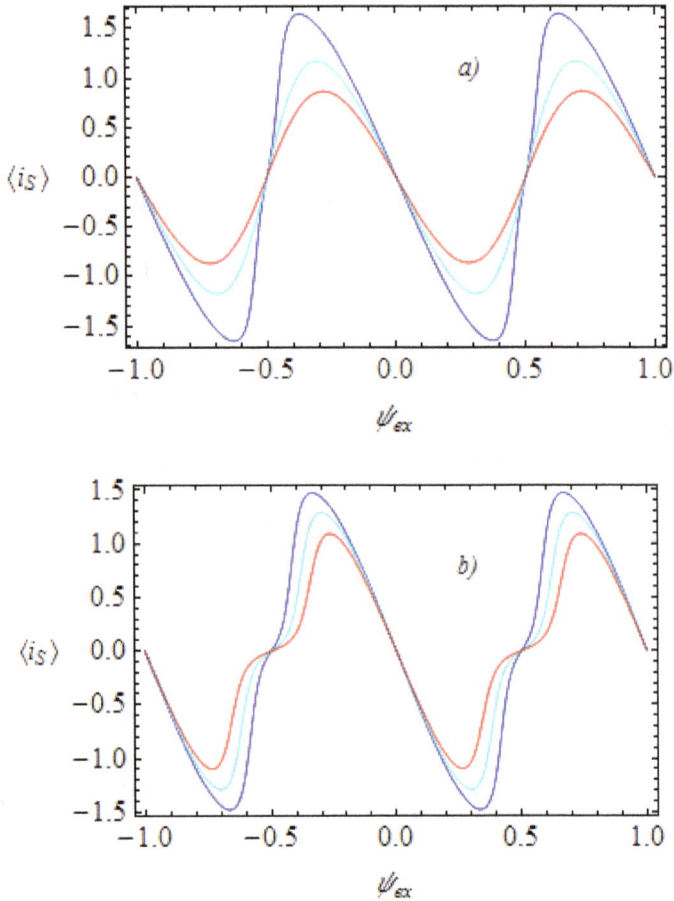

Fig. 4.6. Statistical average $\langle i_S \rangle$ of the circulating currents as function of the normalized externally applied flux ψ_{ex}. In panel a) $i_B = 0.20$ and $\Gamma = 0.10$ (blue curve), $\Gamma = 0.50$ (cyan curve), $\Gamma = 0.90$ (red curve). In panel b) $\Gamma = 0.10$ and $i_B = 0.20$ (blue curve), $i_B = 0.80$ (cyan curve), $i_B = 1.40$ (red curve). In the numerical evaluation of the curves, the series expansion in terms of modified Bessel function has been truncated at the integer $k_{max} = 100$.

4.6.2. *Magnetic susceptibility*

Magnetic susceptibility, as in the case of circulating currents, is defined, by means of Eq.(4.92b) a function of the variable x_2 only. Therefore, we may find its statistical average by means of Eq. (4.109), by setting, according to Eq. (4.92b):

$$f_0 = -(1 - \alpha), \quad f_0' = -\frac{\alpha}{\pi \psi_{ex}}, \quad f_0'' = 0. \qquad (4.118)$$

Therefore, plugging in the values listed in Eq. (4.118) into Eq. (4.109), we have:

$$\langle \chi \rangle = -(1 - \alpha) - 2\pi\beta \frac{\alpha \sin(\pi\psi_{ex})}{\pi\psi_{ex}} \langle \cos x_1 \rangle. \qquad (4.119)$$

Considering now the expression in (4.116) for $\langle \cos x_1 \rangle$, we finally have:

$$\langle \chi \rangle = -(1 - \alpha) - 2\pi\beta\alpha \frac{\sin(\pi\psi_{ex})}{\pi\psi_{ex}} \frac{N_B'}{N_B}. \qquad (4.120)$$

Again, by comparing the above expression with the first-order approximation of the magnetic susceptibility χ given in Eq. (3.35b), we find a perfect agreement of our result, if we consider the cosine term in the $T = 0$ K expression replaced by the statistical average $\langle \cos x_1 \rangle$. When we represent the statistical average $\langle \chi \rangle$ as a function of ψ_{ex}, as done in Fig. 4.7a-c for $\alpha = 0.80$, we may confirm what already noticed for other physical quantities; i.e., that thermal fluctuations tend to make the discontinuities in the χ vs. ψ_{ex} curves obtained at $T = 0$ K (see Fig. 3.4a-b) disappear, in such a way that the curves are smoothly defined on all the ψ_{ex}-axis. In order to show more closely this aspect, in Fig. 4.7a and Fig. 4.7b we fix the value of the parameter Γ to 0.2 . In Fig. 4.7a we take $\beta = 0.04$ and choose the values of the parameter i_B to be the following: 0.00; 0.50; 1.00. In Fig. 4.7b, instead, we fix the bias current, taking $i_B = 0.50$, and choose the values of the parameter β to be the following: 0.01; 0.02; 0.03. Finally, in Fig. 4.7c we fix $\beta = 0.04$ and $i_B = 0.50$, and take: $\Gamma = 0.1$, $\Gamma = 0.4$, $\Gamma = 0.7$. We now spend some words of comment on each figure.

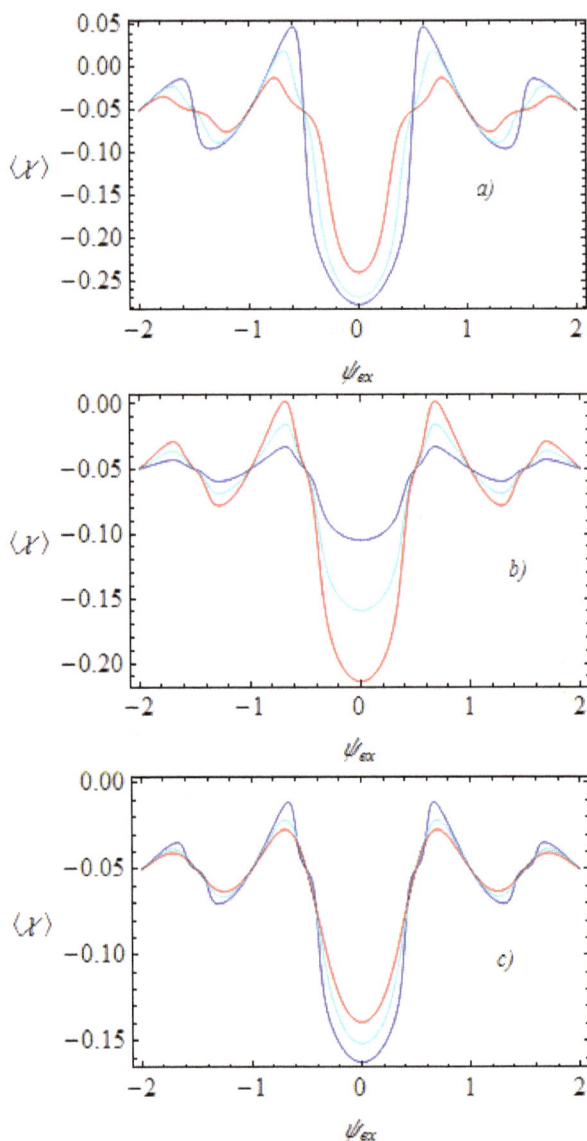

Fig. 4.6. Statistical average $\langle \chi \rangle$ of the magnetic susceptibility as a function of the normalized applied flux ψ_{ex} for $\alpha = 0.95$. In panel (*a*) $\beta = 0.04$, $\Gamma = 0.2$, and $i_B = 0.00$ (blue curve), $i_B = 0.50$ (cyan curve), $i_B = 1.00$ (red curve). In panel (*b*) $i_B = 0.50$, $\Gamma = 0.2$, and $\beta = 0.01$ (blue curve), $\beta = 0.02$ (cyan curve), $\beta = 0.03$ (red curve). Finally, in panel (*c*) $i_B = 0.50$, $\beta = 0.02$, and $\Gamma = 0.1$ (blue curve), $\Gamma = 0.4$ (cyan curve), $\Gamma = 0.7$ (red curve).

As for Fig. 4.7a, notice that, fixing the temperature and the value of β, an increase on the bias current deteriorates the shielding capability of the system at extremely low fields, as it appears from the main lobe of these $\langle \chi \rangle$ vs. ψ_{ex} curves. Furthermore, the paramagnetic response region tends to disappear when i_B grows. As a consequence, as the bias current i_B increases, the curves are compressed in the vertical direction, with plateaus and discontinuities appearing in Fig. 3.4a now replaced by a smooth behavior.

In Fig. 4.7b, on the other hand, where the temperature and the bias current are fixed, an increase in the parameter β makes the shielding capability of the system more effective, so that the curves tend to be stretched in the vertical direction as β grows.

Finally, in Fig. 4.7c the thermal effects for fixed values of β and of the bias current can be detected. Starting from the blue curve obtained for $\beta = 0.04$, $i_B = 0.50$, and $\Gamma = 0.1$, we let thermal fluctuations increase by raising the parameter Γ first to 0.4 and then to 0.7. We thus notice that the wiggles present in the blue curve near the half-integer values of ψ_{ex} tend to fade away in the cyan curve ($\Gamma = 0.4$) and then completely disappear in the red curve ($\Gamma = 0.4$) due to the general smoothing action of noise on the $\langle \chi \rangle$ vs. ψ_{ex} curves, as well as on the $\langle i_S \rangle$ vs. ψ_{ex} curves. Moreover, while the curves are smoothed out by the increase in the parameter Γ, from the main lobe of the curves in Fig. 4.7c a decrease in the shielding capabilities of the system at extremely low fields can be noticed for increasing values of Γ.

Chapter 5

One-dimensional Arrays of Over-damped Josephson Junctions

5.1. A mathematical model

After having considered the basic properties of simple Josephson junction devices, we are now able to consider more complex systems. In the present chapter we analyze the magnetic properties of the one-dimensional Josephson junction array (1D-JJA) represented in Fig. 5.1. The schematic representation of a 1D-JJA given in Fig. 5.1 shows $N + 1$ identical over-damped Josephson junctions connected in parallel. In this figure, the bias current I_B is evenly applied to the two external branches of the array. Other types of bias configurations can be analyzed by following a similar analysis.

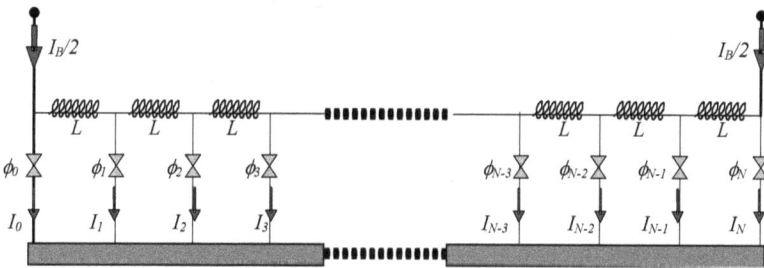

Fig. 5.1. Schematic representation of a one-dimensional array consisting of $N + 1$ equal over-damped Josephson junctions. The bias current I_B is applied to the extremes of the array. Currents flow in the horizontal and vertical branches. An inductance L is associated to each horizontal branch.

The current injected in the 1D-JJA flows in the horizontal and vertical branches in Fig. 5.1 before being collected by a superconducting layer at the bottom. In this way, if we consider the k-th loop (exactly N loops are present in the system), we notice that the incoming current divides into two different branches, according to the scheme in Fig. 5.2.

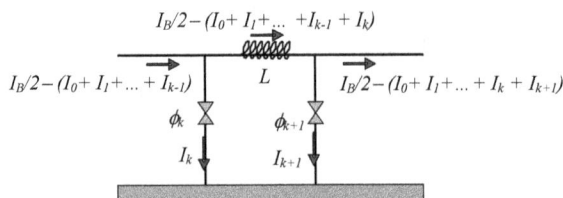

Fig. 5.2. Schematic representation of a single cell in a one-dimensional array consisting of over-damped Josephson junctions. The incoming current $I_B/2 - (I_0 + I_1 + .. + I_{k-1})$ divides into the two branch currents: I_k flowing in the vertical branch on the left; $I_B/2 - (I_0 + I_1 + .. + I_{k-1} + I_k)$ flowing in the horizontal branch inside the cell. By Kirchhoff current law, the same type of splitting occurs at the second upper node of the cell.

By assuming perfectly identical over-damped Josephson junctions with resistive parameter R and maximum Josephson current I_{J0}, we take the inductance L of the horizontal upper branches to be such that the parameter $\beta_0 = LI_{J0}/\Phi_0$ is small; i.e., $\beta_0 \ll 1$. This parameter has been defined as if each single loop could be thought to be a two-junction quantum interference device. By fluxoid quantization, the magnetic flux Φ_{k+1} linked to the cell represented in Fig. 5.2, is related to the superconducting phase differences ϕ_k and ϕ_{k+1} across the two junctions in the cell, so that:

$$2\pi \frac{\Phi_{k+1}}{\Phi_0} + \phi_k - \phi_{k+1} = 2\pi n_{k+1}, \qquad (5.1)$$

where n_{k+1} is an integer, $k = 0, 1, 2, ..., (N-1)$, and the sign of the flux Φ_{k+1} is chosen according to the right-hand rule. By this relation, the superconducting phase difference ϕ_k, for $k = 1, 2, ..., N$, can be written in terms of the homologous quantity ϕ_0, according to the following expression:

$$\phi_k = \phi_0 + 2\pi \sum_{m=1}^{k} \left(\frac{\Phi_m}{\Phi_0} - n_m \right). \tag{5.2}$$

When an external magnetic field $\vec{H}$, orthogonal to the plane of the array, is applied to the system, the geometric applied flux through each cell is $\Phi_{ex} = \mu_0 H A_0$, where A_0 is the cell area. In this way, we may set:

$$\frac{\Phi_{k+1}}{\Phi_0} = \frac{\Phi_{ex}}{\Phi_0} - \beta_0 \left(\frac{i_B}{2} - \sum_{m=0}^{k} i_m \right), \tag{5.3}$$

where $k = 0, 1, 2, \ldots, (N-1)$, $i_B = I_B / I_{J0}$, and $i_k = I_k / I_{J0}$. For short notation, we may further set $\psi_{k+1} = \Phi_{k+1}/\Phi_0$ and $\psi_{ex} = \Phi_{ex}/\Phi_0$.

The dynamic equation for each Josephson junction in the array is written by means of the Resistively Shunted Junction (RSJ) model (Barone, 1982) as follows:

$$\frac{d\phi_k}{d\tau} + \sin \phi_k = i_k, \tag{5.4}$$

where $k = 0, 1, 2, \ldots, N$. Because of Eq. (5.2), and by defining

$$S_k = \sum_{m=1}^{k} (\psi_m - n_m), \tag{5.5}$$

the dynamic equations (5.4) can be written in the following form:

$$\frac{d\phi_0}{d\tau} + \sin \phi_0 = i_0, \tag{5.6a}$$

$$2\pi \frac{dS_k}{d\tau} + \sin(\phi_0 + 2\pi S_k) - \sin \phi_0 = i_k - i_0, \tag{5.6b}$$

with $k = 1, 2, \ldots, N$. For the sake of simplicity, from this point on we take all the loops in the array to be initially magnetically empty; i.e., with no flux inside. In this way, we may set $n_k = 0$, $k = 1, 2, \ldots, N$.

We are interested in getting a set of dynamic equations where we may make use of the assumption $\beta_0 \ll 1$. Therefore, from Eq. (5.3) we write:

$$S_k = k\psi_{ex} - \beta_0 \left(k \frac{i_B}{2} - \sum_{n=1}^{k} \sum_{m=0}^{n-1} i_m \right). \tag{5.7}$$

From Eq. (5.7) we notice that the zero-order approximation for S_k is $S_k^{(0)} = k\psi_{ex}$. The currents in the above equation may be rewritten in terms of the normalized magnetic fluxes ψ_k. In fact, by Eq. (5.3) written for $k + 1$ and k, we can set:

$$i_k = \frac{\psi_{k+1} - \psi_k}{\beta_0}, \tag{5.8}$$

$k = 1, 2, ..., N - 1$. In order to completely define all currents, including i_0 and i_N, we might proceed as follow. We first show that, by evaluating the double summation in Eq. (5.7), we have:

$$\beta_0 \sum_{n=1}^{k} \sum_{m=0}^{n-1} i_m = k[\beta_0 i_0 - \psi_1] + S_k. \tag{5.9}$$

Therefore, by substituting this result into Eq. (5.7), we have:

$$S_k = k \left[\psi_{ex} - \psi_1 - \beta_0 \left(\frac{i_B}{2} - i_0 \right) \right] + S_k. \tag{5.10}$$

In this way, by setting to zero the term in parenthesis in Eq. (5.10), we can define i_0 as follows:

$$i_0 = \frac{i_B}{2} - \frac{\psi_{ex} - \psi_1}{\beta_0}. \tag{5.11}$$

The above expression is consistent with Eq. (5.3) for $k = 0$. Of course, this second route in finding i_0 is more immediate. However, this type of calculation is useful to show that Eq. (5.7) may indeed be used, as noted before, to find a zero-order expression in β_0 for S_k, namely, $S_k^{(0)} = k\psi_{ex}$, but it will not be used to actually calculate S_k. On the same token, by using Eq. (5.3) for $k = N$, we find:

$$i_N = \frac{i_B}{2} - \frac{\psi_N - \psi_{ex}}{\beta_0}. \tag{5.12}$$

Therefore, by virtue of Eq. (5.2), the superconducting phases ϕ_k can be written in terms of the flux variables ψ_k, for $k = 1, 2, ..., N$. Moreover, by means of Eq. (5.8), (5.11), and (5.12), all currents are expressed in terms of the quantities ψ_k, for $k = 1, 2, ... N$. In this way, the dynamic equations (5.6b) can be written in terms of the normalized magnetic fluxes ψ_k and of the variable ϕ_0. We shall explicitly derive these equations in Section 5.3.

5.2. The voltage-flux curves

In the previous section we have considered the general equations of a current-biased one-dimensional array of Josephson junctions. In order to derive the voltage-flux characteristics, at least in the zero-order approximation in β_0, we might take $S_k = k\psi_{ex}$ in Eq. (5.7), so that:

$$\phi_k = \phi_0 + 2\pi k \psi_{ex}. \tag{5.13}$$

By now substituting the above expression in Eq. (5.4) and by summing over k, we have:

$$(N+1)\frac{d\phi_0}{d\tau} + \sum_{k=0}^{N} \sin(\phi_0 + 2\pi k \psi_{ex}) = i_B. \tag{5.14}$$

The summation in Eq. (5.14) can be evaluated as follows. First write:

$$\sigma_N = \sum_{k=0}^{N} \sin(\phi_0 + 2\pi k \psi_{ex}) = Im\{e^{i\phi_0} \sum_{k=0}^{N} e^{i2\pi k \psi_{ex}}\}. \tag{5.15}$$

Therefore, by considering the residual term within the summation sign and by applying the well-known formula of the partial sum of a geometric series, we have:

$$\sigma_N = Im\left\{e^{i\phi_0} \frac{e^{i2\pi(N+1)\psi_{ex}} - 1}{e^{i2\pi\psi_{ex}} - 1}\right\} = \frac{\sin[\pi(N+1)\psi_{ex}]}{\sin(\pi\psi_{ex})} Im\{e^{i(\phi_0 + \pi N \psi_{ex})}\}. \tag{5.16}$$

By now defining

$$\phi = \phi_0 + \pi N \psi_{ex}, \tag{5.17}$$

we may rewrite Eq. (5.14) as follows:

$$\frac{d\phi}{d\tau} + \frac{1}{N+1}\frac{\sin[\pi(N+1)\psi_{ex}]}{\sin(\pi\psi_{ex})}\sin(\phi) = \frac{i_B}{N+1}. \tag{5.18}$$

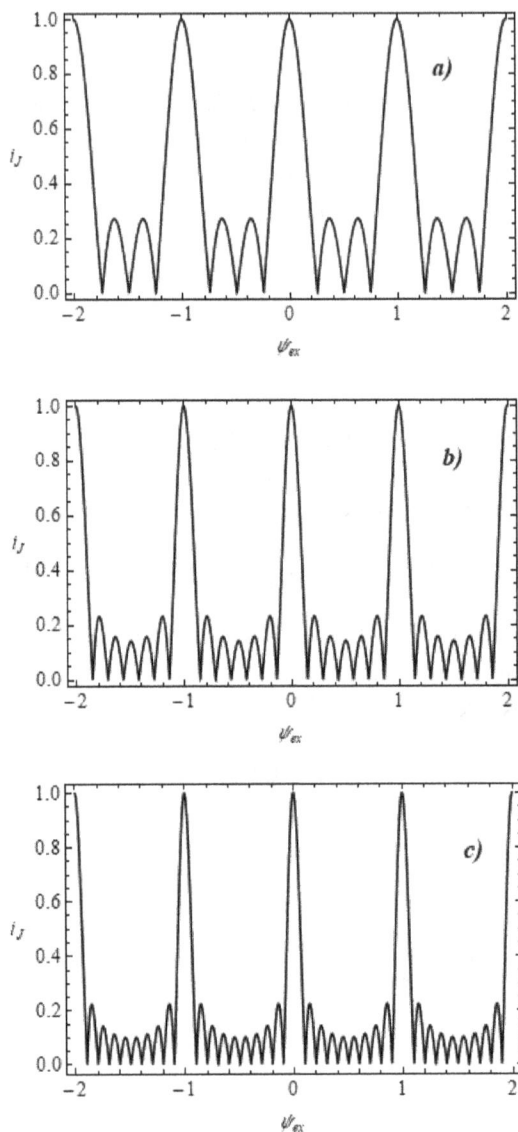

Fig. 5.3. Normalized critical current i_J for a one-dimensional array of $N+1$ over-damped Josephson junctions as a function of the normalized applied flux ψ_{ex}. In panel a) $N = 3$; in b) $N = 6$; in c) $N = 9$. All principal maxima occur at integer values of ψ_{ex}; the number of secondary lobes in between two adjacent principal maxima is equal to $N - 1$. Notice that all curves have the same upper bound $i_J = 1$.

Eq. (5.18) is similar to the dynamic equation for the superconducting phase of a single Josephson junction exhibiting a critical current i_J of the following type:

$$i_J(\psi_{ex}) = \frac{1}{N+1} \left| \frac{\sin[\pi(N+1)\psi_{ex}]}{\sin(\pi\psi_{ex})} \right|. \tag{5.19}$$

A representation of the dependence of i_J on ψ_{ex} is given in Fig. 5.3a–c for various values of N. Recalling the way we proceeded for a two-junction quantum interferometer in Section 3.1, starting from Eq. (3.9), the voltage-current characteristics of a one-dimensional array in the $\beta_0 = 0$ limit can be obtained. In fact, we start by noticing that, for

$$-\left| \frac{\sin[\pi(N+1)\psi_{ex}]}{\sin(\pi\psi_{ex})} \right| \leq i_B \leq +\left| \frac{\sin[\pi(N+1)\psi_{ex}]}{\sin(\pi\psi_{ex})} \right|, \tag{5.20}$$

Eq. (5.18) has a fixed point similar to the one represented in Eq. (3.10), namely:

$$\phi^* = \sin^{-1}\left(\frac{\sin(\pi\psi_{ex})}{\sin[\pi(N+1)\psi_{ex}]} i_B \right). \tag{5.21}$$

Recall now that the voltage V across the device can be normalized as follows:

$$v = \frac{V}{RI_{J0}} = \frac{d\phi}{d\tau}, \tag{5.22}$$

where R is the resistive parameter and I_{J0} is the maximum Josephson current of each junction in the array. Therefore, for bias currents satisfying Eq. (5.20), the time-averaged normalized voltage $\langle v \rangle$ is zero. On the other hand, when Eq. (5.20) is not satisfied, a resistive state with $\langle v \rangle \neq 0$ is realized in the array.

As for the resistive state, by considering the interval of values for i_B complementary to the one specified in Eq. (5.20), we may solve Eq. (5.18) by separation of variables, as done in Section 1.4. In this case the normalized maximum Josephson current of the effective junction is equal to $i_J(\psi_{ex})$, as defined in Eq. (5.19). As done in Chapter 3 for the two-junction quantum interferometer, the solution of the ordinary differential equation (5.18) can be obtained by considering Eq. (1.38), where m_0 is the forcing term, setting

$$m_0 = \frac{i_B}{(N+1)i_J(\psi_{ex})} > 1. \tag{5.23}$$

Let us now divide both members of Eq. (3.9) by $i_J(\psi_{ex})$ and let us define a new time-variable:

$$\tau' = i_J(\psi_{ex})\tau. \tag{5.24}$$

In this way, Eq. (5.18) can be rewritten as follows

$$\frac{d\phi}{d\tau'} + \text{sgn}\left\{\frac{1}{N+1}\frac{\sin[\pi(N+1)\psi_{ex}]}{\sin(\pi\psi_{ex})}\right\}\sin(\phi) = \frac{i_B}{(N+1)i_J(\psi_{ex})}. \tag{5.25}$$

where $\text{sgn}(x)$, we recall, denotes the sign function, which extracts the sign of a real argument x.

For positive values of the multiplicative sign of the sine function in Eq. (5.25), the solution to Eq. (5.18) can be found. In fact, it corresponds to the right hand side of Eq. (1.44) with τ replaced by τ', so that exactly the same expression as in Eq. (3.14) can be written for the function $\phi(\tau')$. As done in Chapter 3, the period T' of the function $\phi(\tau')$ can be seen to be equal to $2\pi/\sqrt{m_0^2 - 1}$. In this way, the time-averaged voltage $\langle v \rangle$ can be written as follows for $i_B > 0$:

$$\langle v \rangle = \frac{2\pi}{T} = \frac{2\pi}{T'}\frac{T'}{T} = i_J(\psi_{ex})\sqrt{\left[\frac{i_B}{(N+1)i_J(\psi_{ex})}\right]^2 - 1}, \tag{5.26}$$

where we used Eq. (5.24) to calculate the ratio T'/T.

For negative values of the multiplicative sign of the sine function in Eq. (5.25), on the other hand, one needs to find solution to the following equation

$$\frac{d\phi}{d\tau'} - \sin\phi = m_0, \tag{5.27}$$

recalling that $m_0 > 1$, as set in Eq. (5.23). Proceeding again as in Section 1.4, we may find the solution in this case is formally equal to Eq. (3.22). Therefore, one argues that the period is again equal to $2\pi/\sqrt{m_0^2 - 1}$, so that Eq. (5.26) gives the correct expression for $\langle v \rangle$ also in this second case. Finally, the case $i_J(\psi_{ex}) = 0$ is immediately solved by noticing that Eq. (5.18) directly gives $\langle v \rangle = i_B/(N + 1)$.

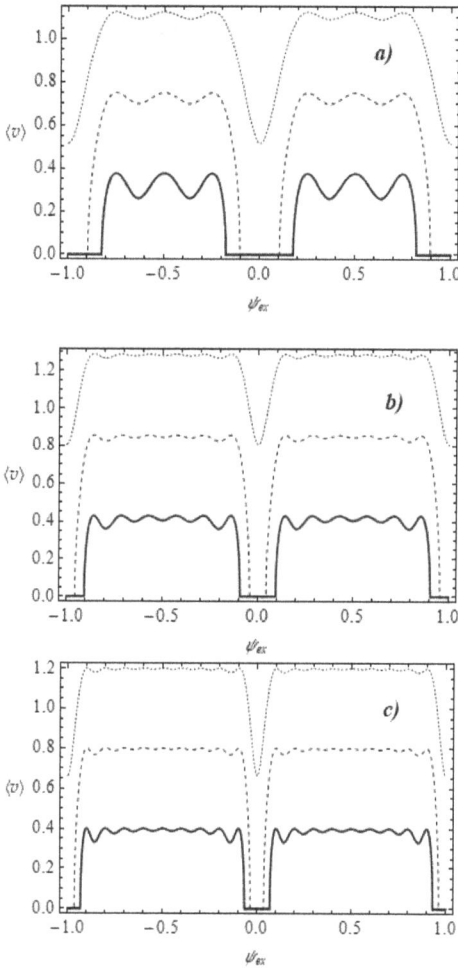

Fig. 5.4. (*a*) Voltage-flux characteristics of one-dimensional arrays of $N + 1$ over-damped Josephson junctions. In panel a) $N = 3$ and $i_B = 1.50$ (full line), $i_B = 3.00$ (dashed line), $i_B = 4.50$ (dotted line). In panel b) $N = 6$ and $i_B = 3.00$ (full line), $i_B = 6.00$ (dashed line), $i_B = 9.00$ (dotted line). In panel c) $N = 9$ and $i_B = 4.00$ (full line), $i_B = 8.00$ (dashed line), $i_B = 12.0$ (dotted line). All curves are periodic with period $\Delta\psi_{ex} = 1.0$.

The voltage-flux characteristics are shown in Fig. 5.4a–c for various numbers of junctions in the array and for three different values of the normalized bias current i_B. In particular, in Fig. 5.4a–c the normalized bias current i_B is taken below (full-line curve), in the vicinity (dashed-line curve) or above (dotted-line curves) the value $N + 1$. In the first two cases $(i_B < N + 1)$ the curves are non-zero only in those intervals in which $i_J(\psi_{ex}) < i_B/(N + 1)$. In the second case $(i_B > N + 1)$ the inequality $i_J(\psi_{ex}) < i_B/(N + 1)$ is satisfied for all values of ψ_{ex}, and the voltage curves lie above the ψ_{ex}-axis. The periodicity of all curves in Fig. 5.4a–c is exactly equal to $\Delta\psi_{ex} = 1.0$, as it can be also argued from Eq. (5.19) and (5.26).

5.3. Static magnetic response

In the previous section we have considered an inductive array of $N + 1$ over-damped Josephson junctions to which both a bias current and a magnetic flux is applied. We have seen that, depending on the value of the normalized applied flux, the array shows a critical current $i_J(\psi_{ex})$, whose expression, in the limit of negligible inductance values $(\beta_0 = 0)$, is given by Eq. (5.19). This system is seen to show richer voltage-flux characteristics than two-junction quantum interference devices.

In the present section we shall consider a similar array with non-negligible value of β_0, in which we have no current bias (i. e., $i_B = 0$). Therefore, by starting from the dynamic equations for the superconducting phase variables ϕ_k given in Eq. (5.6a–b) for $k = 0, 1, 2, \ldots N$, and by setting $i_B = 0$, we shall be able to derive the corresponding equations for the magnetic flux variables ψ_k $(k = 1, 2, \ldots N)$. In this way, the flux and current distribution inside the system can be analyzed in static conditions (absence of flux flow). In addition, the d. c. susceptibility can be found.

Let us then consider the generalization of Eq. (5.8) to all currents flowing in the system by writing again

$$i_k = \frac{\psi_{k+1} - \psi_k}{\beta_0}, \tag{5.28}$$

with the only difference that this this time, by defining $\psi_0 = \psi_{N+1} = \psi_{ex}$, we can extend, for $i_B = 0$, this definition to $k = 0, 1, 2, ... \, N$, according to what was prescribed by Eq. (5.11) and (5.12). In this way, once the flux distribution in the system is known, the current distribution can be readily found by means of Eq. (5.28).

In order to find the equilibrium values of the flux variables ψ_k ($k = 1, 2, ... \, N$) at fixed values of the applied normalized magnetic flux ψ_{ex}, we proceed as follows. By considering Eq. (5.6b), we consider two successive indices and, by noticing that $S_k - S_{k-1} = \psi_k$, we rewrite Eq. (5a–b) as follows:

$$\frac{d\phi_0}{d\tau} + \sin\phi_0 = i_0, \tag{5.29a}$$

$$2\pi \frac{d\psi_1}{d\tau} + \sin(\phi_0 + 2\pi\psi_1) - \sin\phi_0 = i_1 - i_0, \tag{5.29b}$$

$$2\pi \frac{d\psi_k}{d\tau} + \sin(\phi_0 + 2\pi S_k) - \sin(\phi_0 + 2\pi S_{k-1}) = i_k - i_{k-1}, \tag{5.29c}$$

($k = 2, ... \, N$), where the terms S_k sum all flux numbers from 1 to k and where all currents i_k on the right hand side of the above equations can be defined in terms of the normalized quantities ψ_k, ($k = 1, 2, ... \, N$), as specified in Eq. (5.28). Going a step further, we can define the following new set of variables:

$$y_k = \pi\psi_k, \tag{5.30}$$

for $k = 0, 1, ..., N, N + 1$. Therefore, considering the above definition, by using trigonometric identities and by utilizing Eq. (5.28), we can recast Eq. (5.29a–c) in the following form:

$$\frac{d\phi_0}{d\tau} + \sin\phi_0 = \frac{y_1 - y_0}{\pi\beta_0}, \tag{5.31a}$$

$$\frac{dy_k}{d\tau} + \cos(\phi_0 + y_k + 2\sigma_{k-1})\sin y_k = \frac{y_{k+1} - 2y_k + y_{k-1}}{2\pi\beta_0}, \tag{5.31b}$$

where $\sigma_k = \pi S_k$. We notice that coupling between the superconducting phase difference ϕ_0 to the quantities y_k is realized in all expressions in Eq. (5.31a–b). Before attempting to numerically solve the set of differential equations (5.31a–b), we may preliminary notice that, due to the symmetry of the problem, the following relation is satisfied:

$$y_{N-k} = y_{k+1}, \tag{5.32}$$

where $k = 0, 1, 2, \ldots, (N-1)/2$ for N odd or $k = 0, 1, 2, \ldots, N/2$ for N even. In this way, by considering Eq. (5.28) and (5.30), we expect the currents to obey to the following symmetry relation:

$$i_{N-k} = -i_k. \tag{5.33}$$

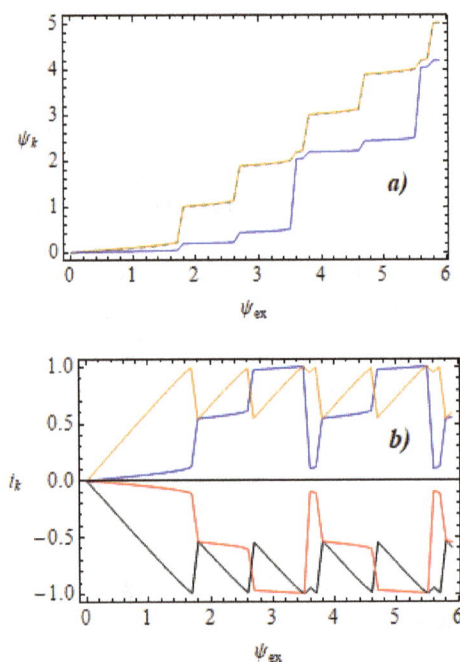

Fig. 5.5. (*a*) Flux distribution in a one-dimensional array of $N + 1 = 4$ over-damped Josephson junctions as a function of ψ_{ex}: the top yellow curve represents the normalized magnetic fluxes ψ_1 and ψ_3 threading the first and third loop, respectively. The lower blue curve represents the flux ψ_2 threading the inner loop. (*b*) Currents flowing in the junctions of a one-dimensional array as a function of the normalized magnetic flux. Black and orange curves are for i_0 and i_3, respectively. Red and blue curves are for i_1 and i_2, respectively. For all curves $\beta_0 = 1.50$.

These features are clearly visible in the numerically evaluated ψ_k vs. ψ_{ex} and in the i_k vs. ψ_{ex} shown in Fig. 5.5a and 5.5b, respectively, for $N = 3$ and $\beta_0 = 1.50$. In fact, in Fig. 5.5a the normalized magnetic flux in the outer loops ($k = 1$ and $k = 3$) are represented by the same curve (in orange), while the flux threading the inner loop is shown as a blue curve. The characteristic features of these graphs are the following. First, of all, the peculiar step-like appearance can be related to the corresponding curves for a single junction interferometer studied in Section 2.4. In fact, in this case both external junctions act as a bridge for flux penetration inside the system. Therefore, as seen for a single-junction interferometer, while the system is able to partially shield the externally applied flux up to a threshold value ψ_{th}, which will be seen to depend on β_0, flux penetration for $\psi_{ex} > \psi_{th}$ occurs in steps of one flux quantum. Fig. 5.5b, on the other hand, clarifies the causes of flux penetration. At extremely low magnetic field, we may set $\psi_2 \approx 0$ and $\psi_1 = \psi_3 \approx \varepsilon \, \psi_{ex}$, with $\varepsilon \ll 1$. In this way, we have

$$i_0 \approx -(1-\varepsilon)\frac{\psi_{ex}}{\beta_0}, \; i_1 = -i_2 \approx -\varepsilon\frac{\psi_{ex}}{\beta_0}, \; i_3 \approx +(1-\varepsilon)\frac{\psi_{ex}}{\beta_0}. \quad (5.34)$$

Therefore, while the peripheral currents increase in intensity for growing values of ψ_{ex}, the absolute values of the two inner currents remain anchored to very low values, when compared to i_0 and i_3. However, as seen in the first chapter, when the normalized current reaches the values ±1 a resistive state sets in and flux penetration occurs. Therefore, by setting $i_3 = 1$ in Eq. (5.34) at $\psi_{ex} = \psi_{th}$, we have:

$$\psi_{th} \approx \frac{\beta_0}{(1-\varepsilon)}. \quad (5.35)$$

By fluxoid quantization on the third loop. we may write:

$$\phi_3 = \phi_2 + 2\pi\psi_3. \quad (5.36)$$

Now, because $i_3 = 1$, one has $\phi_3 = \pi/2$. Therefore, neglecting ϕ_2 in Eq. (5.36), for $\psi_{ex} = \psi_{th}$ we may write:

$$\psi_3 \approx \frac{1}{4} \quad (5.37)$$

In this way, we may set:

$$1 = \frac{\psi_{th} - \psi_3}{\beta_0} \rightarrow \psi_{th} = \beta_0 + \frac{1}{4}, \tag{5.38}$$

so that, by combining Eq. (5.38) and Eq. (5.35), we have

$$\varepsilon \approx \frac{1}{4\beta_0 + 1}. \tag{5.39}$$

Therefore, in order to have $\varepsilon \ll 1$, we need to require at least $\beta_0 > 1$. For $\beta_0 = 1.50$ Eq. (5.38) predicts $\psi_{th} = 1.75$. This figure is reproduced with good agreement, despite the rough approximations made, by the numerical evaluations in Fig. 5.5a–b.

After having explained the basic mechanisms by which flux penetration inside the array occurs, we can compute the magnetic susceptibility χ of the system, by setting

$$\chi = \frac{\psi_{TOT} - \psi_{ex}}{\psi_{ex}}, \tag{5.40}$$

where $\psi_{TOT} = \sigma_N / \pi$ represents the total normalized magnetic flux inside the array. Having already computed the flux and current distribution inside the system at various values of the externally applied normalized flux, the susceptibility χ can be found by the data gathered in the previous numerical analysis. Therefore, in Fig. 5.6 the susceptibility χ is reported for $N = 3$ and for $\beta_0 = 0.30$ (orange curve), $\beta_0 = 0.70$ (cyan curve), and $\beta_0 = 1.10$ (magenta curve). We notice that for the first type of curve ($\beta_0 = 0.30$) the shielding capabilities of the system are weak, so that paramagnetic states at very low fields may appear for specific values of the normalized applied magnetic flux. This is confirmed by the fact that the orange susceptibility curve lies above the ψ_{ex}-axis for non-negligible ranges of the normalized applied magnetic flux. On the other hand, as the parameter β_0 increases, the curves in Fig. 5.6 tend to go toward negative values of the susceptibility χ, as it is evident for the cyan and magenta curves. We finally notice that the discontinuities in the χ vs. ψ_{ex} curves represent irreversible flux penetration in the array.

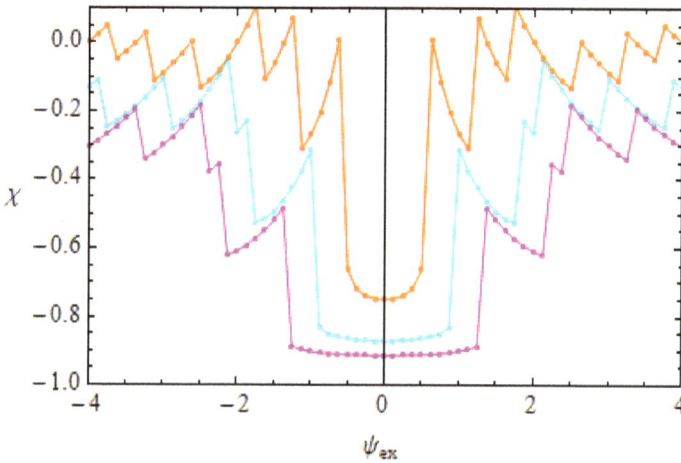

Fig. 5.6. Magnetic susceptibility of a one-dimensional array of $N + 1 = 4$ over-damped Josephson junctions as a function of ψ_{ex} for $\beta_0 = 0.30$ (orange), $\beta_0 = 0.70$ (cyan), and $\beta_0 = 1.10$ (magenta). The numerically computed values are points. The colored lines are guides for the eye.

In order to address more specifically the question of irreversible and reversible magnetic behavior in these systems, we might analyze the ψ_k vs. ψ_{ex} curves for various values of β_0, calculated for a complete sweep of the normalized applied flux. Therefore, starting from zero-field conditions ($\psi_k = 0$ in all loops of the array), we might first let the externally applied Field grow from zero to a maximum value $\Delta\psi_{ex}$. In a second run we let the field decrease from $\Delta\psi_{ex}$ to $-\Delta\psi_{ex}$. Finally, in a third run, the Field is brought again to $\Delta\psi_{ex}$, going through the value $\psi_{ex} = 0$. In the presence of irreversibility, hysteretic behavior of these curves occurs. Therefore, in order to detect the way irreversibility appears in the one-dimensional array considered in Fig. 5.6, we solve the coupled dynamic equations (5.31a–b) for the cyclic variation of described above. The results are shown in Fig. 5.7a–b for two values of the parameter β_0; namely, $\beta_0 = 0.80$ in panel a) and $\beta_0 = 1.20$ in panel b). As we have already noticed, the shielding capability of the system, at very low values of the magnetic field, increases for increasing values of the parameter β_0. By the same token, the hysteretic loops in the ψ_k vs. ψ_{ex} curves are larger for higher values of β_0 as it can be argued from Fig. 5.7a–b. In particular, by looking at the symmetric outer fluxes ψ_1

and ψ_3 in both panels of Fig. 5.7a–b, the dimensions of the hysteretic loops depend principally on the lower threshold field value. The same can be said for the dashed curves, representing the normalized flux ψ_2 in the inner loop of the array.

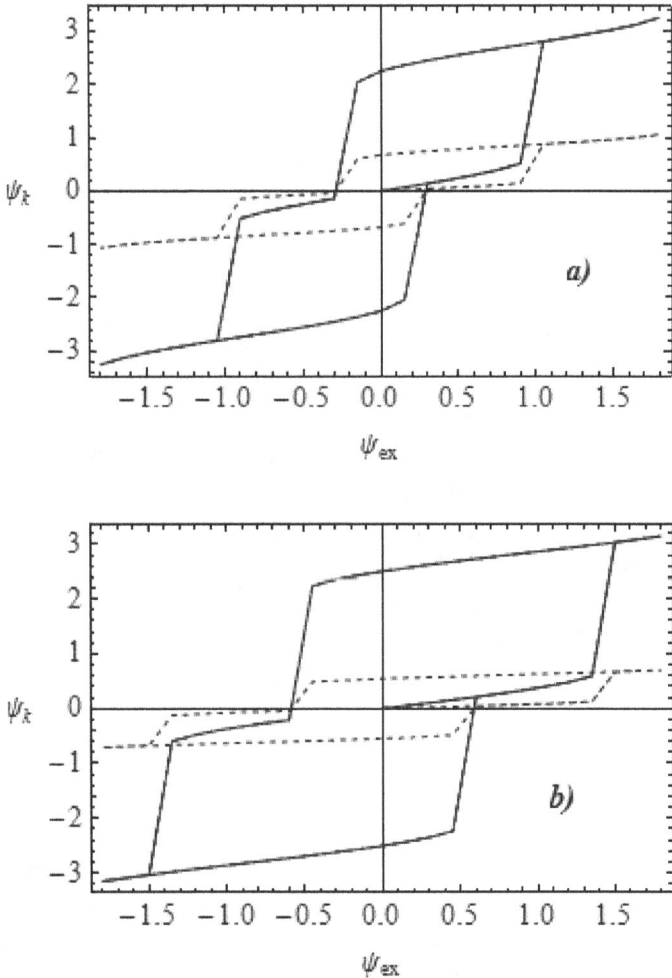

Fig. 5.7. Normalized magnetic flux as a function of ψ_{ex} for a one-dimensional array of $N + 1 = 4$ over-damped Josephson junctions for $\beta_0 = 0.80$ in panel a), and for $\beta_0 = 1.20$ in panel b). In both panels the full lines represent the normalized fluxes in the outer loops ψ_1 and ψ_3. The dashed lines represent the normalized applied flux ψ_2 in the inner loop.

5.4. Non-homogeneous systems and π-junctions

The properties of a two-junction quantum interference device (Barone, 1982; Clarke, 2004) have been already studied in Chapter 3. These systems can be considered to be the simplest one-dimensional Josephson junction array (1D JJA) of over-damped Josephson junctions ($N = 1$). In order to discuss the role of non-homogeneity in 1D JJAs, let us define as 0-JJs those junctions with a conventional Current–Phase Relation (CPR); i. e., those JJs for which the strict Josephson current-phase relation in Eq. (1.4a) applies. The possibility of fabricating π-junctions (π-JJs) (Boulaevskii, 1977; Geshkenbein, 1987; Baselmans, 1999; Ryazanov, 2001) has made possible construction of π-SQUIDs, consisting of a 0-JJs in parallel with a π-JJ. These devices may be realized either by exploiting the symmetry properties of d-wave superconductors (Schultz, 2000) or by utilizing both s-wave and d-wave superconductors (Wollman, 1993; Smilde, 2004).

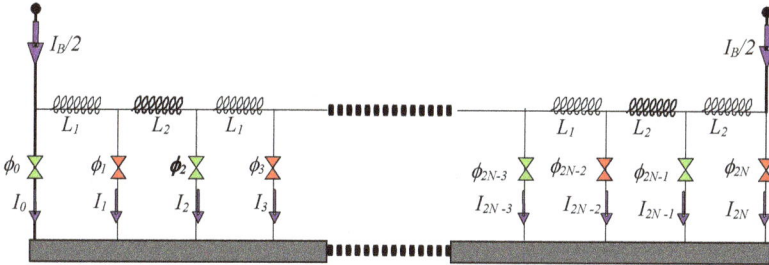

Fig. 5.8. Schematic representation of a one-dimensional array consisting of $2N$ over-damped Josephson junctions and $2N - 1$ superconducting loops. In this scheme in green are represented the 0-junctions and in red the 0-junctions. The bias current I_B is applied to the extremes of the array. Inductances L_1 and L_2 are associated to the horizontal branch in odd-numbered and even-numbered loops, respectively.

As we have previously seen for a 0-SQUID (i.e., a two-junction quantum interference device fabricated with conventional Josephson junctions), a π-SQUID can be viewed as an elementary cell of a $N\times(0-\pi)$ one-dimensional array of overdamped Josephson junctions as shown in Fig. 5.8. Similar systems have been fabricated and tested by Scharinger *et al.* (Scharinger, 2010) using 0-π tunnel junctions with ferromagnetic barrier (Weides, 2006). These devices, in which the critical current density alternates between two opposite values along the junction length, are named multi-facets Josephson junctions (MJJs) (Scharinger, 2010). The experimental properties of MJJs, whose discrete model correspond to a $N \times (0\text{-}\pi)$ one-dimensional array of small Josephson junctions, should therefore show similarities in the magnetic properties of the system shown in Fig. 5.8. However, even though some characteristic features of MJJs can be qualitatively reproduced by $N \times (0\text{-}\pi)$ one-dimensional arrays, one should bear in mind that the latter are, in general, less complex systems than MJJs.

In the previous section it was shown that, for non-negligible values of the characteristic parameter β_0, the magnetic properties of conventional arrays of overdamped Josephson junctions had to be described by means of coupled non-linear ordinary differential equations. Useful information on the quantum interference patterns in these systems could still be obtained in the $\beta_0 = 0$ limit, as done in Section 5.2. In this limit, the current-voltage characteristics for 1D JJAs can be derived by analogy with the well-known results obtained for a single overdamped Josephson junction. Analogously to what was done in Section 5.2, it will be shown that an equivalent single-junction model for a $N \times (0\text{-}\pi)$ one-dimensional array can be found for negligible values of the parameter β_0 defined for a single representative superconducting loop in the array. All junctions in the array will be considered to have the same resistive parameter R. The maximum Josephson current, however, due to the different superconducting coupling (0 or π) between the two electrodes of each junctions, is taken to alternate in sign along the longitudinal direction. Therefore, a π-JJ follows a 0-JJ in the array. Moreover, while in a first subsection we shall consider equal loop areas ($A_k = A_0$, for $k = 1, ..., 2N - 1$) and equal values of the inductances ($L_1 = L_2 = L_0$), in a second subsection we shall consider alternating values of the areas A_1 and A_2 and of the inductances L_1 and L_2.

5.4.1. *N×(0-π) one-dimensional arrays with equal loop areas*

In order to write general equations for a non-homogeneous 1.4.1 $N×(0-\pi)$ one-dimensional arrays with equal loop areas, let us reconsider all expressions written in Section 5.1 for a homogeneous array of 0-JJs. Let us rewrite the fluxoid quantization conditions, linking the superconducting phase differences ϕ_{k-1} and ϕ_k across the two junctions in the k-th cell to the normalized flux ψ_k, so that:

$$2\pi\psi_k + \phi_{k-1} - \phi_k = 2\pi n_k, \tag{5.41}$$

for $k = 1, 2, ..., 2N - 1$, where we take the integers n_k equal to zero for simplicity. By this hypothesis and by recursive use of Eq. (5.41), we may write:

$$\phi_k = \phi_0 + 2\pi \sum_{m=1}^{k} \psi_m. \tag{5.42}$$

If an external magnetic field $\vec{H}$, orthogonal to the plane of the array, is applied to the system, the normalized geometric applied flux through each cell of area A_0 is $\psi_{ex} = \mu_0 H A_0 / \Phi_0$. In this way, for $k = 1, 2, ..., 2N - 1$, we have:

$$\psi_k = \psi_{ex} - \beta_0 \left(\frac{i_B}{2} - \sum_{m=0}^{k-1} i_m \right), \tag{5.43}$$

where $i_B = I_B / I_{J0}$, and $i_k = I_k / I_{J0}$, I_{J0} being the modulus of the maximum Josephson currents of all JJs in the array. One may preliminary notice that Eq. (5.41)–(5.43) are formally identical to Eq. (5.1)–(5.3). The system differs from the 0-JJ array in that the dynamic equation for each Josephson junction have to be written according to the pairing state of device. Therefore, by adopting the Resistively Shunted Junction (RSJ) model (Barone, 1982) and by considering Eq. (5.4) with the substitution

$$\phi_k \rightarrow \phi_k + \frac{1-(-1)^k}{2} \pi, \tag{5.44}$$

we may write:

$$\frac{d\phi_k}{d\tau} + (-1)^k \sin \phi_k = i_k, \tag{5.45}$$

where $k = 0, 1, 2, \ldots, 2N - 1$. By now defining

$$S_k = \sum_{m=1}^{k} \psi_m, \qquad (5.46)$$

the dynamic equations (5.45) can be written in the following form:

$$\frac{d\phi_k}{d\tau} + (-1)^k \sin(\phi_0 + 2\pi S_k) = i_k, \qquad (5.47)$$

with $k = 0, 1, 2, \ldots, 2N - 1$. We are here interested in the quantum interference patterns of the system (i_c vs. ψ_{ex} curves) and in current voltage characteristics ($\langle v \rangle$ vs. i_B curves) for negligible values of the parameter β_0. In this way, we may set $S_k = k\psi_{ex}$, so that from Eq. (5.47) we obtain:

$$2N \frac{d\phi_0}{d\tau} + \sum_{k=0}^{2N-1}(-1)^k \sin(\phi_0 + 2\pi k\psi_{ex}) = i_B, \qquad (5.48)$$

where we have made use of the fact that, in this limit, the derivatives of ϕ_k and ϕ_0 coincide for constant values of ψ_{ex} by (5.42). By carrying out the summation in Eq. (5.48) as in Eq. (5.15) and (5.16), we have:

$$\frac{d\phi_0}{d\tau} + i_P(\psi_{ex}) \sin\left[\phi_0 + \pi(2N - 1)\left(\psi_{ex} + \frac{1}{2}\right)\right] = \frac{i_B}{2N}, \qquad (5.49)$$

where

$$i_P(\psi_{ex}) = \frac{1}{2N} \frac{\sin\left[2\pi N\left(\psi_{ex} + \frac{1}{2}\right)\right]}{\sin\left[\pi\left(\psi_{ex} + \frac{1}{2}\right)\right]}. \qquad (5.50)$$

Therefore, it is remarkable to note that from Eq. (5.48) and (5.50) it can be argued that the properties of $N \times (0\text{-}\pi)$ one-dimensional arrays can be derived from the properties of conventional 1D JJAs by the following simple substitution:

$$\psi_{ex} \to \psi_{ex} + \frac{1}{2}. \qquad (5.51)$$

Therefore, the quantum interference pattern, given by the i_c vs. ψ_{ex} curves, and the voltage-flux characteristics for negligible values of the parameter β_0 can be obtained from Eq. (5.19) and (1.20) and from Eq. (5.26), respectively, by considering the translation given in Eq.

(5.51). In this way, these ψ_{ex}-displaced curves are shown in Fig. 5.9a–c and Fig. 5.10a–c, respectively.

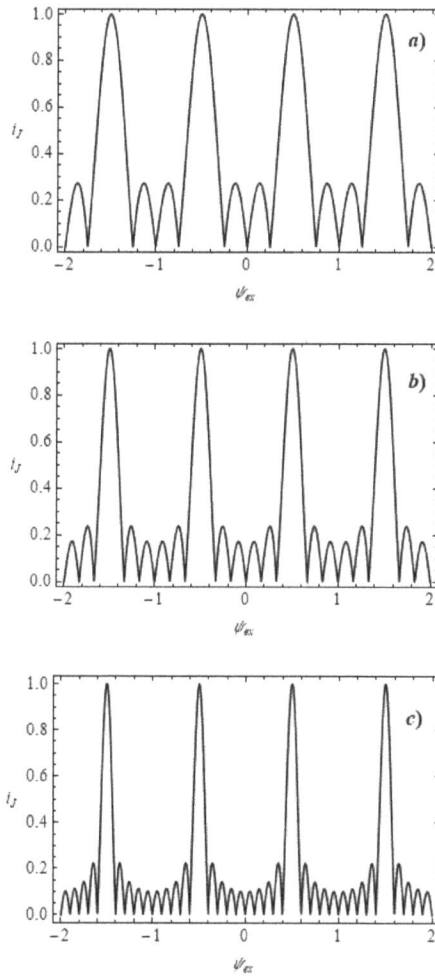

Fig. 5.9. Normalized critical current $i_c = |i_P|$ for of $N\times(0\text{-}\pi)$ one-dimensional arrays as a function of the normalized applied flux ψ_{ex}. In panel a) $N = 2$; in b) $N = 3$; in c) $N = 5$. All principal maxima occur at semi-integer values of ψ_{ex}; the number of secondary lobes in between two adjacent principal maxima is equal to $2N - 2$. Notice that all curves have the same upper bound $i_c = 1$.

In particular, in Fig. 5.9a–c, a representation of the dependence of the critical current $i_c = |i_P|$ on ψ_{ex} is given for the following value of the index N: a) 2; b) 3; c) 5.

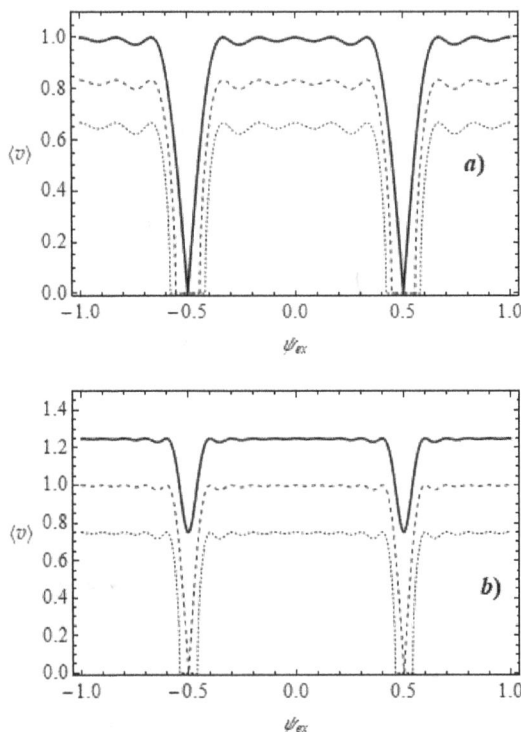

Fig. 5.10. (*a*) Voltage-flux characteristics of $N\times(0\text{-}\pi)$ one-dimensional array. In panel a) $N = 3$ and $i_B = 4.0$ (dotted line), $i_B = 5.0$ (dashed line), $i_B = 6.0$ (full line). In panel b) $N = 5$ and $i_B = 7.5$ (dotted line), $i_B = 10.0$ (dashed line), $i_B = 12.5$ (full line). All curves are periodic with period $\Delta\psi_{ex} = 1.0$.

Notice that zeroes of i_c are now detectable at integer values of the normalized applied flux ψ_{ex}, while the main peaks are located at half-integer values of ψ_{ex}. Notice also that the curves have an upper bound $i_c = 1$.

As for the voltage-flux characteristics of $N\times(0\text{-}\pi)$ one-dimensional arrays, taking account of Eq. (5.26) and of the prescription given in Eq. (5.51), we may write:

$$\langle v \rangle = |i_P(\psi_{ex})| \sqrt{\left[\frac{i_B}{2Ni_P(\psi_{ex})}\right]^2 - 1}. \tag{5.52}$$

Voltage-flux curves of $N\times(0\text{-}\pi)$ one-dimensional arrays are shown in Fig. 5.10a–b for $N = 3$ and $N = 5$ and for three different values of the normalized bias current i_B. In particular, in Fig. 5.10a the number of $0\text{-}\pi$ pairs of junctions is $N = 3$. In this figure the normalized bias current is taken to be $i_B = 4.0$ (dotted-line curve), $i_B = 5.0$ (dashed-line curve), and $i_B = 6.0$ (full-line curves). In the first two cases ($i_B < 2N$) the curves are non-zero only in those intervals of the ψ_{ex}-axis in which $i_J(\psi_{ex}) < i_B/2N$. In the second case ($i_B = 2N$) the inequality $i_J(\psi_{ex}) < i_B/2N$ is satisfied on the entire ψ_{ex}-axis, except for half-integer values of ψ_{ex}. In this way, the full-line voltage curve in Fig. 5.10a lie above the ψ_{ex}-axis, except at half-integer values, where these curves attain a value of zero. A similar behavior can be detected in voltage-flux curves of $N\times(0\text{-}\pi)$ one-dimensional arrays in Fig. 5.10b for $N = 5$ and for the following three values of the normalized bias current: $i_B = 7.5$ (dotted-line curve), $i_B = 10.0$ (dashed-line curve), and $i_B = 12.5$ (full-line curves). Notice that the full-line voltage curve in Fig. 5.10b lie entirely above the ψ_{ex}-axis. Notice also that the periodicity of all curves in Fig. 5.10a–b is exactly equal to $\Delta\psi_{ex} = 1.0$, as it can be also argued from Eq. (5.50) and (5.52).

5.4.2. $N\times(0\text{-}\pi)$ one-dimensional arrays with unequal loop areas

In the previous subsection we have considered the case of equal loop areas ($A_k = A_0$, for $k = 1, ..., 2N - 1$) and equal values of the inductances ($L_1 = L_2 = L_0$) on the upper branch. We have however considered the inductance values to be such can be considered to be negligible. We here consider alternating values of the areas A_1 and A_2 and still consider $\beta_0 = 0$. Let us again reconsider all expressions written in Section 5.1 for a homogeneous array of 0-JJs. Let us thus write down only those expressions which are different from the ones reported in the previous subsection. First of all, we omit writing the explicit expression of the fluxoid quantization conditions, since it is exactly equal to Eq.

(5.41). In this equation we again take the integers n_k equal to zero for simplicity. In this way, the superconducting phase difference across the k-th junction can be written as in Eq. (5.42). The normalized geometric applied flux through the k-th cell ($k = 1, 2, ..., 2N - 1$) of area A_k can be written as follows:

$$\psi_{ex}^{(k)} = \frac{\mu_0 H A_k}{\Phi_0} = \begin{cases} \frac{\mu_0 H A_1}{\Phi_0} = \psi_{ex} & k \text{ odd} \\ \frac{\mu_0 H A_2}{\Phi_0} = \sigma\psi_{ex} & k \text{ even} \end{cases}, \quad (5.53)$$

where $\sigma = A_2/A_1$. As in the previous subsection, the external magnetic field $\vec{H}$ is applied orthogonally to the plane of the array. In this way, we write a new expression for the normalized flux threading the k-th loop ($k = 1, 2, ..., 2N - 1$):

$$\psi_k = \psi_{ex}^{(k)} - \beta_{0k}\left(\frac{i_B}{2} - \sum_{m=0}^{k-1} i_m\right), \quad (5.54)$$

where all quantities, except for the normalized applied fluxes $\psi_{ex}^{(k)}$ and the parameters $\beta_{0k} = L_k I_{J0}/\Phi_0$, are defined as in Eq. (5.43). By adopting the RSJ model (Barone, 1982) and by considering Eq. (5.45), we obtain the same dynamics described by Eq. (5.47), where now

$$S_k = \sum_{m=1}^{k} \psi_{ex}^{(m)}, \quad (5.55)$$

where negligible values of the parameters β_{0k} are assumed. By performing the sum for odd and even values of k, taking account of Eq. (5.53), we have:

$$S_k = \begin{cases} \frac{1+\sigma}{2}\psi_{ex}\left(k + \frac{1-\sigma}{1+\sigma}\right) & k \text{ odd} \\ \frac{1+\sigma}{2}\psi_{ex}k & k \text{ even} \end{cases}, \quad (5.56)$$

By now summing Eq. (5.47) over all k-values, knowing that the derivatives of ϕ_k and ϕ_0 coincide for constant values of ψ_{ex}, we have:

$$\frac{d\phi_0}{d\tau} + \frac{1}{2N}\sum_{k=0}^{2N-1}(-1)^k \sin(\phi_0 + 2\pi S_k) = \frac{i_B}{2N}. \quad (5.57)$$

By considering Eq. (5.56), Eq. (5.57) can be written, after some algebra, as follows:

$$\frac{d\phi}{d\tau} + i_A(\psi_{ex}) \sin \phi = \frac{i_B}{2N}. \tag{5.58}$$

where

$$i_A(\psi_{ex}) = \frac{\sin \pi\psi_{ex}}{N} \frac{\sin[\pi N\psi_{ex}(1+\sigma)]}{\sin[\pi\psi_{ex}(1+\sigma)]}. \tag{5.59a}$$

$$\phi = \phi_0 + \pi(2N - 1)\left(\psi_{ex} + \frac{1}{2}\right). \tag{5.59b}$$

Quantum interference patterns for the system, given by the $i_c = |i_A(\psi_{ex})|$ vs. ψ_{ex} curves, are shown in Fig. 5.11a–c for $N = 5$ and for three values of the area ratio $\sigma = A_2/A_1$; namely, $\sigma = 0.70, 1.05, 1.30$. These curves show that the regularity of the pattern over the short ψ_{ex} range is lost. Nevertheless, from Eq. (5.59), we see that it is still possible to attain periodicity of these curves when σ is a rational number. In fact, it can be shown that the period $\Delta\psi_{ex}$ can be obtained by finding the minimum positive integer value p giving

$$p(1 + \sigma) = M, \tag{5.60}$$

where M is an integer. Therefore, all curves in Fig. 5.11a–c have period $\Delta\psi_{ex} = p = 10$. Naturally, periodicity is lost for irrational values of σ. We finally notice that all curves are modulated by the enveloping term $\sin \pi\psi_{ex}$, reported as a dotted line in Fig. 5.11a–c.

As for the current-voltage characteristics of $N\times(0\text{-}\pi)$ one-dimensional arrays with alternate values of the loop areas, by comparing Eq. (5.58) to Eq. (5.26), we may write:

$$\langle v \rangle = |i_A(\psi_{ex})|\sqrt{\left[\frac{i_B}{2Ni_A(\psi_{ex})}\right]^2 - 1}. \tag{5.61}$$

Voltage-flux curves of $N\times(0\text{-}\pi)$ one-dimensional arrays with alternating loop areas are shown in Fig. 5.12a–b for $N = 3$ and $\sigma = 0.70$ and in Fig. 5.13a–b for $N = 5$ and $\sigma = 1.30$, for three different values of the normalized bias current: $i_B = 4.0$ (dotted-line curve), $i_B = 5.0$ (dashed-line curve), and $i_B = 6.0$ (full-line curve). In order to illustrate the periodicity pattern of these curves we have Fig. 5.12a and Fig. 5.13a are shown in the ψ_{ex}-interval $[-3.0, 3.0]$.

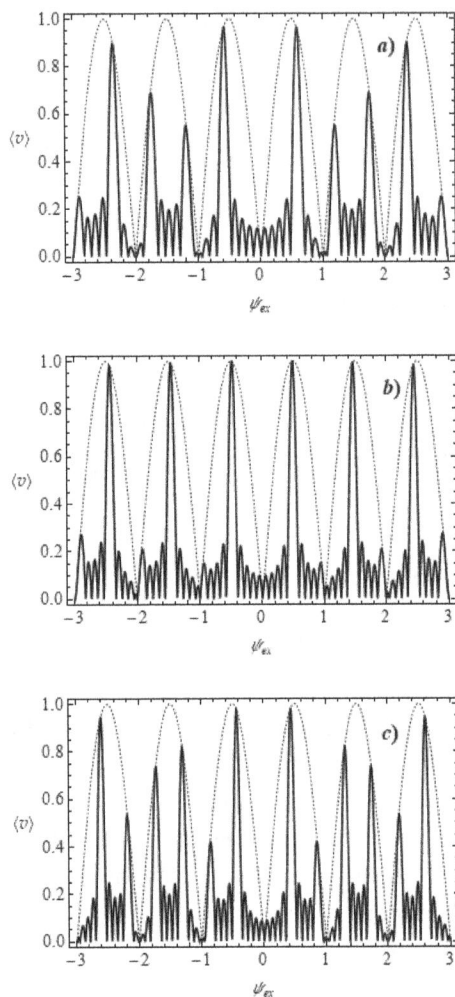

Fig. 5.11. Normalized critical current $i_c = |i_A|$ for of $N \times (0\text{-}\pi)$ one-dimensional array with alternating values of the loop area as a function of the normalized applied flux ψ_{ex}. In all curves the number of loops is $2N - 1 = 9$. The area ratio $\sigma = A_2/A_1$ has been chosen as follows: in panel a) $\sigma = 0.70$; in b) $\sigma = 1.05$; in c) $\sigma = 1.30$. Notice that all curves are modulated by the enveloping term $\sin \pi\psi_{ex}$, reported as a dotted line.

On the other side, the curves in Fig. 5.12b and Fig. 5.13b are shown in the ψ_{ex}-interval $[7.0, 13.0]$. It is noted that homologous curves in Fig. 5.12a and in Fig. 5.12b can be perfectly superimposed. The same is true for homologous curves in Fig. 5.13a and in Fig. 5.13b.

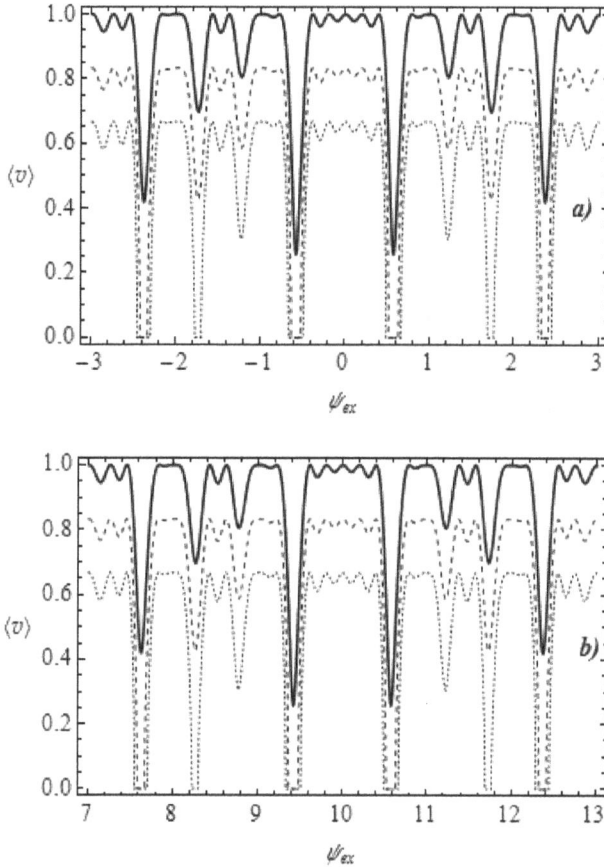

Fig. 5.12. Voltage-flux characteristics of $N\times(0$-$\pi)$ one-dimensional array with alternating loop areas with $\sigma = 0.70$ for $N = 3$ and $i_B = 4.0$ (dotted line), $i_B = 5.0$ (dashed line), $i_B = 6.0$ (full line). In panel a) the $\langle v \rangle$ vs. ψ_{ex} curves are shown in the interval $[-3.0, 3.0]$. In panel b) the same curves are shown in the interval $[7.0, 13.0]$. The homologous curves in panel a) and panel b) can be perfectly superimposed. In this case one can conclude that all curves are periodic with period $\Delta\psi_{ex} = 10$.

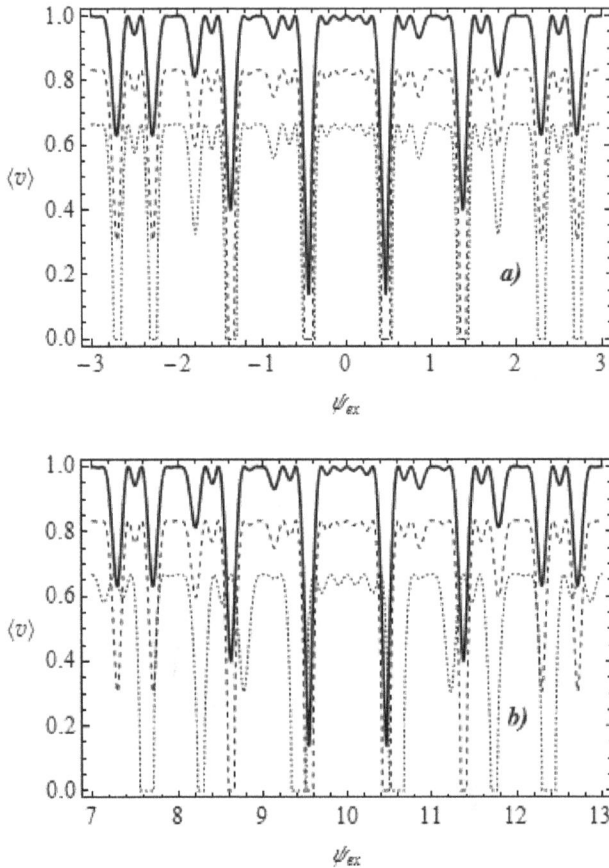

Fig. 5.13. Voltage-flux characteristics of $N\times(0\text{-}\pi)$ one-dimensional array with alternating loop areas with $\sigma = 1.30$ for $N = 3$ and $i_B = 4.0$ (dotted line), $i_B = 5.0$ (dashed line), $i_B = 6.0$ (full line). In panel a) the $\langle v \rangle$ vs. ψ_{ex} curves are shown in the interval $[-3.0, 3.0]$. In panel b) the same curves are shown in the interval $[7.0, 13.0]$. The homologous curves in panel a) and panel b) can be perfectly superimposed. In this case one can conclude that all curves are periodic with period $\Delta\psi_{ex} = 10$.

We may observe that these curves, although preserving the periodicity $\Delta\psi_{ex} = 10$, as specified above, show a more irregular behavior when compared to Fig. 5.10a. Nevertheless, the main features are still preserved. In fact, zero-voltage states are realized for $i_B/2N \leq i_A(\psi_{ex})$, while resistive states are present when this inequality is not satisfied.

5.5. Arrays with unconventional grating structures

In Section 5.1 we have analyzed the properties of one-dimensional Josephson junction arrays (1D JJAs) with conventional coupling and grating structure; i. e., with 0-JJs and with superconducting loops having identical area A_0. 1D JJAs with conventional coupling and unconventional gratings have been extensively studied by Oppenländer *et al.* (Oppenländer J. H., 2000). These systems have lately named SQIFs (Superconducting Quantum Interference Filters) (Oppenländer J. H., 2005). In these devices the voltage vs. applied flux curves present a rather pronounced peak at $\psi_{ex} = 0$. In order to approach the study of these particular arrays, let us consider a 1D JJA with a steadily growing loop area with the index $k = 1, 2, ... N$, so that

$$A_k = A_0(1 + a\,k), \tag{5.62}$$

where a is the relative increment $(A_k - A_{k-1})/A_0$. By immediately stating that we here consider only zero-field-cooling conditions (null values of initially trapped fluxons inside the superconducting loops) and by taking negligible inductance, we write the fluxoid quantization condition for the k-th loop, involving the superconducting phase differences ϕ_k and ϕ_{k-1} , as follows:

$$\phi_k - \phi_{k-1} = 2\pi\psi_{ex}^{(k)}, \tag{5.63}$$

where $k = 1, 2, ..., N$ and where:

$$\psi_{ex}^{(k)} = \frac{\mu_0 H A_k}{\Phi_0} = \psi_{ex}(1 + a\,k), \tag{5.64}$$

H being the amplitude of the external magnetic field $\vec{H}$, applied orthogonally to the plane of the array.

The dynamic equation for each Josephson junction in the array is written as in Eq. (5.4) for $k = 0, 1, 2, ..., N$. By defining

$$S_k = \sum_{m=1}^{k} \psi_{ex}^{(m)} = \psi_{ex} \sum_{m=1}^{k}(1 + a\,m), \tag{5.65}$$

the sum of all dynamic equations can be written in the following form:

$$(N + 1)\frac{d\phi_0}{d\tau} + \sum_{k=0}^{N} \sin(\phi_0 + 2\pi S_k) = i_B. \qquad (5.66)$$

By carrying out the sum in Eq. (5.65), we have:

$$S_k = k\psi_{ex}\left(1 + a\frac{k+1}{2}\right). \qquad (5.67)$$

In this way, Eq. (5.66) can be written as follows:

$$\frac{d\phi_0}{d\tau} + \frac{1}{N+1}\sum_{k=0}^{N} \sin\left[\phi_0 + 2\pi k\psi_{ex}\left(1 + a\frac{k+1}{2}\right)\right] = \frac{i_B}{N+1}. \qquad (5.68)$$

By trigonometric identities, Eq. (5.67) can be expressed as follows:

$$\frac{d\phi_0}{d\tau} + \alpha_N(\psi_{ex})\cos\phi_0 + \beta_N(\psi_{ex})\sin\phi_0 = \frac{i_B}{N+1}, \qquad (5.69)$$

where

$$\alpha_N(\psi_{ex}) = \frac{1}{N+1}\sum_{k=0}^{N} \sin\left[2\pi k\psi_{ex}\left(1 + a\frac{k+1}{2}\right)\right], \qquad (5.70a)$$

$$\beta_N(\psi_{ex}) = \frac{1}{N+1}\sum_{k=0}^{N} \cos\left[2\pi k\psi_{ex}\left(1 + a\frac{k+1}{2}\right)\right]. \qquad (5.70b)$$

Before solving numerically Eq. (5.66) or, alternatively, Eq. (5.69), let us consider the properties of the pre-factors $\alpha_N(\psi_{ex})$ and $\beta_N(\psi_{ex})$. In Fig. 5.14a–b we report the graphs of $\alpha_N(\psi_{ex})$ for $N = 20$ and for $a = 0.001$ (panel a) and $a = 0.005$ (panel b). In Fig. 5.15a–b, on the other hand, the graphs of $\beta_N(\psi_{ex})$ for $N = 20$ and for $a = 0.001$ (panel a) and $a = 0.005$ (panel b) are shown.

In Fig. 5.14a–b we notice the presence of a pair of peaks of opposite sign in correspondence of integer values of ψ_{ex}. However, when the value of the parameter a is rather small, as in Fig. 5.14a, these couples of counter-peaks repeat without significant variations, starting from $\psi_{ex} = 0$, at finite field values. The same is not true, however, when the parameter a is slightly increased, as in Fig. 5.14b. The same type of feature can be observed in the β_N vs. ψ_{ex} curves in Fig. 5.15a–b. In fact, the somewhat regular peaks present at integer values of ψ_{ex} for $a =$

0.001 in Fig. 5.15a, suffer a significant variation in shape and in amplitude when the parameter a is increased to 0.005 as in Fig. 5.15b.

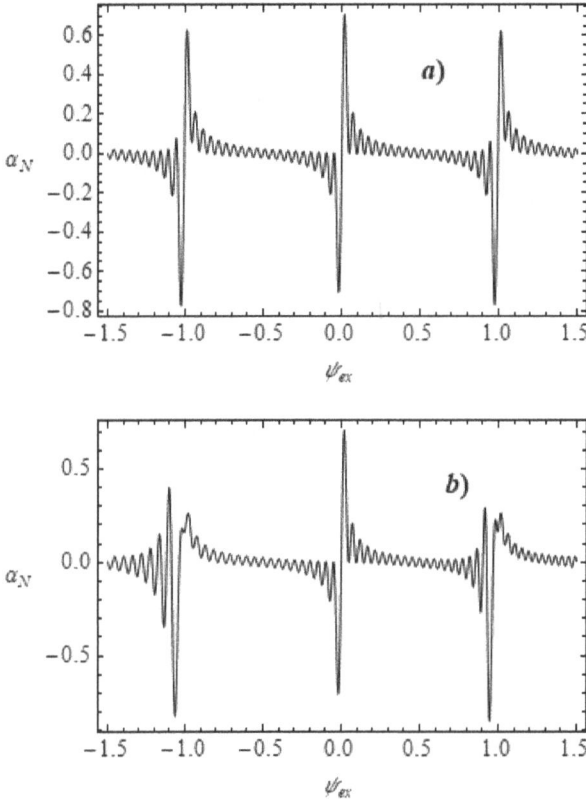

Fig. 5.14 Pre-factors α_N of the cosine and sine terms, respectively, in the effective single-junction model of a one-dimensional array with unconventional grating. For both curves $N = 20$, while $a = 0.001$ in panel a) and $a = 0.005$ in panel b).

We therefore argue that the parameter a acts as a sort of de-coherence factor in a 1D JJA with unconventional grating in the presence of a magnetic field. Therefore, one may think that the only properties of 1D JJAs fabricated with conventional grating preserved by 1D JJAs with unconventional grating structures would be those in the vicinity of $\psi_{ex} = 0$. As a consequence, one would expect that the characteristic quantum interference properties observed in conventional devices (see, for example, Fig. 5.3) get degraded in the critical current of 1D JJAs with unconventional grating structures at finite values of ψ_{ex}. This argument can be supported by the direct calculation of the maximum of the following function:

$$j(x) = \frac{i_B}{N+1} = \alpha_N(\psi_{ex}) \cos x + \beta_N(\psi_{ex}) \sin x, \qquad (5.71)$$

which represents the maximum normalized current i_B that can be injected in the array before a resistive state sets in, as deduced from Eq. (5.69), when we take $\phi_0 = x$.

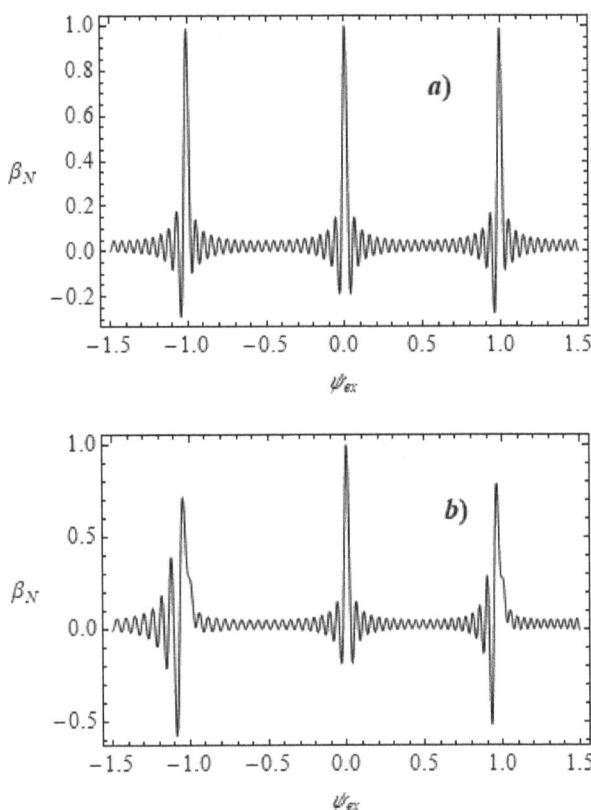

Fig. 5.15 Pre-factors β_N of the cosine and sine terms, respectively, in the effective single-junction model of a one-dimensional array with unconventional grating. For both curves $N = 20$, while $a = 0.001$ in panel a) and $a = 0.005$ in panel b).

By therefore setting $j'(x) = 0$, where the prime stands for "derivative with respect to x", one finds the following expression for the critical current of the device:

$$i_c(\psi_{ex}) = (N + 1)\sqrt{\alpha_N^2(\psi_{ex}) + \beta_N^2(\psi_{ex})}, \qquad (5.72)$$

In Fig. 5.16a–c the critical current i_c, normalized to the number of junctions, is represented as a function of ψ_{ex} in the vicinity of $\psi_{ex} = 0$ ($-1.50 \le \psi_{ex} \le 1.50$) for $N = 10$. In particular, in Fig. 5.16a the relative increase in the loop areas is taken to be $a = 0.01$, in Fig. 5.16b $a = 0.05$, and in Fig. 5.16c $a = 0.10$.

The previous analysis was useful in deriving some general properties of 1D JJAs fabricated with unconventional grating structures. In order to find the $\langle v \rangle$ vs. ψ_{ex} curves of the system, we may proceed in solving the non-linear differential equation (5.66) by numerical means. Therefore, after having numerically integrated Eq. (5.58), by averaging the time derivative of the superconducting phase difference ϕ_0 over sufficiently long intervals of time, we obtain the voltage response of the system as a function of the externally applied magnetic flux ψ_{ex} for a well-defined bias current i_B. The results are reported, in the range $-6.00 \le \psi_{ex} \le 6.00$, in Fig. 5.17a–c for $a = 0.10$ and for various values of N and in Fig. 5.18a–c for $N = 20$ and for various values of the parameter a.

By keeping the parameter a fixed, the $\langle v \rangle$ vs. ψ_{ex} curves shown in Fig. 5.17a–c for increasing values of N ($N = 10, 20, 30$) and of i_B ($i_B = 15.0, 25.0, 35.0$) present the following properties. For all values of N a central *delta-like* peak, as defined by Oppenländer *et al.* (Oppenländer J. H., 2000), at $\psi_{ex} = 0$ is present. These peaks become sharper and sharper as N increases. Furthermore, the lateral structures of the $\langle v \rangle$ vs. ψ_{ex} curves tend to flatten for increasing values of N.

Next, by keeping fixed the number of junctions ($N + 1 = 21$) and the current bias ($i_B = 25.0$), the $\langle v \rangle$ vs. ψ_{ex} curves shown in Fig. 5.18a–c for increasing values of a ($a = 0.02, 0.04, 0.06$) suggest the following remark: The central *delta-like* peak at $\psi_{ex} = 0$ is not appreciably affected, in this range of variation of a, by the increase of this parameter.

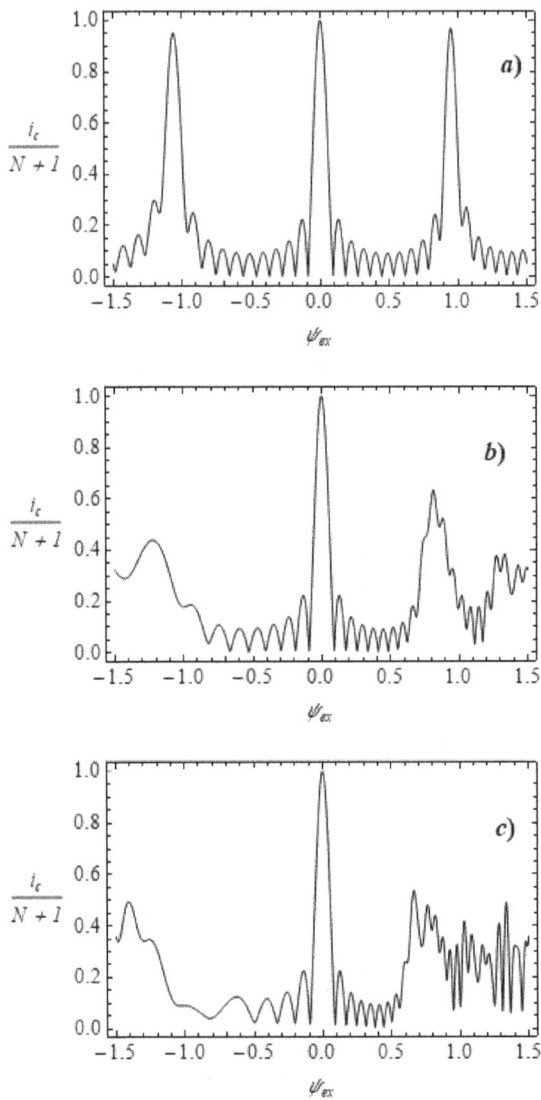

Fig. 5.16. Time-averaged voltage $\langle v \rangle$ as a function of the externally applied magnetic flux ψ_{ex}. For all curves $a = 0.1$. In panel a) $N = 10$ and $i_B = 15$; in panel b) $N = 20$ and $i_B = 25$; in panel c) $N = 30$ and $i_B = 35$.

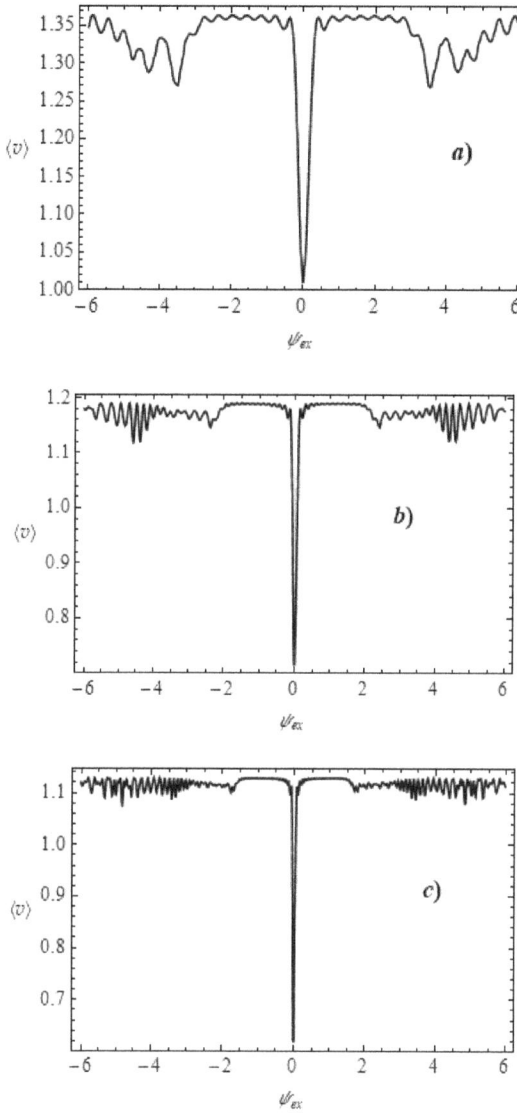

Fig. 5.17. Time-averaged voltage $\langle v \rangle$ as a function of the externally applied magnetic flux ψ_{ex}. For all curves $a = 0.1$. In panel a) $N = 10$ and $i_B = 15$; in panel b) $N = 20$ and $i_B = 25$; in panel c) $N = 30$ and $i_B = 35$.

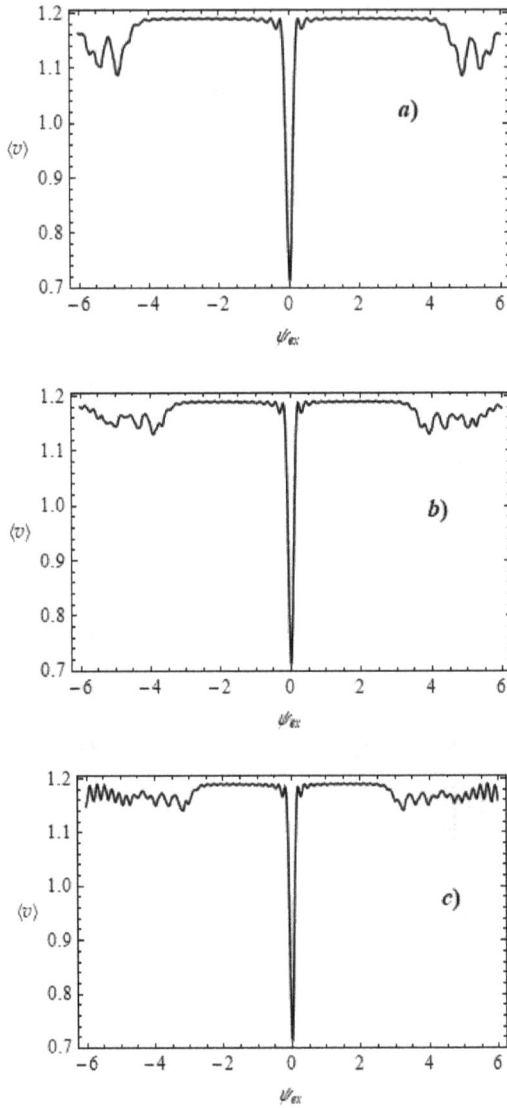

Fig. 5.18. Time-averaged voltage $\langle v \rangle$ as a function of the externally applied magnetic flux ψ_{ex}. For all curves $N = 20$ and $i_B = 25$. In panel a) $a = 0.02$; in panel b) $a = 0.04$; in panel c) $a = 0.06$.

However, as far as the lateral structures of the $\langle v \rangle$ vs. ψ_{ex} curves are concerned, one may detect a contraction of the slightly undulated plateau in the vicinity of $\psi_{ex} = 0$. In fact, while for $a = 0.02$ this plateau extends on the right up to $\psi_{ex} = 4.0$, for higher values of the parameter a it tends to recede. For $a = 0.06$, in fact, its farthest extension only reaches the value of $\psi_{ex} = 2.5$.

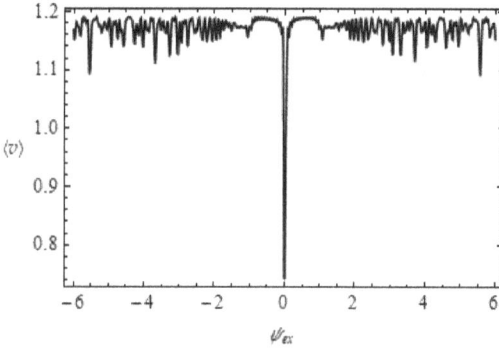

Fig. 5.19. Time-averaged voltage $\langle v \rangle$ as a function of the externally applied magnetic flux ψ_{ex} for $N = 20$, $i_B = 25$, and $a = \sqrt{2}/5$.

For further increase of the structural parameter a, this plateau contracts further, as shown in Fig. 5.19, where the $\langle v \rangle$ vs. ψ_{ex} curve is reported for $N = 10$, $i_B = 25$, and $a = \sqrt{2}/5$. This figure is significant in that it clarifies that the amplitude of the delta-like central peak increases for increasing values of a. Furthermore, it was purposely calculated for an irrational value of a, in order to clarify that the de-coherence effects on the interference pattern only depend upon the structural configuration of the device, regardless of the fact that the loop areas may be commensurate or not. Finally, from Fig. 5.19 one can argue that the central peak amplitude increases when N is kept constant and the parameter a is increased.

Chapter 6

Two-dimensional Networks of Over-damped Josephson Junctions

6.1. Square arrays

A rather obvious generalization of one-dimensional Josephson junction arrays (1D JJA) consists in extending these systems to a two-dimensional networks of Josephson junctions (JJs) on a square lattice. We shall see that this type of generalization, *per se* interesting from the theoretical and the experimental point of view, can be used to describe, in a rather idealized way, the low-field magnetic properties of granular superconducting films. In fact, when one considers the general morphology of granular superconductors, Josephson junction networks (JJNs) arise quite naturally from the fact that each individual superconducting grain is weakly coupled to all next-neighboring grains. Before tackling the study of granular superconductors in the next chapter, we shall here lay a rather useful theoretical basis upon which description of the magnetic properties of granular superconducting systems can be made to rest.

In Fig. 6.1 a schematic representation of a single superconducting loop of a network of Josephson junctions in two dimensions is shown. This loop represents an elementary cell (a *plaquette*) of a square lattice of $N \times M$ superconducting nodes giving rise to the distribution of JJs shown in the same figure. In order to correctly obtain the flux quantization conditions and the flux-current relations, we specify the direction of the current flowing through each JJ coherently with the phase difference across the junction and the positive flux threading the loop according to the right-hand rule. A positive externally applied flux is assumed to be applied to each loop.

We may therefore proceed to write the general equations for the system. As done for the 1D JJAs, we here assume perfectly identical over-damped Josephson junctions with resistive parameter R and maximum Josephson current I_{J0} and we take the inductances $L/4$ of the horizontal and vertical branches, so that a parameter $\beta_L = LI_{J0}/\Phi_0$ can be associated to each loop.

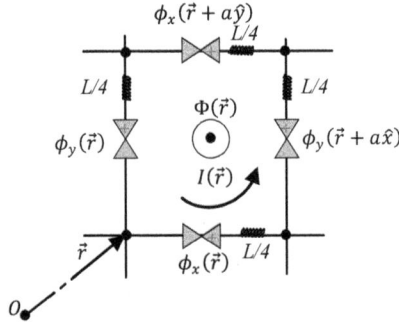

Fig. 6.1. Schematic representation of the physical quantities associated to an elementary cell of a two-dimensional Josephson network. The lower left vertex of the cell is at a position $\vec{r} = a(i\hat{x} + j\hat{y})$, a being the length of the cell side. An inductance L is associated to each loop of the network.

Consider the single plaquette represented in Fig. 6.1, whose lower left vertex is at a position $\vec{r} = a(i\hat{x} + j\hat{y})$, with respect to the origin 0, a being the length of the cell side, i and j being integers. We may write the fluxoid quantization conditions in terms of the indexed phase differences ϕ_x and ϕ_y reported in this same figure as follows:

$$2\pi\psi(\vec{r}) + \phi_y(\vec{r} + a\hat{x}) - \phi_y(\vec{r}) + \phi_x(\vec{r}) - \phi_x(\vec{r} + a\hat{y}) = 2\pi n(\vec{r}), \quad (6.1)$$

where $\psi(\vec{r}) = \Phi(\vec{r})/\Phi_0$ is the normalized flux threading the plaquette and $n(\vec{r})$ is an integer. Because of the presence of a positive external magnetic flux $\Phi_{ex} = \mu_0 H A_0$, where A_0 is the cell area, the flux through the elementary cell in Fig. 6.1 is the following:

$$\psi(\vec{r}) = \psi_{ex} + \beta_L i(\vec{r}) + \sum_{\vec{r}'} m(\vec{r}, \vec{r}')i(\vec{r}'), \quad (6.2)$$

where $\psi_{ex} = \Phi_{ex}/\Phi_0$, $i(\vec{r}) = I(\vec{r})/I_{J0}$ is the normalized current circulating in the cell, and $m(\vec{r}, \vec{r}')$ is the normalized coefficient of mutual inductance between the loop at position $\vec{r}$ and that at position $\vec{r}'$. In what follows we shall consider the mutual inductance coefficient to be negligible. The dynamic equation for each Josephson junction in the network can be written, as already seen, by means of the Resistively Shunted Junction (RSJ) model (Barone, 1982). By taking account of the current flowing in the JJ at position $\vec{r}$ indexed with the subscript ξ (which may be either x or y), and by making use of the usual normalization in time, we may write:

$$\frac{d\phi_\xi(\vec{r})}{d\tau} + \sin \phi_\xi(\vec{r}) = \varepsilon_{\xi\eta z} [i(\vec{r}) - i(\vec{r} - a\hat{\eta})], \tag{6.3}$$

where η, on its turn, can be either x or y, and where $\varepsilon_{\xi\eta z}$ is the Levi-Civita symbol, which excludes the possibility of setting $\xi = \eta$. The other two equations for the junctions in the loop can be obtained by applying translation operators, acting on a scalar function $f(\vec{r})$, representing a physical variable related to the network, as follows:

$$\widehat{T}_\eta f(\vec{r}) = f(\vec{r} + a\hat{\eta}), \tag{6.4}$$

where η can be either x or y. The relations in Eq. (6.1), (6.2), and (6.3) thus represent the constitutive equations of two-dimensional Josephson junction networks (2D JJNs), where the vector $\vec{r}$ may vary over all possible positions in the network. We remark that the validity of these equations may be extended also to junctions lying on the boundary of the network. In fact, in this case, one simply does not consider the loop currents which are not present in the network. For instance, when considering the JJ at position $\vec{r} = 0$, we have to write:

$$\frac{d\phi_x(0)}{d\tau} + \sin \phi_x(0) = i(0), \tag{6.5a}$$

$$\frac{d\phi_y(0)}{d\tau} + \sin \phi_y(0) = - i(0). \tag{6.5b}$$

We notice that, although the currents $i(-a\hat{y})$ and $i(-a\hat{x})$ are not present in the network, so that they are set to zero in Eq. (6.3), their fictitious use is useful in determining the sign of the right-hand side of Eq. (6.5a) and

(6.5b). We finally recall that the assumptions made in writing Eq. (6.5a) and (6.5b) are that all JJs are identical and the lattice consists of regular square cells of side a.

In order to write the above equations in a more compact form, we may introduce the forward and backward difference operators Δ_η and $\Delta_{-\eta}$ defined as follows:

$$\Delta_\eta f(\vec{r}) = f(\vec{r} + a\hat{\eta}) - f(\vec{r}), \qquad (6.6a)$$

$$\Delta_{-\eta} f(\vec{r}) = f(\vec{r}) - f(\vec{r} - a\hat{\eta}), \qquad (6.6b)$$

where $f(\vec{r})$ is any scalar physical variable related to the network. Therefore, we rewrite all constitutive equations for the network, neglecting mutual inductance between loops and taking $n(\vec{r}) = 0$ (zero-field cooling conditions), as follows:

$$2\pi\psi(\vec{r}) + \Delta_x \phi_y(\vec{r}) - \Delta_y \phi_x(\vec{r}) = 0, \qquad (6.7a)$$

$$\psi(\vec{r}) = \psi_{ex} + \beta_L\, i(\vec{r}), \qquad (6.7b)$$

$$\frac{d\phi_\xi(\vec{r})}{d\tau} + \sin \phi_\xi(\vec{r}) = \varepsilon_{\xi\eta z}\, \Delta_{-\eta} i(\vec{r}). \qquad (6.7c)$$

Since the translator operator in Eq. (6.4) commutes with the differential operator in Eq. (6.7c) and with the forward and backward difference operators Δ_η and $\Delta_{-\eta}$, we may consider the above set of equations as representing the overall behavior of the whole 2D JJN.

6.2. Simple networks

In this section we analyze the properties of rather simple networks, for which an analytic perturbation approach is possible for small values of the parameter β_L. We shall perform numerical calculations for values of β_L which do not allow this type of analysis.

6.2.1. *A single plaquette*

The simplest two-dimensional square network of Josephson junctions consists of four junctions distributed on the sides of a square. In order to illustrate the way one proceeds in writing down equations (6.7a–c) for this specific case, we explicitly set:

$$2\pi\psi(0) + \phi_y(a\hat{x}) - \phi_y(0) - \phi_x(a\hat{y}) + \phi_x(0) = 0, \quad (6.8a)$$

$$\psi(0) = \psi_{ex} + \beta_L\, i(0), \quad (6.8b)$$

$$\frac{d\phi_x(0)}{d\tau} + \sin\phi_x(0) = i(0), \quad (6.8c)$$

$$\frac{d\phi_y(0)}{d\tau} + \sin\phi_y(0) = -i(0), \quad (6.8d)$$

$$\frac{d\phi_y(a\hat{x})}{d\tau} + \sin\phi_y(a\hat{x}) = i(0), \quad (6.8e)$$

$$\frac{d\phi_x(a\hat{y})}{d\tau} + \sin\phi_x(a\hat{y}) = -i(0). \quad (6.8f)$$

Therefore, by inspection, one has that the solutions of Eq. (6.8c–f) obey to the following relations:

$$\phi_y(a\hat{x}) = \phi_x(0) = -\phi_y(0) = -\phi_x(a\hat{y}) = \phi. \quad (6.9)$$

In this way, one can write down the following reduced system of equations, in which only the phase difference ϕ is involved:

$$\pi\psi(0) + 2\phi = 0, \quad (6.10a)$$

$$\psi(0) = \psi_{ex} + \beta_L\, i(0), \quad (6.10b)$$

$$\frac{d\phi}{d\tau} + \sin\phi = i(0). \quad (6.10c)$$

After having expressed $\psi(0)$ in terms of ϕ in Eq. (6.10a) and having solved for $i(0)$ in Eq. (6.10b), Eq. (6.10c) can be written as follows:

$$\frac{d\phi}{d\tau} + \sin\phi + \frac{2}{\pi\beta_L}\phi = -\frac{\psi_{ex}}{\beta_L}. \quad (6.11)$$

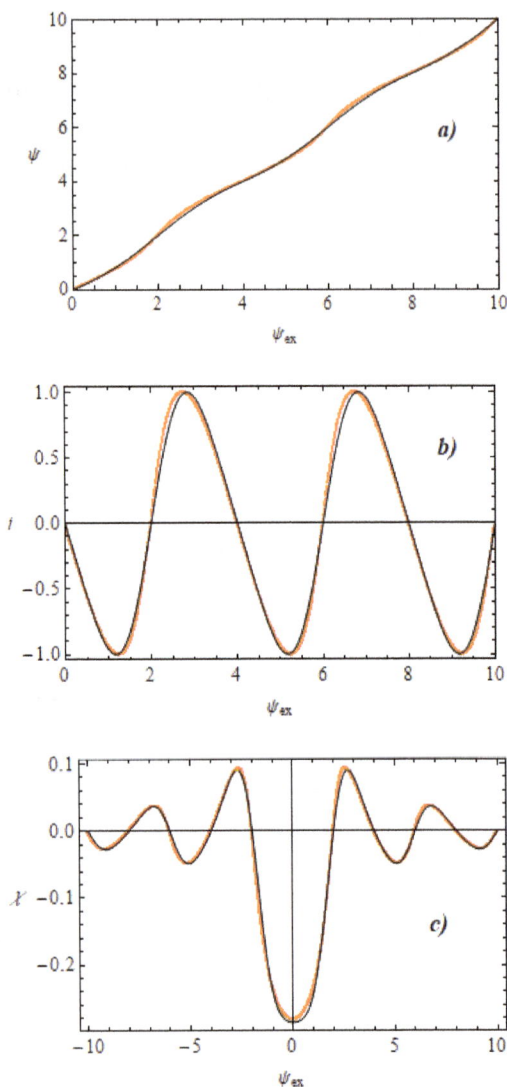

Fig. 6.2. Physical quantities related to the simplest two-dimensional Josephson network: a single plaquette with four junctions. Panel a): normalized magnetic flux $\psi(0)$ threading the loop; panel b): superconducting current $i(0)$; panel c): d. c. susceptibility χ. In all panels $\beta_L = 0.25$. The black curves in panels a), b), and c), respectively, represent the analytically determined functions $\psi(0)$, $i(0)$, and χ in terms of the normalized externally applied flux ψ_{ex}. The orange curves represent the same quantities determined by numerically integrating the non-linear dynamic equation for the system.

By considering the stationary solutions of (6.11), we set:

$$\pi \beta_L \sin \phi + 2\phi = -\pi \psi_{ex}. \tag{6.12}$$

The above non-linear equation can be solved by a perturbation approach for small values of β_L, giving the following approximate solution to second order in β_L:

$$\phi = -\frac{\pi}{2}\psi_{ex} + \frac{\pi}{2}\beta_L \sin\left(\frac{\pi}{2}\psi_{ex}\right)\left[1 - \frac{\pi}{2}\beta_L \cos\left(\frac{\pi}{2}\psi_{ex}\right)\right]. \tag{6.13}$$

The normalized magnetic flux $\psi(0)$ threading the loop at various values of ψ_{ex} can thus be obtained by Eq. (6.10a), so that:

$$\psi(0) = \psi_{ex} - \beta_L \sin\left(\frac{\pi}{2}\psi_{ex}\right)\left[1 - \frac{\pi}{2}\beta_L \cos\left(\frac{\pi}{2}\psi_{ex}\right)\right]. \tag{6.14}$$

On the same token, the stationary current $i(0)$ can be obtained by Eq. (6.10c) and (6.14):

$$i(0) = -\sin\left[\frac{\pi}{2}\psi(0)\right]. \tag{6.15}$$

Finally, the d. c. susceptibility χ of the network can be obtained by setting:

$$\chi = \frac{\psi(0)-\psi_{ex}}{\psi_{ex}} = -\beta_L \frac{\sin\left[\frac{\pi}{2}\psi(0)\right]}{\psi_{ex}}. \tag{6.16}$$

In Fig. 6.2a–c the curves for the physical quantities in Eq. (6.14), (6.15) and (6.16) are shown for $\beta_L = 0.25$. The analytically determined values obtained by the perturbation approach described above are reported in black, the numerically determined values of the same quantities are reported in orange. In particular, in Fig. 6.2a the $\psi(0)$ vs. ψ_{ex} curves are shown. These curves present a steady increase of the magnetic flux $\psi(0)$ for increasing values of ψ_{ex}. In Fig. 6.2b the $i(0)$ vs. ψ_{ex} curves are represented. We remark the similarity between the circulating currents in this simple 2D JJN and in a single superconducting loop interrupted by one JJ. In this latter case, however, one should keep in mind that the different number of JJs in the loop give different periodicities of the curves and different cross-over value of β_L for irreversible magnetic behavior. In this case, in fact, we have a period $\Delta\psi_{ex} = 4$ and a crossover value of β_L from reversible to irreversible magnetic behavior equal to $\beta_c = 2/\pi$. As for the susceptibility curves in Fig. 6.2c, we notice that it still reproduces the already known phenomenon of paramagnetic behavior in well-defined magnetic flux intervals, the intensity of the paramagnetic signal depending on β_L, as it will be shown by considering higher values of this parameter. Furthermore, the minimum of the χ vs. ψ_{ex} curve at $\psi_{ex} = 0$ can be calculated from Eq. (6.14) and (6.16) to be approximately equal to

$$\chi_0 \approx -\frac{\pi}{2}\beta_L \left(1 - \frac{\pi}{2}\beta_L\right). \tag{6.17}$$

The approximated minimum value of the χ vs. ψ_{ex} curve at $\psi_{ex} = 0$ for $\beta_L = 0.25$ is therefore $\chi_0 \approx -0.24$. This value of χ_0 agrees with the numerically determined true value, as one can see from Fig. 6.2c. However, as we shall see in what follows, higher order terms are needed to obtain a more refined result.

Naturally, for higher values of β_L the perturbation approach cannot be used and the results are to be computed by means of numerical integration of Eq. (6.11). Therefore, in Fig. 6.3a–c we show the numerical results obtained for the same physical quantities in Fig. 6.2a–c, this time for $\beta_L = 0.6$ (black curves) and $\beta_L = 1.2$ (gray curves). As a general remark, we may notice that the black curve still maintain a reversible character, while the increase in β_L in the gray curves produces an overall sharpening of the transitions of flux quanta inside the system, because of irreversible magnetic behavior. Furthermore, the currents in Fig. 6.3b show the same period $\Delta\psi_{ex} = 4$ already detected for small values of the parameter β_L. Another important point to be noticed for the gray curves in Fig. 6.3b is the first irreversible flux transition inside the network, occurring at the value $\psi_{ex} = \psi_1$, where the normalized current in Eq. (6.15) reaches its minimum value of -1. In this way, by setting $\phi = -\pi/2$ in Eq. (6.12), we have $\psi_1 = 1 + \beta_L$. This value can be readily detected in the first sharp increase of the current in the gray curve in Fig. 6.3b.

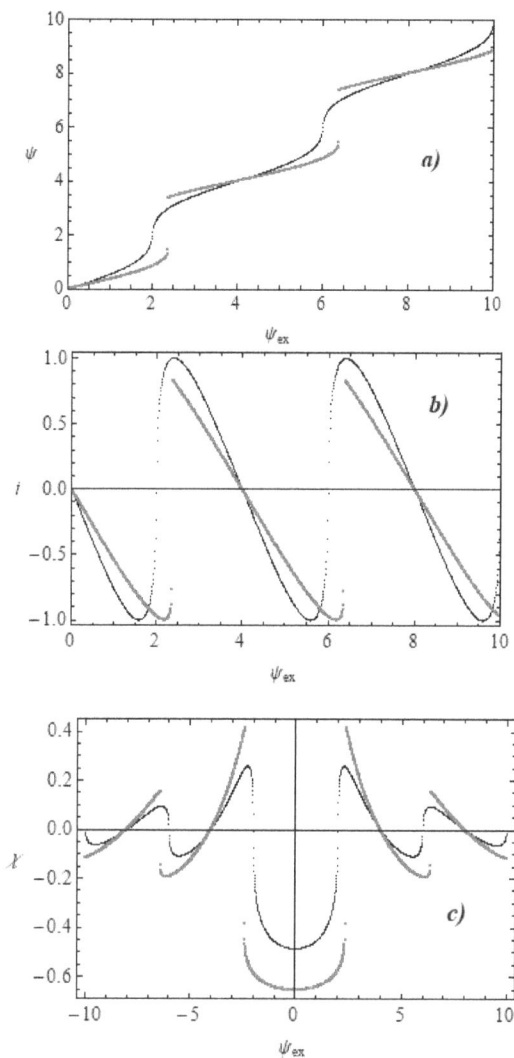

Fig. 6.3. Physical quantities related to the simplest two-dimensional Josephson network: a single plaquette with four junctions for $\beta_L = 0.6$ (black curves) and $\beta_L = 1.2$ (gray curves). Panel a): normalized magnetic flux $\psi(0)$ threading the loop; panel b): superconducting current $i(0)$; panel c): d. c. susceptibility χ. The curves in all panels represent the numerically determined quantities $\psi(0)$, $i(0)$, and χ in terms of the normalized externally applied flux ψ_{ex}.

Finally, in Fig. 6.3c we can distinctly notice how the amplitude of the diamagnetic signal present at extremely low values of the normalized externally applied flux is more pronounced as β_L increases. As for the approximated minimum value of the χ vs. ψ_{ex} curve at $\psi_{ex} = 0$, the approximate result in Eq. (6.17) has to be replaced by a more refined estimate. Therefore, by considering higher order terms for $\psi(0)$ in determining the approximate solution of Eq. (6.12), we have:

$$\chi_0 = -\frac{\frac{\pi}{2}\beta_L}{1+\frac{\pi}{2}\beta_L}. \tag{6.18}$$

The above result, consistent with Eq. (6.17) for small values of β_L, can be obtained by retaining higher order terms in the Eq. (6.12) and by recognizing that the series expansion obtained for χ_0 is consistent with the expression in Eq. (6.18). Alternatively, a recursive solution of Eq. (6.10a–c) in the stationary regime, illustrated in details in the following subsection, can be used. Therefore, by means of Eq. (6.18), we have $\chi_0 \approx -0.49$ for $\beta_L = 0.6$ (black curve in Fig. 6.3c) and $\chi_0 \approx -0.65$ for $\beta_L = 1.2$ (gray curve in Fig. 6.3c). These values match the numerically determined quantities reported.

6.2.2. *A 2×2 network*

After having taken a closer look at the simplest possible 2D JJN consisting of a single superconducting square loop with one JJ on each side, we next consider a 2D JJN consisting of 4 plaquettes. We shall see that this system, as far as magnetic response is concerned, may be compared to the case studied in the previous subsection. In fact, by symmetry reasons, starting from zero-field cooling conditions and applying uniformly the normalized flux ψ_{ex} to the network, the system presents the flux and current distribution shown in Fig. 6.4. In this figure the peripheral junctions in the network are seen to be active, showing a superconducting phase difference of either $+\phi_0$ or $-\phi_0$, depending on the direction in which the loop current i_0 flows in them. On the other hand, the four inner junctions are not active, since the loop currents in adjacent loops flow in opposite direction, giving a net current equal to zero in the four inner branches.

Equations (6.7a–c) in this case reduce to the following set:

$$\pi\psi_0 + \phi_0 = 0, \tag{6.19a}$$

$$\psi_0 = \psi_{ex} + \beta_L\, i_0, \tag{6.19b}$$

$$\frac{d\phi_0}{d\tau} + \sin\phi_0 = i_0, \tag{6.19c}$$

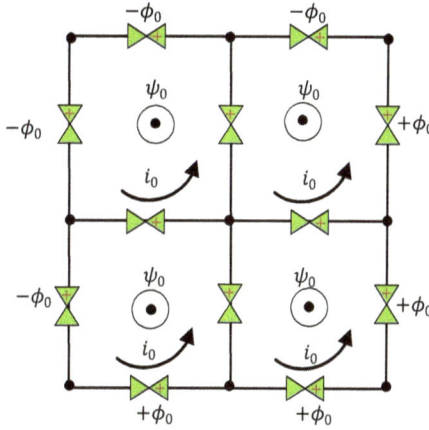

Fig. 6.4. Schematic representation of the physical quantities associated to a 2×2 two-dimensional Josephson network. The inductance L, still associated to each loop of the network, has been omitted in this picture. The peripheral junctions in the network are active, presenting a superconducting phase difference of either $+\phi_0$ or $-\phi_0$, while the four inner junctions are not active, since null net currents flow in them.

where ψ_0 is the flux linked to each plaquette. Therefore, proceeding as in the previous subsection, the stationary equation for the relevant phase difference ϕ_0 can be written as follows:

$$\pi\beta_L \sin\phi_0 + \phi_0 = -\pi\psi_{ex}. \tag{6.20}$$

For small values of β_L, the above equation could be solved by means of a perturbation approach, giving, to second order in the parameter β_L, the following solution for ϕ_0:

$$\phi_0 = -\pi\psi_{ex} + \pi\beta_L \sin(\pi\psi_{ex})\left[1 - \pi\beta_L \cos(\pi\psi_{ex})\right]. \quad (6.21)$$

In this subsection, we shall also present an alternative simple recursive procedure leading to the solution of Eq. (6.20), showing that the solution obtained by this approach coincides to the second order approximation given in Eq. (6.21). In fact, if we considered β_L negligible in Eq. (6.20), we could immediately write the solution as follows: $\phi_0 = -\pi\psi_{ex}$. We could thus substitute this expression in the left-hand-side sine-term in Eq. (6.20), obtaining:

$$\phi_0 = -\pi\psi_{ex} + \pi\beta_L \sin(\pi\psi_{ex}). \quad (6.22)$$

The above expression coincides, to first order in β_L, to the solution (6.21). By iterating the procedure once again; i.e., by substituting Eq. (6.22) in the left-hand-side sine-term in Eq. (6.20), we have:

$$\phi_0 = -\pi\psi_{ex} + \pi\beta_L \sin[\pi\psi_{ex} - \pi\beta_L \sin(\pi\psi_{ex})]. \quad (6.23)$$

By simple trigonometric identities and by considering only terms up to second-order in β_L, we arrive at the same expression for ϕ_0 given in (6.21). This alternative procedure could be more useful when higher order terms in β_L are to be considered.

The normalized magnetic flux ψ_0 threading each loop in the network can thus be obtained by Eq. (6.19a), so that:

$$\psi_0 = \psi_{ex} - \beta_L \sin(\pi\psi_{ex})\left[1 - \pi\beta_L \cos(\pi\psi_{ex})\right]. \quad (6.24)$$

The stationary current i_0, circulating in each loop of the network, can be obtained by Eq. (6.19c):

$$i_0 = -\sin(\pi\psi_0). \quad (6.25)$$

The d. c. susceptibility χ of the network can be finally found to be:

$$\chi = -\beta_L \frac{\sin(\pi\psi_0)}{\psi_{ex}}. \quad (6.26)$$

In Fig. 6.5a–c the approximated values of the physical quantities given by Eq. (6.24), (6.25) and (6.26) for $\beta_L = 0.10$ are shown as black lines, while the numerically determined values, obtained by numerical integration of Eq. (6.19c), are shown as orange curves. In particular, in Fig. 6.5a the ψ_0 vs. ψ_{ex} curves are reported. These curves are continuous and show a steady increase of the magnetic flux ψ_0 for increasing values of ψ_{ex}. In Fig. 6.5b the i_0 vs. ψ_{ex} curves are represented. We may notice that these curves are similar to the previously obtained ones shown in Fig. 6.3b. In the latter figures, however, the cross-over value of β_L for irreversible magnetic behavior is $\beta_c = 2/\pi$, while in the case of the 2×2 network in Fig. 6.4 we have $\beta_c = 1/\pi$. Furthermore, the period $\Delta\psi_{ex} = 2$ is halved with respect to that of a single plaquette. As for the susceptibility curves in Fig. 6.5c, the same qualitative feature in Fig. 6.3c are present. However, the minimum of the χ vs. ψ_{ex} curve at $\psi_{ex} = 0$ is now calculated from Eq. (6.19a–c), to be:

$$\chi_0 = -\frac{\pi\beta_L}{1+\pi\beta_L}. \tag{6.27}$$

The approximated minimum value of the χ vs. ψ_{ex} curve in Fig. 6.5c at $\psi_{ex} = 0$ for $\beta_L = 0.10$ is therefore $\chi_0 \approx -0.24$, which is very close to the value calculated by numerical integration.

As in the previous subsection, the perturbation approach cannot be used for higher values of β_L and the physical properties of the system need to be computed by means of numerical integration of Eq. (6.19c). Therefore, in Fig. 6.6a–c we show the numerical results obtained for the same physical quantities in Fig. 6.5a–c, this time for $\beta_L = 0.3$ (black curves) and $\beta_L = 0.6$ (gray curves). In Fig. 6.6a–c we notice that the black curves still maintain a reversible character, while the increase in β_L in the gray curves produces an overall sharpening of the transitions of flux quanta inside the system, because of irreversible magnetic behavior. Furthermore, the currents in Fig. 6.6b show the same period $\Delta\psi_{ex} = 2$, already detected for smaller values of the parameter β_L in Fig. 6.5b. In the gray curves in Fig. 6.6b the first irreversible flux transition inside the network, occurs at $\psi_{ex} = \psi_1 = 1/2 + \beta_L$, as it can be seen by setting $\phi_0 = -\pi/2$ in Eq. (6.20).

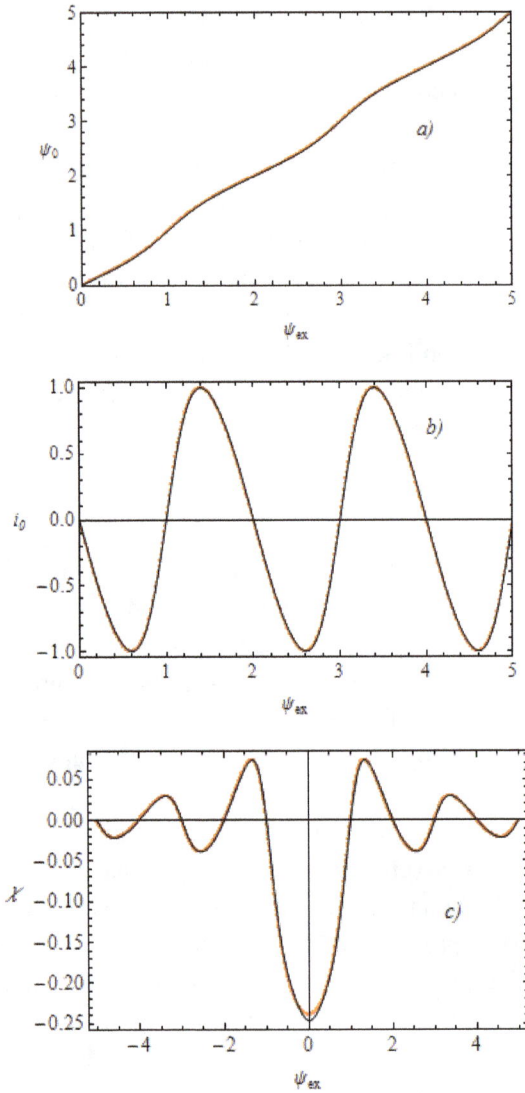

Fig. 6.5. Physical quantities related to a 2×2 two-dimensional Josephson network for $\beta_L = 0.10$: a) normalized magnetic flux ψ_0 threading each loop of the network; b) superconducting loop current i_0; c) d. c. susceptibility χ. The black curves in panels a), b), and c), respectively, represent the analytically determined functions ψ_0, i_0, and χ in terms of the normalized externally applied flux ψ_{ex}. The orange curves represent the same quantities determined by numerically integrating the non-linear dynamic equation for the system.

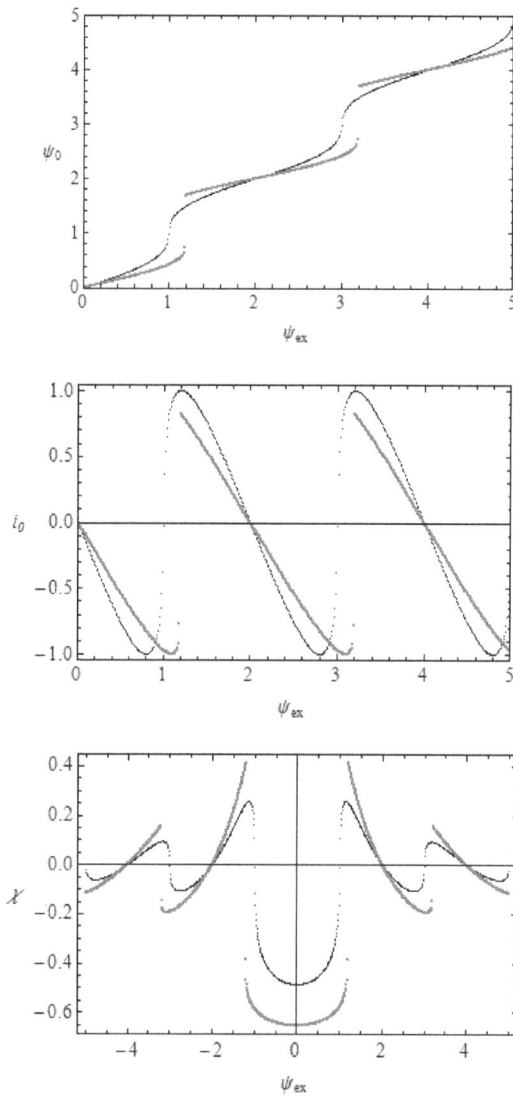

Fig. 6.6. Numerical evaluation of the physical quantities related to a 2×2 two-dimensional Josephson network for $\beta_L = 0.30$ (black curves) and $\beta_L = 0.60$ (gray curves): a) normalized magnetic flux ψ_0 threading each loop of the network; b) superconducting loop current i_0; c) d. c. susceptibility χ.

Finally, in Fig. 6.6c we can notice that the amplitude of the diamagnetic signal present at extremely low values of the normalized externally applied flux is more pronounced as β_L increases. As for the minimum value of the χ vs. ψ_{ex} curve at $\psi_{ex} = 0$, by Eq. (6.27) we have $\chi_0 \approx -0.48$ for $\beta_L = 0.3$ (black curve in Fig. 6.6c) and $\chi_0 \approx -0.65$ for $\beta_L = 0.6$ (gray curve in Fig. 6.6c). These values match the numerically determined quantities.

6.3. A 3×3 square array

The 3×3 network of Josephson junctions we shall consider in the present section is represented in Fig. 6.7.

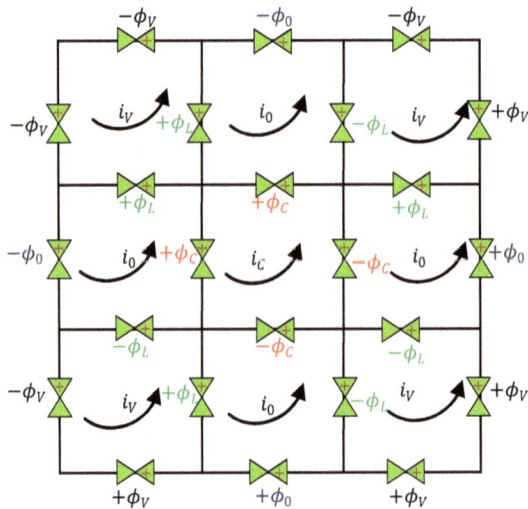

Fig. 6.7. Schematic representation of the physical quantities associated to a 3×3 two-dimensional Josephson network. The inductance L, still associated to each loop of the network, has been omitted in this picture. The peripheral junctions in the network are of two types, presenting a superconducting phase difference either equal to $\pm\phi_0$ or equal to $\pm\phi_V$. The next-to-peripheral JJs are characterized by a phase difference $\pm\phi_L$. Finally, the central junctions possess a phase difference $\pm\phi_C$.

By what already seen in the subsections 6.2.1 and 6.2.2, we may treat this more complex case by writing down the following quantization conditions:

$$\pi\psi_V + \phi_V + \phi_L = 0, \tag{6.28a}$$

$$2\pi\psi_0 + \phi_0 - 2\phi_L - \phi_C = 0, \tag{6.28b}$$

$$\pi\psi_C + 2\phi_C = 0, \tag{6.28c}$$

where ψ_V, ψ_0, and ψ_C are the normalized magnetic fluxes threading the superconducting loops in which the currents i_V, i_0, and i_C circulate, respectively. For these same loops one may write the following relations:

$$\psi_V = \psi_{ex} + \beta_L i_V, \tag{6.29a}$$

$$\psi_0 = \psi_{ex} + \beta_L i_0, \tag{6.29b}$$

$$\psi_C = \psi_{ex} + \beta_L i_C. \tag{6.29c}$$

The dynamic equations for the four types of the 24 JJs in the network are the following:

$$\frac{d\phi_V}{d\tau} + \sin\phi_V = i_V, \tag{6.30a}$$

$$\frac{d\phi_0}{d\tau} + \sin\phi_0 = i_0, \tag{6.30b}$$

$$\frac{d\phi_L}{d\tau} + \sin\phi_L = i_V - i_0, \tag{6.30c}$$

$$\frac{d\phi_C}{d\tau} + \sin\phi_C = i_C - i_0. \tag{6.30d}$$

Having specified the relevant equations for the system, we now proceed to finding a stationary solution for the superconducting phase in terms of the normalized magnetic flux ψ_{ex}. In order to do this, we proceed by numerical integration in the normalized time τ of Eq. (6.30a–d), until a stationary solution is reached. Therefore, after having expressed the currents in terms of the superconducting phase differences by means of (6.28a–c) and (6.29a–c), by substituting in Eq. (6.30a–d), we write:

$$\frac{d\phi_V}{d\tau} + \sin \phi_V = -\frac{\pi \psi_{ex} + \phi_V + \phi_L}{\pi \beta_L}, \qquad (6.31a)$$

$$\frac{d\phi_0}{d\tau} + \sin \phi_0 = -\frac{2\pi \psi_{ex} - 2\phi_L + \phi_0 - \phi_C}{2\pi \beta_L}, \qquad (6.31b)$$

$$\frac{d\phi_L}{d\tau} + \sin \phi_L = -\frac{2\phi_V + 4\phi_L - \phi_0 + \phi_C}{2\pi \beta_L}, \qquad (6.31c)$$

$$\frac{d\phi_C}{d\tau} + \sin \phi_C = -\frac{2\phi_L - \phi_0 + 5\phi_C}{2\pi \beta_L}. \qquad (6.31d)$$

Finding the stationary solutions for small values of β_L by means of the same procedure adopted in in subsections 6.2.1 and 6.2.2 now becomes a rather cumbersome task. Therefore, even though the analytic approach could be followed also in this case for small values of the parameter β_L, we shall here numerically solve the non-linear ordinary differential equations 6.31a–d.

By letting the normalized applied flux ψ_{ex} vary gradually from zero to some upper value, we find the superconducting phase differences by standard use of the software application Mathematica, obtaining the normalized applied fluxes shown in Fig. 6.8a–b. In Fig. 6.8a the normalized fluxes ψ_V (magenta), ψ_0 (cyan), and ψ_C (orange) are reported for $\beta_L = 0.1$. The same is done in Fig. 6.8b for $\beta_L = 0.2$. From the ψ vs. ψ_{ex} curves represented in Fig. 6.8b we may notice the presence of clear discontinuities, not evident in Fig. 6.8a, for which $\beta_L = 0.1$. From the presence of these clear discontinuities at $\beta_L = 0.2$, we may argue that the system, at least for this and higher values of β_L, behaves irreversibly, as far as its magnetic properties are concerned. We shall analyse the penetration mechanisms in the system by means of the normalized currents i vs. ψ_{ex} curves shown in Fig. 6.9a–b. In particular, the normalized currents i_V (magenta), i_0 (cyan), and i_C (orange) are reported for $\beta_L = 0.1$ in Fig. 6.9a. The same is done in Fig. 6.9b for $\beta_L = 0.2$.

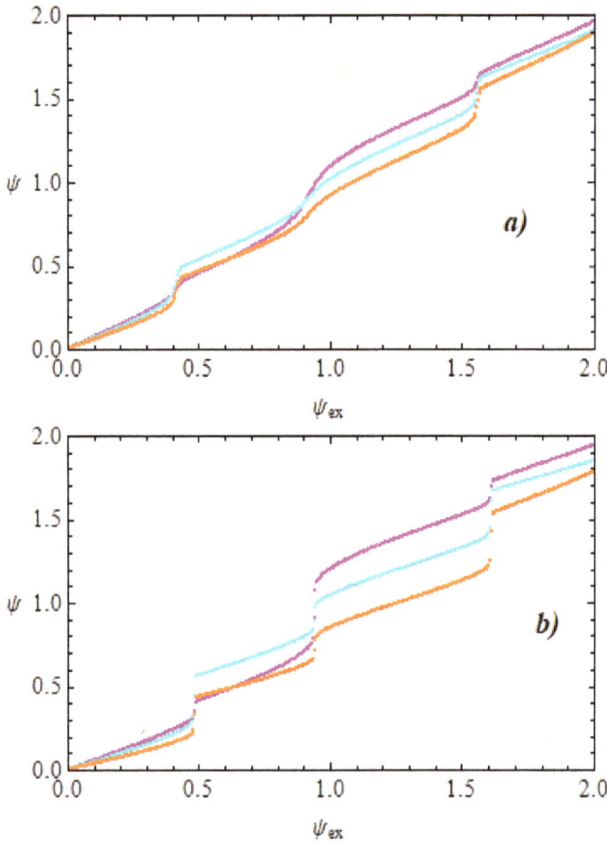

Fig. 6.8. Normalized magnetic fluxes ψ threading the superconducting loops in a 3×3 two-dimensional network of Josephson junctions for $\beta_L = 0.1$ (panel a) and $\beta_L = 0.2$ (panel b). The curves in magenta are for ψ_V (superconducting loops at the edge of the network). The curves in cyan are for ψ_0 (peripheral superconducting loops in between edge loops). Finally, the curves in orange are for ψ_C (central superconducting loop). In panel b a first irreversible penetration of flux quanta is recognizable by the discontinuities in all three curves at $\psi_{ex} \approx 0.5$.

Analysing the clearly irreversible magnetic behaviour of the i vs. ψ_{ex} curves in Fig. 6.9b, we notice that a first shielding region is present in these curves for $0 < \psi_{ex} < \psi_1$, where ψ_1 is the value of the normalized applied flux at which a first irreversible flux penetration occurs. In this region the normalized loop currents vary almost linearly with ψ_{ex}. The curve with the lowest slope is the one for the current i_V.

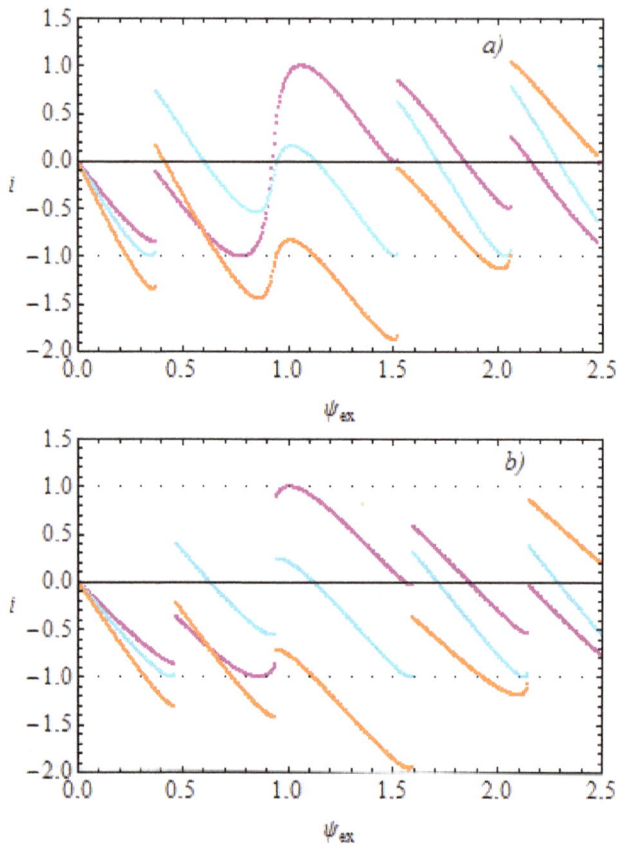

Fig. 6.9. Normalized loop currents in a 3×3 two-dimensional network of Josephson junctions for $\beta_L = 0.1$ (panel a) and $\beta_L = 0.2$ (panel b). The curves in magenta are for i_V (superconducting loops at the edge of the network). The curves in cyan are for i_0 (peripheral superconducting loops in between edge loops). Finally, the curves in orange are for i_C (central superconducting loop). In panel b a first irreversible penetration of flux quanta is recognizable by the discontinuities in all three curves at $\psi_{ex} \approx 0.5$. As discussed in the text, the flux penetration mechanism inside the network can be reconstructed by means of the oscillating character of these curves.

Noticing that the maximum possible value for the absolute value of the current i_C is +2, given that the effective current flowing in the central JJs is $i_C - i_0$, as shown in Fig. 6.7, we focus our attention on the current i_0 in Fig. 6.9b. The absolute value of this current reaches its maximum possible value (+1) because of a 2π phase slip of the superconducting phase difference ϕ_0 at $\psi_{ex} = \psi_1$. After first flux penetration, a change in sign (from negative to positive) of the current i_0 is accompanied by a decrease of the absolute value of the currents i_V and i_C. A second irreversible flux penetration event occurs by means of a 2π phase slip of the superconducting phase difference ϕ_V at the value of ψ_{ex} at which $i_V = -1$. A third irreversible flux penetration is present, in the interval reported in Fig. 6.9a–b, at the value of ψ_{ex} at which the current i_0 reaches again the value of -1. This qualitative description is useful to interpret correctly the flux penetration mechanisms inside the network and, at the same time, it clarifies the role of the loop currents in determining flux distribution in the system. As far as the role of the parameter β_L is concerned, from Fig. 6.8a–b and Fig. 6.9a–b we may argue that the flux transitions inside the superconducting loops are sharper for higher values of β_L and that the value of ψ_1, at which a first irreversible flux penetration occurs, increases for increasing values of this parameter, as it has already been noticed in subsections 6.2.1 and 6.2.2. This is also in line with what already seen in one-junction and two-junction interferometers and in one-dimensional arrays of Josephson junctions, where the parameter β_0 was specifically used in the latter case.

Further insight in the magnetic properties of a 3×3 two-dimensional network of Josephson junctions may be attained by considering the susceptibility χ calculated by the following expression:

$$\chi = \frac{\psi_{TOT} - 9\psi_{ex}}{9\psi_{ex}}, \tag{6.32}$$

where $\psi_{TOT} = 4\psi_V + 4\psi_0 + \psi_C$ is the total normalized magnetic flux inside the network. Two χ vs. ψ_{ex} curves after zero-field cooling are represented in Fig. 6.10, one for $\beta_L = 0.1$, one for $\beta_L = 0.2$.

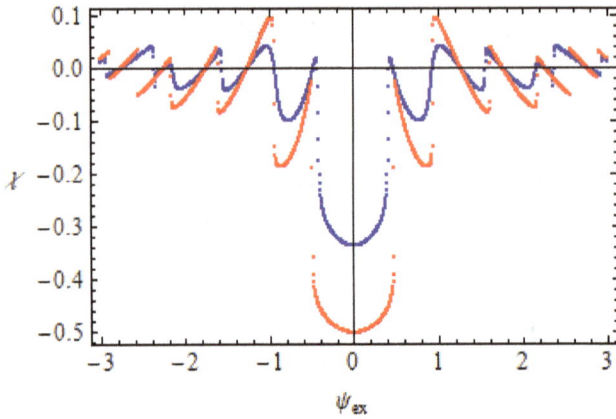

Fig. 6.10. Magnetic susceptibility χ of a 3×3 two-dimensional network of Josephson junctions as a function of the normalized applied flux ψ_{ex} for $\beta_L = 0.1$ (blue points) and $\beta_L = 0.2$ (red points).

As already noticed in the previous subsections, even in the more complex case of a 3×3 two-dimensional network of Josephson junctions these curves present a first diamagnetic region for $-\psi_1 < \psi_{ex} < \psi_1$, in which the loop currents are able to screen the externally applied flux. In this way, the response of the system is markedly diamagnetic in this interval. The bottom of the reversed lobe centred about the absolute minimum $\psi_{ex} = 0$ depends on the parameter β_L: the minimum reaches values closer and closer to $\chi = -1$ as β_L increases. Moreover, oscillations between diamagnetic and paramagnetic stationary states are present for absolute values of ψ_{ex} higher than ψ_1; the same behavior was also detected in the previous section.

In order to understand flux penetration mechanisms from the outer loops to the inner ones in more details, we analyze the case $\beta_L = 1.0$, for which the discontinuities in the flux and current distributions inside the system are expected to be rather sharp.

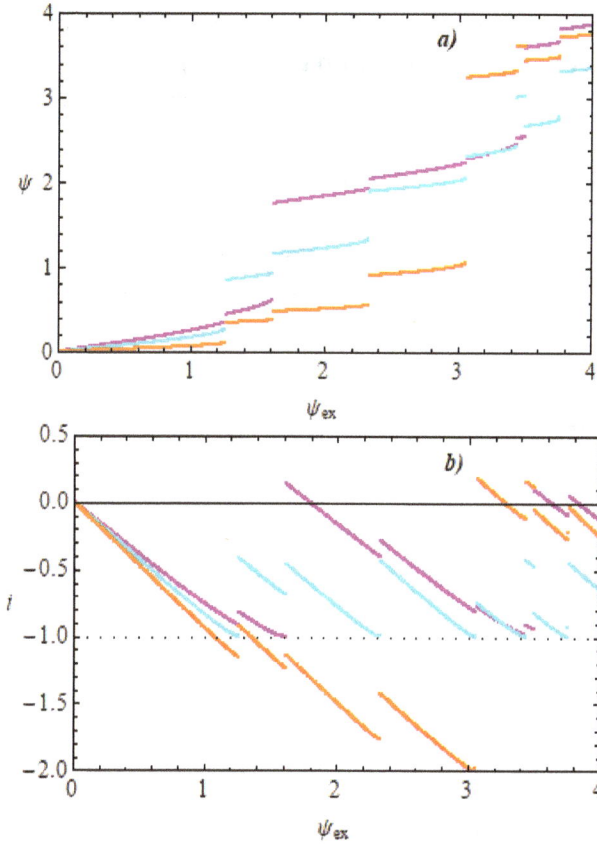

Fig. 6.11. In panel a the normalized magnetic fluxes ψ threading the superconducting loops in a 3×3 two-dimensional network of Josephson junctions are shown for $\beta_L = 1.0$. In panel b the normalized loop currents i are reported for the same network. The curves in magenta are for the superconducting loops at the edge of the network. The curves in cyan are for the peripheral superconducting loops in between edge loops. Finally, the curves in orange are for the central superconducting loop. A first irreversible penetration of flux quanta is recognizable by the discontinuities in all three curves at $\psi_{ex} \approx 1.25$.

Therefore, we report the normalized fluxes ψ_V (magenta), ψ_O (cyan), and ψ_C (orange) in Fig. 6.11a for $\beta_L = 1.0$. In Fig. 6.11b the normalized loops currents i_V (magenta), i_O (cyan), and i_C (orange) are shown for the same value of β_L. From the ψ vs. ψ_{ex} curves represented in Fig. 6.11a we may notice that first irreversible flux penetration occurs at $\psi_{ex} = \psi_1 \approx 1.25$. In fact, from the normalized currents vs. ψ_{ex} curves shown in

Fig. 6.11b a first shielding region is present in these curves for $0 <$
$\psi_{ex} < \psi_1$. By considering the current i_0 in Fig. 6.11b, we again notice
that it reaches a lower value of -1 at $\psi_{ex} = \psi_1$. A second irreversible
flux penetration event occurs at $\psi_{ex} = \psi_2$, because the current i_V reaches
the value of -1, and so on. In these figures we need to notice the rather
large discontinuity given by the penetration of flux in the central loop at
$\psi_1 \approx 3.05$. After a penetration of about two flux quanta in the central
loop, the field gradient (from the peripheral to the central loop) is
relieved and the values of the curves for the normalized fluxes ψ_V
(magenta) and ψ_0 (cyan) tend to coincide.

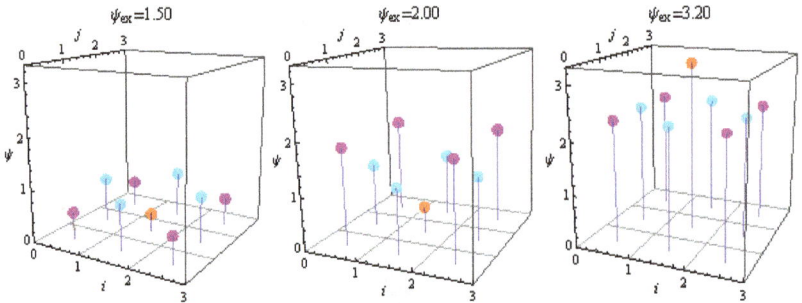

Fig. 6.12. Flux distribution in a 3×3 two-dimensional network of Josephson junctions for
$\beta_L = 1.0$ and for various values of the normalized applied flux ψ_{ex} (reported on top of
each figure). The points in magenta are for the superconducting loops at the edge of the
network. The points in cyan are for the peripheral superconducting loops in between edge
loops. Finally, the points in orange are for the central superconducting loop.

In order to better visualize these penetration mechanisms, let us recur to
a flux-distribution diagram, as the one reported in Fig. 6.12, where the
heights of the magenta, cyan, and orange points represent the values of
the normalized fluxes ψ_V, ψ_0, and ψ_C, respectively. Three different
values of the externally applied flux have been chosen; namely, $\psi_{ex} =$
1.50, $\psi_{ex} = 2.00$, and $\psi_{ex} = 3.20$. From Fig. 6.12 we may notice that at
$\psi_{ex} = 1.50$, after first irreversible flux penetration, the flux ψ_0 is
predominant with respect to ψ_V and ψ_C. This is because the
superconducting phase difference ϕ_0 suffers a 2π phase shift at $\psi_{ex} =$
ψ_1. On the other hand, for $\psi_{ex} = 2.00$, the plateaus in Fig. 6.9a are
distributed in such a way that $\psi_V > \psi_0 > \psi_C$, so that the magenta points

are on top, the cyan points in the middle, and the orange point in the lowest part of the graph. Finally, for $\psi_{ex} = 3.20$, after penetration of two fluxons in the central loop, we have $\psi_C > \psi_V \approx \psi_0$, so that the orange point is on top of the remaining points. Therefore, one argues that the inner part of the superconducting network can be partially shielded from the externally applied field only for rather low values of the forcing parameter ψ_{ex}.

6.4. Concentric circular arrays: the discrete model

One of the possible ways of extending the simple networks studied in the previous sections of this chapter is to consider cylindrically symmetric systems. We have already noticed that, in some cases, junctions in the network would remain inactive, allowing a simplification of the analysis.

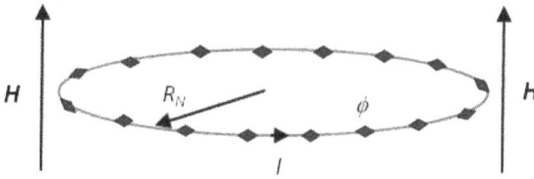

Fig. 6.13. Schematic representation of a circular array of N identical Josephson junctions. When an orthogonal external magnetic field is applied to the system, a current I flows in all JJs and, as a consequence, the same superconducting phase difference ϕ appears across each junction. The radius of the superconducting loop interrupted by the N JJs is R_N.

This occurred, for example, in the 2×2 network where only the peripheral junctions in the network were active, while the four inner junctions played no role. In fact, in this simple network the loop currents in adjacent loops flow in opposite directions (see Fig. 6.4), giving a net current equal to zero in the four inner branches. In this case, we could have as well analyzed the response of a single superconducting loop interrupted by eight Josephson junctions, in which one has to consider the size of the network, formed by four plaquettes network. Therefore, in the present section we shall first analyze the behavior of a circular array of N Josephson junctions in series. Successively, we shall analyze the response of concentric circular arrays of Josephson junctions.

6.4.1. *A circular array of Josephson junctions*

Let us start by recalling, in a rather general way, the response of the circular array of N identical JJs shown in Fig. 6.13, as already studied in Section 3.4.1. In this array the parameter β_N is taken to be proportional to the area $S_N = \pi R_N^2$ enclosed in the superconducting loop, of inductance L_N, interrupted by the N JJs having maximum Josephson current I_{J0} and resistive parameter R. Therefore, by assuming a fixed linear density in the distribution of the JJs along the loop, we may set $\beta_N = L_N I_{J0}/\Phi_0 = N^2 \beta_0$, where β_0 is a rescaled characteristic parameter of the system. When an orthogonal external magnetic field H is applied to the superconducting loop as shown in Fig. 6.13, a current I flows in all junctions. In this way, the same superconducting phase difference ϕ appears across each JJ of the circular array. Recalling now the results in Section 3.4.1 by means of fluxoid quantization and by defining $\Psi = \Phi/\Phi_0$, where Φ is the magnetic flux linked to the superconducting loop, we have

$$\phi = -\frac{2\pi}{N}\Psi. \tag{6.33}$$

The dynamic equation in terms of the variable Ψ and of the appropriate normalized time $\tau = NRt/L_N$ can be written as follows:

$$\frac{d\Psi}{d\tau} + \beta_N \sin\left(2\pi\frac{\Psi}{N}\right) + \Psi = \Psi_{ex}, \tag{6.34}$$

where $\Psi_{ex} = \mu_0 H S_N/\Phi_0$. Let us now rescale the fluxes Ψ and Ψ_{ex}, so that the new variables ψ and ψ_{ex} are defined as follows:

$$\psi = \frac{\Psi}{N}; \ \psi_{ex} = \frac{\Psi_{ex}}{N}. \tag{6.35}$$

Therefore, Eq. (6.34) can be rewritten as follows:

$$\frac{d}{d\tau}\psi + N\beta_0 \sin(2\pi\psi) + \psi = \psi_{ex}. \tag{6.36}$$

As remarked in Section 3.4.1, the multi-junction loop behaves like a superconducting loop interrupted by a single JJ. In this way, the system shows reversible magnetic behavior for $\beta_0 < 1/2\pi N$, and its general magnetic properties can be deduced from section 2.4.

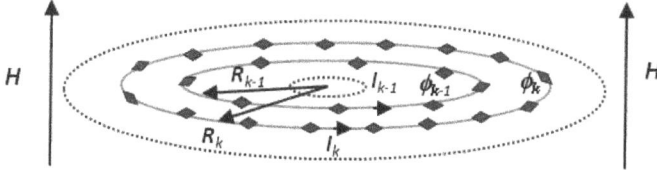

Fig. 6.14. Schematic representation of a concentric circular array of identical Josephson junctions. An orthogonal external magnetic field is applied to the system, so that the current I_k flows in all n_k JJs of the k-th concentric loop; as a consequence, a superconducting phase difference ϕ_k appears across each junction in the loop. The radius of the k-th superconducting concentric loop interrupted by the n_k JJs is R_k.

6.4.2. *Concentric circular arrays of Josephson junctions*

Let us now consider the system shown in Fig. 6.14, where identical Josephson junctions with maximum Josephson current I_{J0} and resistive parameter R is seen to form a concentric circular array. When an orthogonal external magnetic field is applied to this system, we may notice that a current I_k flows in all n_k JJs interrupting the k-th concentric superconducting ring. We take $k = 1, 2, \dots, N$. Therefore, a superconducting phase difference ϕ_k appears across each junction of the loop. The radius of the k-th superconducting concentric ring interrupted by the n_k JJs can be expressed as $R_k = n_k \delta/2\pi$, δ being the constant distance between adjacent junctions in each loop. The area enclosed in the k-th superconducting ring is thus proportional to n_k^2, so that we may set $\beta_k = n_k^2 \beta_0$, where β_0 will be defined later. We here take $n_k = k$, for simplicity. By now repeating the analysis done in the previous subsection for each loop in the network, we define $\Psi_k = \Phi_k/\Phi_0$, where Φ_k is the magnetic flux linked to the k-th superconducting ring. By fluxoid quantization, we may thus write:

$$\phi_k = -\frac{2\pi}{n_k}\Psi_k. \qquad (6.37)$$

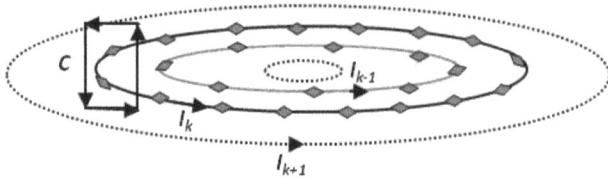

Fig. 6.15. Integration path C embracing the current I_k.

The dynamic equations for the superconducting phase differences ϕ_k can now be written, for $k = 1, 2, ..., N$, as follows:

$$\frac{\Phi_0}{2\pi R I_{J0}}\frac{d\phi_k}{dt} + \sin\phi_k = i_k, \tag{6.38}$$

where $i_k = I_k/I_{J0}$. We now need to express the currents i_k in terms of the fluxes Ψ_k. In order to do this, let us consider Ampere's law for the rectangular path shown in Fig. 6.15, so that:

$$\frac{\Phi_k - \Phi_{k-1}}{S_k - S_{k-1}}d - \frac{\Phi_{k+1} - \Phi_k}{S_{k+1} - S_k}d = \mu_0 I_k, \tag{6.39}$$

where d is the height of the ring and the multiplicative terms of d represent the local magnetic field h_k, defined as follows:

$$\mu_0 h_k = \frac{\Phi_k - \Phi_{k-1}}{S_k - S_{k-1}}. \tag{6.40}$$

In Eq. (6.39) and (6.40) the index k can well go from 1 to N, provided that we set:

$$\frac{\Phi_{N+1} - \Phi_N}{S_{N+1} - S_N} = \mu_0 H, \tag{6.41a}$$

$$\frac{\Phi_1}{S_1} = \mu_0 h_1. \tag{6.41b}$$

In Eq. (6.41b) we have thus considered equal to zero the quantities Φ_k and S_k labeled with index $k = 0$. Naturally, the latter positions need not to be confused with the definition of the elementary flux quantum and by other successive definitions.

By now normalizing the fluxes Φ_k to the elementary flux quantum Φ_0 in Eq. (6.39), by multiplying both sides of the same expression for S_k, setting $S_k = k^2 S_0$, with $S_0 = \delta^2/4\pi$, and by setting $L_k = \mu_0 S_k/d = k^2(\mu_0 S_0/d) = k^2 L_0$, we have:

$$\beta_0 i_k = \frac{\Psi_k - \Psi_{k-1}}{2k-1} - \frac{\Psi_{k+1} - \Psi_k}{2k+1}, \tag{6.42}$$

By which we can now finally define $\beta_0 = L_0 I_{J0}/\Phi_0$. By substituting Eq. (6.37) and (6.42) in Eq. (6.38), we obtain the dynamic equation for the normalized fluxes for $k = 1, 2, \dots, N$:

$$\frac{d\Psi_k}{d\tau} + k\beta_0 \sin\left(\frac{2\pi\Psi_k}{k}\right) + \frac{4k^2}{4k^2-1}\Psi_k = \frac{k}{2k-1}\Psi_{k-1} + \frac{k}{2k+1}\Psi_{k+1}, \tag{6.43}$$

where $\tau = Rt/L_0$, $\Psi_0 = 0$, and where, by Eq. (6.41a), we set:

$$\frac{\Psi_{N+1} - \Psi_N}{S_{N+1} - S_N} = \frac{\mu_0 H S_N}{\Phi_0}\frac{1}{S_N} = \frac{\Psi_{ex}}{S_N}. \tag{6.44}$$

In this way, the dynamic equation for $k = N$ can be written as follows:

$$\frac{d\Psi_N}{d\tau} + N\beta_0 \sin\left(\frac{2\pi\Psi_N}{N}\right) + \frac{N}{2N-1}\Psi_N = \frac{N}{2N-1}\Psi_{N-1} + \frac{\Psi_{ex}}{N}. \tag{6.45}$$

By the stationary solutions of Eq. (6.43), the flux distribution inside the system can be determined. On the other hand, by knowledge of the Ψ_k vs. Ψ_{ex} curves the currents i_k can be found by Eq. (6.42). Next, by defining the normalized magnetization m as follows

$$m = \Psi_N - \Psi_{ex}, \tag{6.46}$$

we may compute this quantity over a complete cycle in the quantity Ψ_{ex}.

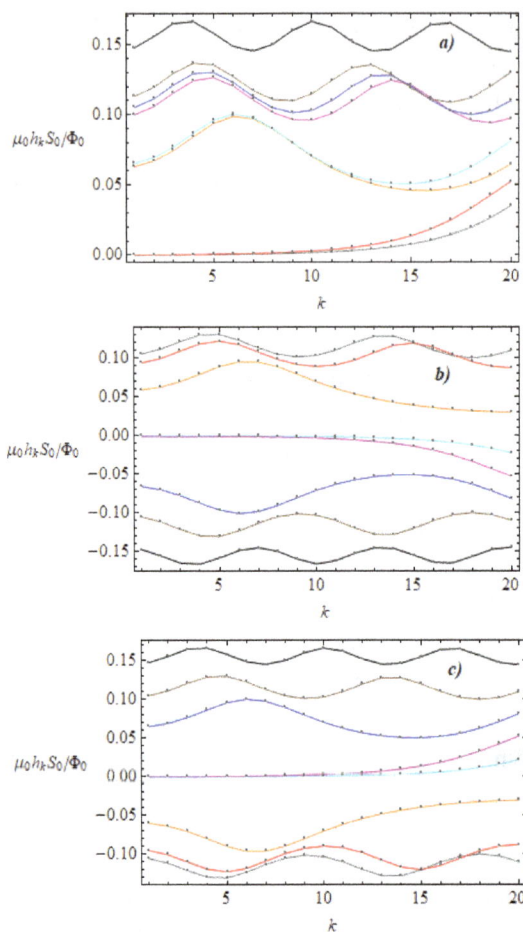

Fig. 6.16. Magnetic field distribution in the concentric circular array model of Josephson junctions for $N = 20$ and $\beta_0 = 0.01$. The index k ranges from 1 (inner shell) to 20 (external shell). Colored lines are used to distinguish the values of the normalized fluxes Ψ_k (grey points) at various values of Ψ_{ex}. In panel a, the increasing Ψ_{ex} values are: $+18.0$ (gray); $+24.0$ (red); $+30.0$ (orange); $+36.0$ (cyan); $+42.0$ (magenta); $+48.0$ (blue); $+54.0$ (brown); $+60.0$ (black). In panel b, the decreasing Ψ_{ex} values are: $+48.0$ (gray); $+36.0$ (red); $+12.0$ (orange); -12.0 (cyan); -24.0 (magenta); -36.0 (blue); -48.0 (brown); -60.0 (black). In panel c, the increasing Ψ_{ex} values are: -48.0 (gray); -24.0 (red); -12.0 (orange); $+12.0$ (cyan); $+24.0$ (magenta); $+36.0$ (blue); $+48.0$ (brown); $+60.0$ (black).

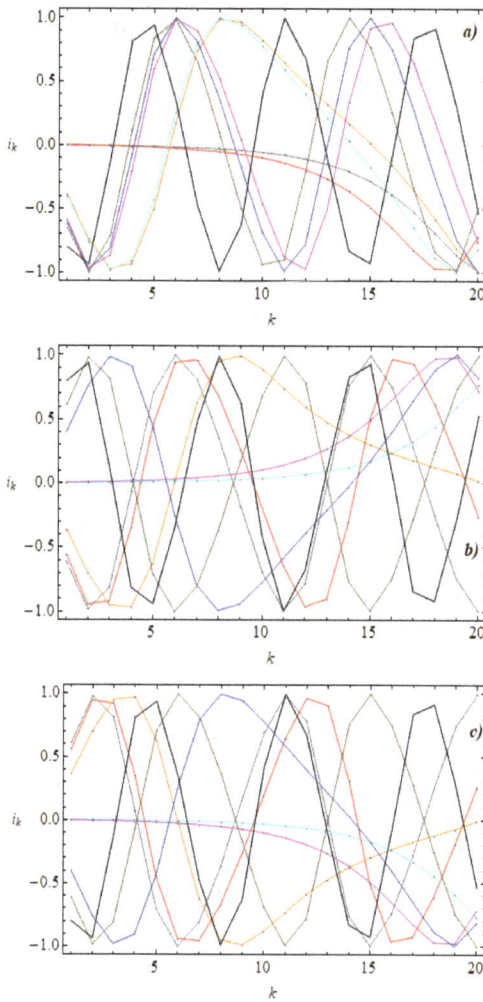

Fig. 6.17. Current distribution in the concentric circular array model of Josephson junctions for $N = 20$ and $\beta_0 = 0.01$. The index k ranges from 1 (inner shell) to 20 (external shell). Colored lines are used to distinguish the values of the normalized currents i_k (grey points) at various forcing terms Ψ_{ex}. In panel a, the increasing Ψ_{ex} values are: $+18.0$ (gray); $+24.0$ (red); $+30.0$ (orange); $+36.0$ (cyan); $+42.0$ (magenta); $+48.0$ (blue); $+54.0$ (brown); $+60.0$ (black). In panel b, the decreasing Ψ_{ex} values are: $+48.0$ (gray); $+36.0$ (red); $+12.0$ (orange); -12.0 (cyan); -24.0 (magenta); -36.0 (blue); -48.0 (brown); -60.0 (black). In panel c, the increasing Ψ_{ex} values are: -48.0 (gray); -24.0 (red); -12.0 (orange); $+12.0$ (cyan); $+24.0$ (magenta); $+36.0$ (blue); $+48.0$ (brown); $+60.0$ (black).

In Fig. 6.16a–c the normalized local field distribution ξ_k inside the system, defined as follows

$$\xi_k = \mu_0 h_k S_0 / \Phi_0,\qquad\qquad(6.47)$$

is shown for $\beta_0 = 0.01$ and for various values of the normalized applied magnetic flux Ψ_{ex}. In particular, in panel a) the quantity ξ_k is shown for increasing values of Ψ_{ex} in the interval $[0, +60]$; in panel b) for decreasing values of Ψ_{ex} in the interval $[-60, +60]$; in panel c) for increasing values of Ψ_{ex} in the interval $[-60, +60]$. Different colors for all values of Ψ_{ex} are used.

In order to correctly interpret the process of flux penetration inside the circular array of JJs in Fig. 6.14, it is necessary to represent the current distribution inside this system. Therefore, in Fig. 6.17a–c the i_k vs. k curves are shown for different values of the normalized applied flux Ψ_{ex}.

By analyzing the curves in Fig. 6.16a and in Fig. 6.17a in details, we see that, for normalized applied flux values $\Psi_{\text{ex}} = 18.0$ (gray curves) and $\Psi_{\text{ex}} = 24.0$ (orange curves) only the outermost superconducting rings are penetrated by flux quanta, the inner ones being undisturbed by the external forcing Ψ_{ex}. In fact, a strong diamagnetic response of the network, visible in the current distribution in Fig. 6.17a, prevents flux penetration deep inside the system. However, when Ψ_{ex} is raised to the value of 30.0 (orange curves) and 36.0 (cyan curves), a first alternating behavior in the ξ_k vs. k curves with a single positive hump, and alternating signs of the current in the i_k vs. k curves appear inside the circular array of JJs. When the externally applied magnetic flux is further increased to $\Psi_{\text{ex}} = 42.0$ (magenta curves), $\Psi_{\text{ex}} = 48.0$ (blue curves) and $\Psi_{\text{ex}} = 54.0$ (brown curves), the positive humps in the magnetic field distribution become two. Even though the qualitative behavior of the system at these values of Ψ_{ex} is similar, the amplitude of $\xi_k = \mu_0 h_k S_0 / \Phi_0$ at a given value of k grows with increasing values of Ψ_{ex}. Correspondingly, in the i_k vs. k curves a clear double oscillation can be noticed. The two positive peaks of the latter curves tend to be shifted to the left (in the inner part of the circular array) as Ψ_{ex} increases. The positive peaks are located in the regions where the gradient of the local field is minimum. In fact, according to Eq. (6.40) and (6.42), the currents i_k can be written, in terms of the normalized local field variables ξ_k, as follows:

$$i_k = -\frac{1}{\beta_0}(\xi_{k+1} - \xi_k). \qquad (6.48)$$

Finally, when $\Psi_{ex} = 60.0$ (black curves), the number of positive humps in the ξ_k vs. k curves becomes equal to three. In the same way, three positive peaks appear in the i_k vs. k curve.

Let us now consider the curves in Fig. 6.16b and in Fig. 6.17b. We may notice that decreasing the externally applied flux from $\Psi_{ex} = 60.0$ to $\Psi_{ex} = 48.0$ (gray curves) and successively to $\Psi_{ex} = 36.0$ (red curves), the number of positive humps present in the ξ_k vs. k curves lowers from three to two. Notice, however that, for $\Psi_{ex} = 36.0$, the magnetic response is different from what can be argued from the homologous curve in Fig. 6.16a obtained for values of Ψ_{ex} increasing from zero. By further lowering the value of the externally applied flux to $\Psi_{ex} = 12.0$ (orange curves) only one positive hump is detectable in the ξ_k vs. k curve and a corresponding single positive peak in the current is present in the i_k vs. k curve. Letting the normalized externally magnetic flux decrease to $\Psi_{ex} = -12.0$ (cyan curves) and successively to $\Psi_{ex} = -24.0$ (magenta curves), a negative, but rather flat, flux distribution inside the system is obtained and corresponding positive currents circulate in the outermost superconducting rings. A first reversed (negative) hump in the magnetic field distribution and one negative peak in the currents appear for $\Psi_{ex} = -36.0$ (blue curves). On the other hand, two reversed humps in the magnetic flux distribution and two negative peaks in the currents appear for $\Psi_{ex} = -48.0$ (brown curves). Finally, three reversed humps and three negative current peaks are present for $\Psi_{ex} = -60.0$ (black curves).

Let us now consider the curves in Fig. 6.16c and in Fig. 6.17c. For a first increase of the externally applied flux from $\Psi_{ex} = -60.0$ to $\Psi_{ex} = -48.0$ (gray curves) and successively to $\Psi_{ex} = -24.0$ (red curves), the number of reversed humps in the ξ_k vs. k curves lowers from three to two. Correspondingly, two negative peaks are present in the current distribution for $\Psi_{ex} = -48.0$ (gray curves) and $\Psi_{ex} = -24.0$ (red curves). By further increasing the value of the externally applied flux to $\Psi_{ex} = -12.0$ (orange curves) only one reversed hump is detectable in the ξ_k vs. k curve. For normalized external flux values of $\Psi_{ex} = 12.0$ (cyan curves) and $\Psi_{ex} = 24.0$ (magenta curves) flat flux and current distributions are obtained, except for the outermost superconducting loop, inside the concentric circular array. One positive hump and one positive current peak appear in the ξ_k vs. k blue curve and in the i_k vs. k blue curve for $\Psi_{ex} = 36.0$. Two positive humps and two positive current peaks are instead present in the ξ_k vs. k brown curve and in the i_k vs. k brown curve for $\Psi_{ex} = 48.0$. Finally, when the value of the externally applied flux is raised to $\Psi_{ex} = 60.0$ three positive humps and three positive peaks are detectable in the ξ_k vs. k black curve and in the ξ_k vs. k black curve, respectively.

This rather detailed description of the flux and current distributions in the concentric circular array for the rather low value of $\beta_0 = 0.01$ can be seen as propaedeutic to the analysis of the magnetization curve obtained by means of Eq. (6.46).

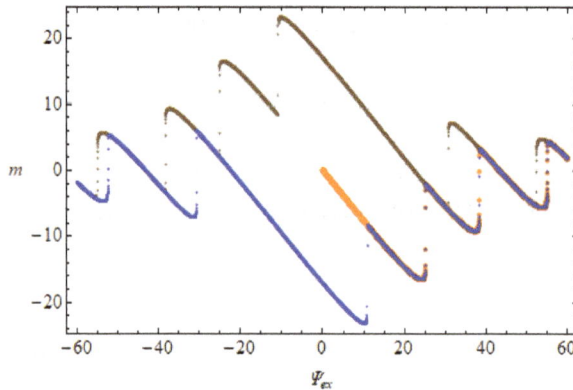

Fig. 6.18. Magnetization curve of a concentric circular array for $N = 20$ and $\beta_0 = 0.01$. The points in orange are for the first magnetization part of the cycle ($0 < \Psi_{ex} < 60$). The brown point represent the portion of the cycle for which Ψ_{ex} decreases from 60 to -60. Finally, the blue points are for the portion of the magnetization cycle obtained by letting Ψ_{ex} increase from -60 to 60.

In fact, by considering the magnetization cycle reported in Fig. 6.18, we may notice that the initial diamagnetic behavior in the first magnetization part (orange points) recedes after two flux transitions occurring at $\Psi_{ex} \approx$ 25 and $\Psi_{ex} \approx 40$. In fact, considering Fig. 6.16a, we observe that for values of Ψ_{ex} lower than 25 (gray and red curves), only the outermost superconducting rings host magnetic fluxes in the interstices between loops. On the other hand, the hump in the ξ_k vs. k orange curve ($\Psi_{ex} =$ 30.0) signifies a penetration of flux quanta, detectable in the discontinuity in the magnetization curve at $\Psi_{ex} \approx 25$. Furthermore, we see that, in the vicinity of $\Psi_{ex} = 40$, the system shows, even in ZFC conditions, paramagnetic behavior. This character is accentuated when the second cycling portion in Ψ_{ex} is considered (brown points), in which the externally applied flux decreases from $+60.0$ to -60.0. In fact, reverse flux transitions appear at different values with respect to the first magnetization portion.

As a consequence, positive magnetization Ψ_{ex} intervals grow in size, the maximum interval being approximately $(-12.5, 22.5)$. The third portion of the magnetization curve (blue points) complete the cycle, by making the magnetization curve symmetric with respect to the origin.

In the following we shall explore the role of the parameter β_0 in determining flux penetration inside the system, by briefly commenting on the results obtained for ξ_k vs. k curves for $\beta_0 = 0.1$.

In Fig. 6.19a–c the normalized local field distribution $\xi_k = \mu_0 h_k S_0 / \Phi_0$ is shown for $N = 20$, $\beta_0 = 0.1$, and for various values of the normalized applied magnetic flux Ψ_{ex}. In particular, in panel a) the quantity ξ_k is shown for increasing values of Ψ_{ex} in the interval $[0, +300]$; in panel b) for decreasing values of Ψ_{ex} in the interval $[-300, +300]$; in panel c) for increasing values of Ψ_{ex} in the interval $[-300, +300]$. In Fig. 6.20 the magnetization curve for $N = 20$, $\beta_0 = 0.1$ in the Ψ_{ex} interval $[-300, +300]$ is shown.

When we compare Fig. 6.19a–c to Fig. 16.16a–c and Fig. 6.20 to Fig. 6.18, we notice that the first irreversible penetration, starting from ZFC conditions, occurs at $\Psi_{ex} = \Psi_{ex}^{(1)} \approx 25$ for $\beta_0 = 0.01$ and at $\Psi_{ex} = \Psi_{ex}^{(1)} \approx 80$ for $\beta_0 = 0.1$. In order to investigate deeper on the dependence of $\Psi_{ex}^{(1)}$ on the parameter β_0, let us proceed as follows.

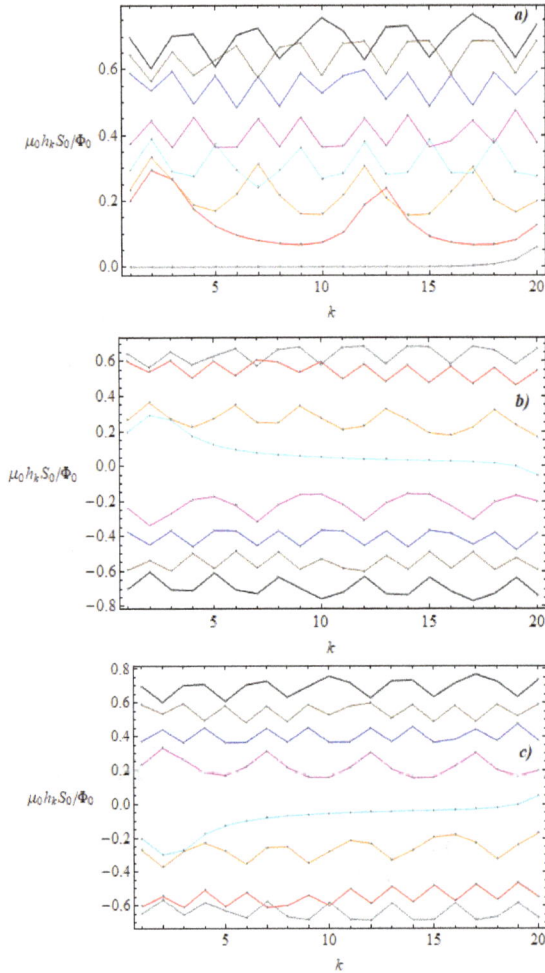

Fig. 6.19. Magnetic field distribution in the concentric circular array model of Josephson junctions for $N = 20$ and $\beta_0 = 0.1$. The index k ranges from 1 (inner shell) to 20 (external shell). Colored lines are used to distinguish the values of the normalized fluxes Ψ_k (grey points) at various values of Ψ_{ex}. In panel a, the increasing Ψ_{ex} values are: +600 (gray); +90.0 (red); +120.0 (orange); +150.0 (cyan); +180.0 (magenta); +240.0 (blue); +280.0 (brown); +300.0 (black). In panel b, the decreasing Ψ_{ex} values are: +240.0 (gray); +180.0 (red); +60.0 (orange); −60.0 (cyan); −120.0 (magenta); −180.0 (blue);−240.0 (brown); −300.0 (black). In panel c, the increasing Ψ_{ex} values are: −240.0 (gray); −180.0 (red); −60.0 (orange); +60.0 (cyan); +120.0 (magenta); +180.0 (blue); +240.0 (brown); +300.0 (black).

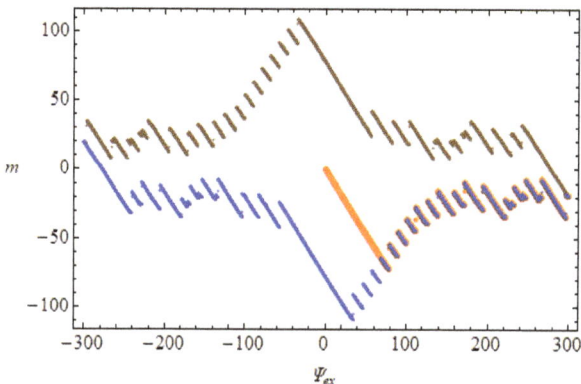

Fig. 6.20. Magnetization curve of a concentric circular array for $N = 20$ and $\beta_0 = 0.1$. The points in orange are for the first magnetization part of the cycle $(0 < \Psi_{ex} < 300)$. The brown points represent the portion of the cycle for which Ψ_{ex} decreases from 300 to -300. Finally, the blue points are for the portion of the magnetization cycle obtained by letting Ψ_{ex} increase from -300 to 300.

Starting from Eq. (6.43), we first write the following approximate stationary equations:

$$k\beta_0 \sin\left(\frac{2\pi\Psi_k}{k}\right) + \Psi_k = \frac{\Psi_{k-1}+\Psi_{k+1}}{2},\tag{6.49}$$

valid for $k \gg 1$. By expanding the sine term in Eq. (6.49) to first order in the argument, we have:

$$(1 + 2\pi\beta_0)\Psi_k = \frac{\Psi_{k-1}+\Psi_{k+1}}{2}.\tag{6.50}$$

By taking

$$\Psi_k = s^{(N-k)}\Psi_N,\tag{6.51}$$

we find that the above expression satisfies Eq. (6.50) provided that

$$s^2 - 2(1 + 2\pi\beta_0)s + 1 = 0.\tag{6.52}$$

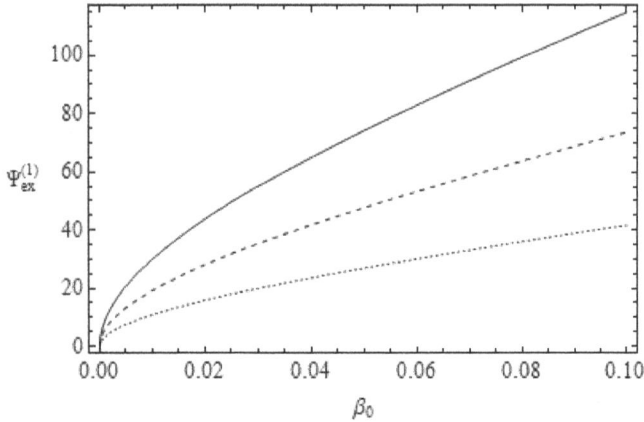

Fig. 6.21. Dependence on β_0 of the value of Ψ_{ex} at which first irreversible flux penetration in a concentric circular array occurs, for three values of N: $N = 15$ (dotted line), $N = 20$ (dashed line), $N = 25$ (dotted line).

Among the two positive solutions we choose the one giving an increase in k, so that:

$$s(\beta_0) = 1 - 2\pi\beta_0 \left(\sqrt{1 + \frac{1}{\pi\beta_0}} - 1 \right). \tag{6.53}$$

Substituting now Eq. (6.51) in the stationary part of Eq. (6.45), where we impose that the argument of the sine is $\pi/2$ and $\Psi_{ex} = \Psi_{ex}^{(1)}$, we have:

$$\Psi_{ex}^{(1)} \approx N^2 \left[\beta_0 + \frac{N(1-s(\beta_0))}{4(2N-1)} \right]. \tag{6.54}$$

We need to recall that Eq. (6.54) provides only a rough estimate of the true values of $\Psi_{ex}^{(1)}$. However, Eq. (6.53) and (6.54) reproduce fairly well the overall qualitative dependence of $\Psi_{ex}^{(1)}$ on β_0 and N. In Fig. 6.21 the $\Psi_{ex}^{(1)}$ vs. β_0 curve, obtained from Eq. (6.54), is reported for three values of N: $15, 20, 25$. One notices that, apart from a non-linear increase of $\Psi_{ex}^{(1)}$ for very small values of β_0, the $\Psi_{ex}^{(1)}$ vs. β_0 curves approach an oblique asymptote for $\beta_0 > 0.1$.

As the number of loops in the array increases, the discrete model can be replaced by a continuous one, as it will be shown in the following section.

6.5. Concentric circular arrays: The continuous model

In the present section we would like to specify the continuous form of the discrete model given in the previous section. In other words, we would like to consider the concentric circular array model for a very small distance a between adjacent superconducting loops.

Therefore, by recalling Eq. (6.48), we specify its continuous version by writing:

$$\beta_0 i(r,t) = -\frac{\partial \xi(r,t)}{\partial r} a, \qquad (6.55)$$

where $\xi = \xi(r,t) = \mu_0 h(r,t)S_0/\Phi_0$ is the time-dependent normalized local field at distance r from the center of the concentric loops and $\beta_0 = L_0 I_{J0}/\Phi_0$ is defined as in the previous section. By introducing the normalization $\varrho = r/a$ for the radial distance, we rewrite Eq. (6.55) as follows:

$$\beta_0 i(\rho,t) = -\frac{\partial \xi(\rho,t)}{\partial \rho}. \qquad (6.56)$$

The above equation can be also obtained by applying Ampere's law to a path inside the system enclosing the normalized current i. For each junction in a loop carrying the current i, we can write the following dynamic equation:

$$\frac{\Phi_0}{2\pi R I_{J0}} \frac{\partial \phi(r,t)}{\partial t} + \sin \phi(r,t) = i(r), \qquad (6.57)$$

where R and I_{J0} are the resistive parameter and the maximum Josephson current of the junctions in the network. By now taking the number of junctions in a current loop at distance r from the center of the concentric array to be $n(r)$, we may rewrite the quantization condition in Eq. (6.33) as follows:

$$\phi(r,t) = -\frac{2\pi}{n(r)} \psi(r,t). \qquad (6.58)$$

By considering Eq. (6.40), defining the local magnetic field in terms of the magnetic fluxes, we may write:

$$\xi(r,t) = \frac{\psi(r,t) - \psi(r-a,t)}{n^2(r) - n^2(r-a)} = \frac{\frac{\partial \psi(r,t)}{\partial r}}{2n(r)\frac{dn}{dr}}. \tag{6.59}$$

Therefore, by taking account of Eq. (6.56), (6.58), and (6.59), considering the normalization $\varrho = r/a$, Eq. (6.57) can be recast in the following form:

$$\frac{1}{n(\rho)}\frac{\partial \psi(\rho,\tau)}{\partial \tau} + \beta_0 \sin\left(\frac{2\pi\psi(\rho,\tau)}{n(\rho)}\right) = \frac{\partial}{\partial \rho}\left[\frac{\frac{\partial \psi(\rho,\tau)}{\partial \rho}}{2n(\rho)\frac{dn}{d\rho}}\right], \tag{6.60}$$

where $\tau = Rt/L_0$. We now write

$$n(\rho) = 2\pi\rho/\delta = \kappa\rho, \tag{6.61}$$

κ being a constant and δ the normalized distance between the junctions along the circumference having normalized radius ρ. Therefore, the constant κ can be expressed as follows:

$$\kappa = \frac{2\pi}{\delta} = \frac{dn}{d\rho}. \tag{6.62}$$

We thus notice that this constant defines the rate at which the number of junctions varies upon a variation $d\rho$ in the normalized radius. Because of Eq. (6.61) and (6.62), Eq. (6.60) becomes:

$$\frac{\partial \psi(\rho,\tau)}{\partial \tau} + \kappa\rho\beta_0 \sin\left(\frac{2\pi\psi(\rho,\tau)}{\kappa\rho}\right) = \frac{\rho}{2\kappa}\frac{\partial}{\partial \rho}\left[\frac{1}{\rho}\frac{\partial \psi(\rho,\tau)}{\partial \rho}\right], \tag{6.63}$$

While in the previous section we considered an unitary increase of the number of Josephson junctions hosted on each superconducting loop for an increase of the radius of a, here we leave this number more generally equal to κ. The above differential equation will be solved under the ZFC initial condition

$$\psi(\rho, 0) = 0, \tag{6.64}$$

and under boundary conditions, which shall here define. The first can be inferred from Eq. (6.44):

$$\frac{\partial \psi(\rho,\tau)}{\partial \rho}\bigg|_{\rho=\rho_{max}} = \frac{2\psi_{ex}}{\rho_{max}}, \tag{6.65}$$

where ρ_{max} is the normalized radial distance of the junctions in the outermost loop in the concentric network and ψ_{ex} is the normalized maximum value of the normalized applied flux, defined, according to Eq. (6.44), as follows:

$$\psi_{ex} = \frac{\mu_0 H(\pi \rho_{max}^2)}{\Phi_0},$$ (6.66)

where H is the amplitude of the applied field orthogonal to the system of concentric loops. The second is just a way of saying that no magnetic flux can be present in a loop of vanishing area, so that:

$$\lim_{\rho \to 0^+} \psi(\rho, \tau) = 0.$$ (6.67)

Having specified the initial condition and the boundary conditions, in the following section we proceed to solve the partial differential equation (6.63) in the limit of very small values of the parameter β_0.

6.5.1. *Analytic approach for small values of β_0*

We start noticing that, in the limit of very small values of the parameter β_0, we might try to solve the non-linear partial differential equation (PDE) reported in Eq. (6.63) by means of a perturbation approach to first order in β_0, by setting:

$$\psi(\rho, \tau) \approx \psi_0(\rho, \tau) + \beta_0 \psi_1(\rho, \tau).$$ (6.68)

By substituting the above approximated expression in Eq. (6.63) and by equating terms of similar order in the parameter β_0 on the right-hand side and on the left-hand side of the PDE, we have:

$$\frac{\partial \psi_0(\rho,\tau)}{\partial \tau} = \frac{\rho}{2\kappa} \frac{\partial}{\partial \rho} \left[\frac{1}{\rho} \frac{\partial \psi_0(\rho,\tau)}{\partial \rho} \right],$$ (6.69a)

$$\frac{\partial \psi_1(\rho,\tau)}{\partial \tau} - \frac{\rho}{2\kappa} \frac{\partial}{\partial \rho} \left[\frac{1}{\rho} \frac{\partial \psi_1(\rho,\tau)}{\partial \rho} \right] = -\kappa \rho \sin \left(\frac{2\pi \psi_0(\rho,\tau)}{\kappa \rho} \right).$$ (6.69b)

Despite the fact that the nonlinear term in Eq. (6.63) becomes a forcing term in Eq. (6.69b), where the argument of the sine can be found by solving Eq. (6.69a), we still face a rather challenging task in solving

Eq. (6.69a–b). We shall here attempt to give an analytic solution only of Eq. (6.69a), subject to the following initial and boundary conditions:

$$\psi_0(\rho, 0) = 0, \tag{6.70a}$$

$$\lim_{\rho \to 0^+} \psi_0(\rho, \tau) = 0, \tag{6.70b}$$

$$\left. \frac{\partial \psi_0(\rho,\tau)}{\partial \rho} \right|_{\rho=\rho_{max}} = \frac{2\psi_{ex}}{\rho_{max}}. \tag{6.70c}$$

In order to transform the above nonhomogeneous boundary conditions into homogeneous ones, we set:

$$\Gamma_0(\rho, \tau) = \psi_0(\rho, \tau) - \frac{2\psi_{ex}}{\rho_{max}}\rho. \tag{6.71}$$

By substituting the expression for $\psi_0(\rho, \tau)$ arguable from Eq. (6.71) into Eq. (6.69a), we have:

$$\rho\frac{\partial \Gamma_0(\rho,\tau)}{\partial \tau} - \frac{\rho^2}{2\kappa}\frac{\partial}{\partial \rho}\left[\frac{1}{\rho}\frac{\partial \Gamma_0(\rho,\tau)}{\partial \rho}\right] = -\frac{\psi_{ex}}{\kappa\rho_{max}}, \tag{6.72}$$

which is valid in the same interval $0 \leq \rho \leq \rho_{max}$ of Eq. (6.63) and is now subject to the following initial conditions and boundary conditions:

$$\Gamma_0(\rho, 0) = -\frac{2\psi_{ex}}{\rho_{max}}\rho, \tag{6.73a}$$

$$\Gamma_0(0, \tau) = 0, \tag{6.73b}$$

$$\left. \frac{\partial \Gamma_0(\rho,\tau)}{\partial \rho} \right|_{\rho=\rho_{max}} = 0, \tag{6.73c}$$

where, for brevity, we have replaced the limit in Eq. (6.70b) by the standard boundary condition at $\rho = 0$ in Eq. (6.73b). Naturally, having specified the interval $0 \leq \rho \leq \rho_{max}$, over which the function is defined, the expression in Eq. (6.73b) is now unambiguous. Being now the boundary condition of Eq. (6.72) homogeneous, the overall problem can be treated by means of Sturm-Liouville theory (Boyce, 1976).

Therefore, in order to obtain an analytical solution for $\Gamma_0(\rho, \tau)$, we could proceed as follows. We would first obtain a solution for the homogeneous problem by separation of variables. Therefore, we set

$$\Gamma_0(\rho, \tau) = T(\tau)R(\rho). \tag{6.74}$$

In this way, Eq. (6.69a) could be written as follows:

$$2\kappa \left[\frac{dT}{d\tau}/T(\tau)\right] = \frac{\rho}{R(\rho)} \frac{d}{d\rho} \left[\frac{1}{\rho} \frac{dR}{d\rho}\right] = -\lambda, \tag{6.75}$$

where we take $\lambda > 0$, in order to get finite solutions of $T(\tau)$ for $\tau \to +\infty$. By considering the differential equation for the radial part, for which we preserve homogenous boundary conditions, we write:

$$\rho \frac{d}{d\rho} \left[\frac{1}{\rho} \frac{dR}{d\rho}\right] = -\lambda R(\rho). \tag{6.76}$$

This equation leads to a Bessel equation of order one if we make the further substitution

$$y(\rho) = \frac{R(\rho)}{\rho}, \tag{6.77}$$

For which we still require boundedness in the interval $0 \le \rho \le \rho_{max}$. Because of the above definition, by taking into account the boundary conditions in (6.73b) and (6.73c) the problem to be solved is now the following:

$$\rho^2 y'' + \rho y' + (\lambda \rho^2 - 1)y = 0. \tag{6.78}$$

with boundary conditions

$$y(0) = 0, \tag{6.79a}$$

$$\rho_{max} y'(\rho_{max}) + y(\rho_{max}) = 0. \tag{6.79b}$$

By the hypothesis of boundedness of the function y over the interval $0 \le \rho \le \rho_{max}$, we can express the solution of Eq. (6.78) in terms of the Bessel function of order one $J_1(\sqrt{\lambda}\rho)$. Therefore, the boundary condition (6.79a) can be satisfied and Eq. (6.79b) requires that

$$\bar{x} J_1'(\bar{x}) + J_1(\bar{x}) = 0. \tag{6.80}$$

where $\bar{x} = \sqrt{\lambda}\rho_{max}$ and $J_1'(x) = dJ_1/dx$. Eq. (6.80) has infinitely many solutions $\bar{x}_n = \sqrt{\lambda_n}\rho_{max}$, as it can be readily verified, for example, by plotting the functions $x J_1'(x)$ and $-J_1(x)$ and by looking for the intersections between the graphs of the two functions. A more refined argument will be given in the following section, where these zeroes are actually identified better. Therefore, the solution Eq. (6.78) can be expressed in terms of the orthonormal eigenfunctions $\phi_n(\rho)$, defined as follows:

$$\phi_n(\rho) = c_n J_1\left(\sqrt{\lambda_n}\rho\right) = \frac{\sqrt{2}}{\rho_{max}} \frac{J_1\left(\sqrt{\lambda_n}\rho\right)}{|J_1\left(\sqrt{\lambda_n}\rho_{max}\right)|}, \qquad (6.81)$$

with the implicit definition of the normalization coefficients c_n. The corresponding solutions in the radial function R can thus be written as follows:

$$R_n(\rho) = \rho\phi_n(\rho). \qquad (6.82)$$

Therefore, we might express the solution for the function $\Gamma_0(\rho, \tau)$ as follows:

$$\Gamma_0(\rho, \tau) = \sum_{n=1}^{\infty} b_n(\tau) R_n(\rho) = \rho \sum_{n=1}^{\infty} b_n(\tau)\phi_n(\rho), \qquad (6.83)$$

where the time-dependent coefficients $b_n(\tau)$ are to be evaluated by recurring to Eq. (6.72) which, upon substitution of Eq. (6.83), can be written as follows:

$$\sum_{n=1}^{\infty} \left[b_n'(\tau) + \frac{\lambda_n}{2\kappa} b_n(\tau) \right] \phi_n(\rho) = -\frac{\psi_{ex}}{\kappa\rho_{max}} \frac{1}{\rho^2}. \qquad (6.84)$$

By now expressing the function $1/\rho^2$ in terms of the eigenfunctions $\phi_n(\rho)$ in the open interval $0 < \rho \leq \rho_{max}$, we set:

$$\frac{1}{\rho^2} = \sum_{n=1}^{\infty} \gamma_n \, \phi_n(\rho), \qquad (6.85)$$

where the constant coefficients γ_n can be found by means of the orthogonality property to be:

$$\gamma_n = \lim_{\epsilon \to 0} \int_\epsilon^{\rho_{max}} \frac{\phi_n(\rho)}{\rho} d\rho = c_n \lim_{\epsilon \to 0} \int_\epsilon^{\rho_{max}} \frac{J_1\left(\sqrt{\lambda_n}\rho\right)}{\rho} d\rho. \qquad (6.86)$$

Use of the recursion relations of Bessel functions (Gradshteyn, 1979) allows us to express the integrand in the right hand side of Eq. (6.86) as follows:

$$\frac{J_1(x)}{x} = J_0(x) - \frac{d}{dx} J_1(x). \tag{6.87}$$

Where we set $x = \sqrt{\lambda_n} \rho$. By further making use of the integral relation

$$\int_0^{\bar{x}} J_0(x) dx = 2 \sum_{j=0}^{\infty} J_{2j+1}(\bar{x}), \tag{6.88}$$

we may finally write:

$$\gamma_n = c_n \left[2 \sum_{j=1}^{\infty} J_{2j+1}(\sqrt{\lambda_n} \rho_{max}) - J_1(\sqrt{\lambda_n} \rho_{max}) \right]. \tag{6.89}$$

We can now write Eq. (6.84) as follows:

$$\sum_{n=1}^{\infty} \left[b_n'(\tau) + \frac{\lambda_n}{2\kappa} b_n(\tau) + \frac{\psi_{ex}}{\kappa \rho_{max}} \gamma_n \right] \phi_n(\rho) = 0. \tag{6.90}$$

Because of the orthogonality of the eigenfunctions ϕ_n, from the above expression we may set

$$b_n'(\tau) + \frac{\lambda_n}{2\kappa} b_n(\tau) = -\frac{\psi_{ex}}{\kappa \rho_{max}} \gamma_n. \tag{6.91}$$

for each integer $n = 1, 2, \ldots$ The above equation has the following immediate solution:

$$b_n(\tau) = \beta_n e^{-\frac{\lambda_n}{2\kappa} \tau} - \frac{2\psi_{ex}}{\lambda_n \rho_{max}} \gamma_n, \tag{6.92}$$

where the constants β_n are to be determined by means of the initial condition (6.73a). Let us then first express the solution $\Gamma_0(\rho, \tau)$, as we wrote in Eq. (6.83), in terms of all known quantities, except for the coefficients β_n:

$$\Gamma_0(\rho, \tau) = \rho \sum_{n=1}^{\infty} \left(\beta_n e^{-\frac{\lambda_n}{2\kappa} \tau} - \frac{2\psi_{ex}}{\lambda_n \rho_{max}} \gamma_n \right) \phi_n(\rho). \tag{6.93}$$

In order to find the constants β_n, we recall the initial condition in Eq. (6.73a). Therefore, by setting $\tau = 0$ in Eq. (6.93), we obtain:

$$\sum_{n=1}^{\infty} \left[\beta_n - \frac{2\psi_{ex}}{\lambda_n \rho_{max}} \gamma_n \right] \phi_n(\rho) = -\frac{2\psi_{ex}}{\rho_{max}}. \tag{6.94}$$

The term in brackets on the left hand side of Eq. (6.94), which represent the expansion coefficients of the constant function on the right hand side, is thus given by the following expression:

$$\beta_n - \frac{2\psi_{ex}}{\lambda_n \rho_{max}} \gamma_n = -\frac{2\psi_{ex}}{\rho_{max}} \int_0^{\rho_{max}} \rho \, \phi_n(\rho) \, d\rho. \qquad (6.95)$$

Therefore:

$$\beta_n = \frac{2\psi_{ex}}{\rho_{max}\lambda_n} \left[\gamma_n - c_n \int_0^{\bar{x}_n} x J_1(x) \, dx \right], \qquad (6.96)$$

where $x = \sqrt{\lambda_n}\rho$ and $\bar{x}_n = \sqrt{\lambda_n}\rho_{max}$. By means of the relation

$$J_1(x) = -\frac{d}{dx} J_0(x), \qquad (6.97)$$

after having integrated by parts the integral on the right hand side of Eq. (6.96), recalling Eq. (6.88) and (6.89) we have:

$$\beta_n = \frac{2\psi_{ex} c_n}{\sqrt{\lambda_n}} \left[J_1(\bar{x}_n) - \frac{J_1(\bar{x}_n)}{\bar{x}_n} \right], \qquad (6.98)$$

or

$$\beta_n = \frac{2\psi_{ex} c_n}{\sqrt{\lambda_n}} J_1'(\sqrt{\lambda_n}\rho_{max}) = \frac{2\psi_{ex} c_n \bar{x}_n}{\lambda_n} J_1'(\bar{x}_n). \qquad (6.99)$$

In this way, setting again $\bar{x}_n = \sqrt{\lambda_n}\rho_{max}$, we write the solution of Eq. (6.72) as follows:

$$\Gamma_0(\rho, \tau) = -\frac{2\psi_{ex}\rho}{\rho_{max}} \sum_{n=1}^{\infty} \left[c_n J_1(\bar{x}_n) e^{-\frac{\lambda_n}{2\kappa}\tau} + \gamma_n \right] \frac{\phi_n(\rho)}{\lambda_n}. \qquad (6.100)$$

Having treated in details the initial condition (6.73a), we now make a very rapid check on the boundary conditions (6.73b) and (6.73c). The boundary condition (6.73b) is readily satisfied by the above solution, because of the convergence of the series and of the overall multiplication by the variable ρ. As for the condition (6.73c), we proceed as follows. First define the solution in Eq. (6.100) as follows:

$$\Gamma_0(\rho, \tau) = -\frac{2\psi_{ex}}{\rho_{max}} \rho S_0(\rho, \tau), \qquad (6.101)$$

where $S_0(\rho, \tau)$ represents the infinite sum in Eq. (6.100). By now evaluating the first derivative with respect to the radial variable, we have

$$\frac{\partial \Gamma_0(\rho, \tau)}{\partial \rho} = -\frac{2\psi_{ex}}{\rho_{max}}\left[S_0(\rho, \tau) + \rho\frac{\partial S_0(\rho, \tau)}{\partial \rho}\right]. \tag{6.102}$$

We now realize that the term inside the bracket in the infinite sum of $\Gamma_0(\rho, \tau)$ in Eq. (6.100) is left unaltered by the derivative with respect to ρ, so that we have:

$$\frac{\partial \Gamma_0(\rho, \tau)}{\partial \rho} = -\frac{2\psi_{ex}}{\rho_{max}}\sum_{n=1}^{\infty}\left[c_n J_1(\bar{x}_n)e^{-\frac{\lambda_n}{2\kappa}\tau} + \gamma_n\right]\frac{\phi_n(\rho) + \rho\frac{d\phi_n(\rho)}{d\rho}}{\lambda_n}. \tag{6.103}$$

It now suffices to evaluate the above derivative in $\rho = \rho_{max}$ and to show that

$$\phi_n(\rho_{max}) + \rho_{max}\frac{d\phi_n(\rho)}{d\rho}\bigg|_{\rho=\rho_{max}} = 0. \tag{6.104}$$

This can be done by means of Eq. (6.80), after having expressed the eigenfunctions $\phi_n(\rho)$ in terms of the Bessel function J_1 as in Eq. (6.81). Finally, by recalling Eq. (6.71), the solution of Eq. (6.69a) can be found:

$$\Psi_0(\rho, \tau) = \frac{2\psi_{ex}\rho}{\rho_{max}}\left\{1 - \sum_{n=1}^{\infty}\left[c_n J_1(\bar{x}_n)e^{-\frac{\lambda_n}{2\kappa}\tau} + \gamma_n\right]\frac{\phi_n(\rho)}{\lambda_n}\right\}. \tag{6.105}$$

As stated before, the above solution can be used to solve Eq. (6.69b). However, the analytic procedure in solving for the function $\Psi_1(\rho, \tau)$, although being similar to what was already done, is rather long, so that we will content ourselves with the leading-order solution given by Eq. (6.105).

Another important point to be made is that the solution in Eq. (6.105) can be used when we start from ZFC initial conditions, as specified in Eq. (6.70a). Had we chosen a different magnetic configuration, with initially trapped flux inside the system, a different solution would be expected. In the following subsection we shall consider some immediate results of the solution found.

6.5.2. *Some results for very small values of β_0*

Having found a leading-order analytic expression to the problem for small values of the parameter β_0, we are in a position to explore in which way the analytic model reproduces, at least qualitatively, the numerical results reported in the previous section for $\beta_0 = 0.01$. First of all, we should evaluate the infinitely many roots of Eq. (6.80), in order to obtain the quantities $\bar{x}_n$ ($n = 1, 2, ...$). We thus proceed as follows. First, we recognize, by means of Eq. (6.87), that we can write:

$$\bar{x} J_1'(\bar{x}) + J_1(\bar{x}) = \bar{x} J_0(\bar{x}). \tag{6.106}$$

Therefore, for $\bar{x} > 0$, the roots $\bar{x}_n$ of the expression in Eq. (6.80) coincide with the roots of $J_0(\bar{x})$, which are tabulated. For example, for the first four values of $\bar{x}_n$, retaining four significant digits, we can set (Gradshteyn, 1979):

$$\bar{x}_1 = 2.405, \bar{x}_2 = 5.520, \bar{x}_3 = 8.654, \bar{x}_4 = 11.79. \tag{6.107}$$

Let us now define, in terms of the stationary part of the leading order solution $\Psi_0(\rho, \tau)$, the stationary values of the normalized local magnetic field and current, namely, of the quantities $\xi_S(\rho)$ and $i_S(\rho)$. Therefore, first define:

$$\Psi_S(\rho) = \lim_{\tau \to \infty} \Psi_0(\rho, \tau). \tag{6.108}$$

By calculating the above limit, we have:

$$\Psi_S(\rho) = \frac{2\Psi_{ex}\rho}{\rho_{max}} \left[1 - \sum_{n=1}^{\infty} \frac{\gamma_n}{\lambda_n} \phi_n(\rho)\right]. \tag{6.109}$$

By Eq. (6.59), recalling Eq. (6.61), we first set

$$\xi_S(\rho) = \frac{1}{2\kappa^2\rho} \frac{d\Psi_S(\rho)}{d\rho}. \tag{6.110}$$

By Eq. (6.55) and Eq. (6.110) we can thus write:

$$i_S(\rho) = -\frac{1}{\beta_0} \frac{d\xi_S(\rho)}{d\rho} = -\frac{1}{2\kappa^2\beta_0} \frac{d}{d\rho}\left[\frac{1}{\rho} \frac{d\Psi_S(\rho)}{d\rho}\right]. \tag{6.111}$$

The above expression can be compared with the stationary part of Eq. (6.63), to obtain the following more compact expression for the normalized current:

$$i_S(\rho) = -\sin\left(\frac{2\pi\Psi_S(\rho)}{\kappa\rho}\right). \tag{6.112}$$

Let us now compare the results obtained for the magnetic field and current distributions within the discrete system obtained for $\beta_0 = 0.01$ and shown in Fig. 6.16a and 6.17a, respectively, for increasing values of the normalized applied magnetic flux ψ_{ex}, to the analogous quantity given by the analytical expression in Eq. (6.110). In Fig. 6.22a, 6.22b, and 6.22c we show the stationary solution $\Psi_S(\rho)$, the normalized local magnetic field $\xi_S(\rho)$, and the normalized current $i_S(\rho)$, respectively. In these figures, the same parameters chosen for Fig. 6.16a and Fig. 6.17a have been used, and the normalized applied magnetic flux ψ_{ex} is taken to increase from 10 to 60, with step 10.

From Fig. 6.22a we notice the expected steadily increasing behavior of the function $\Psi_S(\rho)$. We recall that this quantity indicates the normalized flux linked to the circle of radius ρ. Therefore, in the presence of a constant field distribution, this quantity would increase as ρ^2. From Fig. 6.22b a rather flat field distribution inside the system can be indeed detected, with plateau levels increasing in height with the value of the externally applied flux. When comparing the field distribution in Fig. 6.22b with the results obtained in Fig. 6.16a, we need to recall that in the latter case we had used a finite value of β_0. In fact, having we computed $\xi_S(\rho)$, starting from the leading order solution Ψ_0, we should not expect a strict quantitative agreement between the curves in Fig. 6.22b and those in Fig. 6.16a. However, we can detect the same qualitative behavior between these two curves. The first characteristic point is the rather flat behavior of these functions, apart from undulating effects in Fig. 6.16a due to the discreteness of the model. The second feature is a regular increase of the plateau levels with the increase of ψ_{ex}, as it can be noticed both from Fig. 6.16a and Fig. 6.22b.

The oscillating behavior of the current distribution in Fig. 6.22c can also be discussed in terms of the behavior of the discrete model represented in Fig. 6.17a.

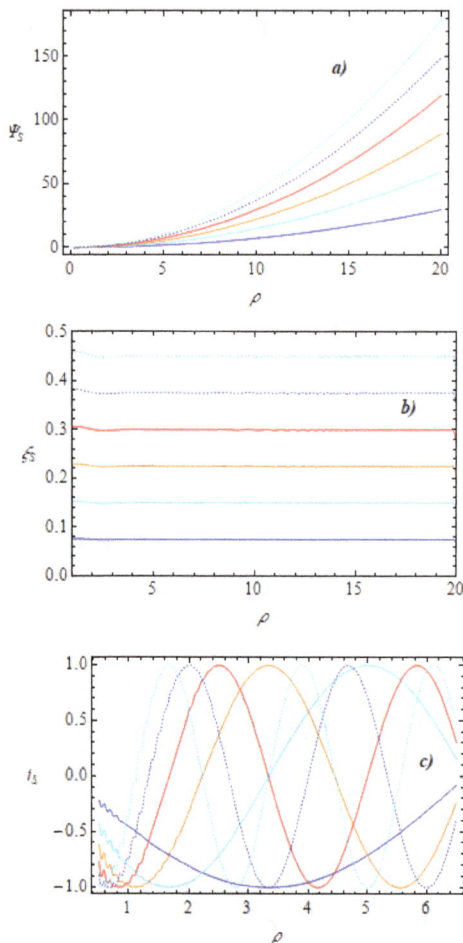

Fig. 6.22. a) Stationary solution Ψ_S of the continuous model of circular arrays of Josephson junctions; b) normalized field distribution ξ_S and c) current distribution i_S inside the model system for $\beta_0 = 0$. The normalized radial dimension of the system is $\rho_{max} = 20$. For all panels colored lines are used to distinguish the normalized quantities, calculated at increasing values of the forcing terms ψ_{ex}: 10.0 (blue full line); 20.0 (cyan full line); 30.0 (orange full line); 40.0 (red full line); 50.0 (blue dotted line); 60.0 (cyan dotted line). In panel c) i_S is shown only in a limited range of the interval $[0, \rho_{max}]$.

In this respect, the incidence of the discrete character of the system and of the finiteness of the value of β_0 on the form of the i_S vs. ρ curves in Fig. 6.17a needs to be considered more carefully. In fact, in the case of

Fig. 6.22c, we obtain the expression of the leading-order value of the current i_S from Eq. (6.112). Therefore, the current in Fig. 6.22c is represented by a sine function with an argument which depends on the radial distance ρ in a nearly linear way. On the other hand, this is not the case for the curves at relatively low values of ψ_{ex} in Fig. 6.17a (see, for instance, the gray and the red curves in the latter figure). In fact, the finiteness of the parameter β_0 implies, as we have seen in Section 6.4.2, the existence of a lower threshold value $\psi_{ex}^{(1)}$ of the externally applied flux, above which flux penetration is allowed. This feature, together with the discrete character of the model, gives a flux-penetrated region close to the external portion of the system ($\rho \approx \rho_{max}$) for ψ_{ex} slightly above $\psi_{ex}^{(1)}$. In other words, a field-dependent effective penetration length arises in these systems for finite values of β_0. In this way, for increasing values of ψ_{ex}, a deformed oscillating character of the current is obtained for $\beta_0 > 0$, differently from Fig. 6.22c, where the sine function distribution spreads over all the circle of radius ρ_{max}. Nevertheless, we can still detect and interpret the characteristic feature of the decrease of the effective wavelength of the oscillating pattern as the externally applied flux ψ_{ex} is increased. This feature is also detectable in fig. 6.17a and can be understood as follows. Let us consider Eq. (6.110), where the factor ψ_{ex} represents a scaling term in the expression for $\Psi_S(\rho)$. Notice now the latter function is present in the argument of the sine term in Eq. (6.112). By assuming a nearly-quadratic dependence of $\Psi_S(\rho)$ on the variable ρ and by noticing that ψ_{ex} is a scaling term also for the nearly-linear dependence of the argument of the sine in Eq. (6.112), we may justify the regular decrease of the effective wavelength of the function $i_S(\rho)$ with increasing values of ψ_{ex} in Fig. 6.22c. In fact, the effective wavelength of the oscillating pattern in Fig. 6.22c can be seen to be inversely proportional to ψ_{ex}.

6.5.3. *Numerical analysis*

In the previous subsections we have noticed that, for small values of the parameter β_0, it is possible to exhibit an analytic solution to the dynamic equation describing magnetic flux penetration inside concentric circular arrays of Josephson junctions. In fact, in the limit of vanishing values of β_0, a perturbation analysis allows a closed analytic solution of Eq. (6.63). After having found this solution in Section 6.5.1 some results have been analyzed in Section 6.5.2. In the present section we investigate the magnetic properties of concentric circular arrays of Josephson junctions for $\beta_0 > 0$ by means of numerical analysis.

In order to take into account the magnetic history of the system, we need to consider the complete time evolution given in Eq. (6.63). We therefore start by solving this equation numerically for $\beta_0 = 0.01$. This is done by standard numerical routines and produces the results shown in Fig. 6.23a–c. In Fig. 6.23a, in particular, the stationary values $\Psi_S(\rho)$ of the total normalized flux within a circle of radius ρ are reported for five different values of the externally applied normalized magnetic flux ψ_{ex} ($\psi_{ex} = 400, 800, 1200, 1600, 2000$). When comparing these curves to the ones shown in Fig. 6.22a, we detect a similar qualitative behavior. In Fig. 6.23b normalized field distribution curves $\xi_S(\rho)$ are shown for the same five values of the forcing term ψ_{ex}. The flat magnetic field distribution in Fig. 6.22b for $\beta_0 = 0$ is still present for $\beta_0 = 0.01$. Finally, in Fig. 6.23c current distributions in the system are shown for the same five values of the forcing term ψ_{ex} used in the previous two graphs. The oscillating character of the sine term is only approximately reproduced in the latter figure, due to the fact that a discrete grid has been adopted for the radial variable, in order to obtain more readily the solution of Eq. (6.63) by a numerical approach. In Fig. 6.23c the feature already noticed in Fig. 6.22c of the regular decrease of the effective wavelength of the function $i_S(\rho)$ with increasing values of ψ_{ex} is still present. Before tackling the analysis of the behavior of the system under a complete ψ_{ex} cycle, we might try to understand in which way an increase of the parameter β_0 affects the magnetic field and current distributions in concentric circular arrays of Josephson junctions.

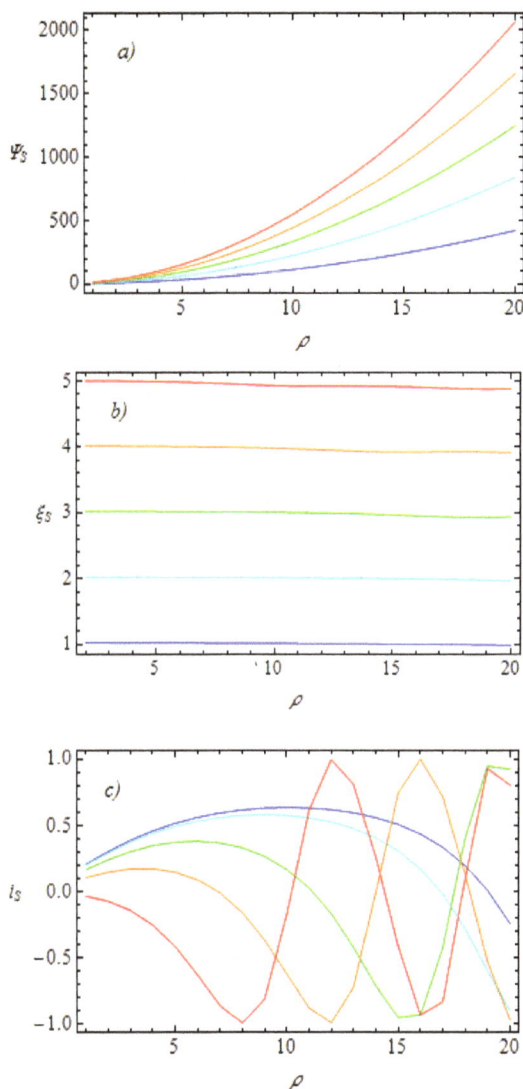

Fig. 6.23. a) Stationary solution Ψ_S of the continuous model of circular arrays of Josephson junctions; b) normalized field distribution ξ_S and c) current distribution i_S inside the model system for $\beta_0 = 0.01$. The normalized radial dimension of the system is $\rho_{max} = 20$. For all panels colored lines are used to distinguish the normalized quantities, calculated at increasing values of the forcing terms ψ_{ex}: 400 (blue); 800 (cyan); 1200 (green); 1600 (orange); 2000 (red).

Therefore, in Fig. 6.24a–c we report results obtained by means of the same type of numerical analysis carried out to attain the results shown in Fig. 6.23a–c, this time for $\beta_0 = 0.05$ and for the following five values of the externally applied normalized magnetic flux ψ_{ex}: 800, 1600, 2400, 3200, 4000. In this way, the stationary quantities $\Psi_S(\rho)$, $\xi_S(\rho)$, and $i_S(\rho)$ are shown in Fig. 6.24a, Fig. 6.24b and Fig. 6.24c, respectively. These curves are rather similar, from the qualitative point of view, to the corresponding ones in Fig. 6.23a–c. From the quantitative side, we might notice that at $\psi_{ex} = 1600$, for example, the current in Fig. 6.23c (orange curve) has one positive and one negative peak, in Fig. 6.24c (cyan curve), $i_S(\rho)$ shows only one peak. In fact, with increasing value of β_0 the system tends, as we have already seen in other cases, to shield in a more effective way the external magnetic field. In order to support this statement, let us look at the value of the total normalized flux $\Psi_S(\rho)$ at $\rho = \rho_{max}/2$. Extracting this information from our numerical analysis or, more roughly, from the graphs in Fig. 6.23b and 6.24b, we have that, for $\beta_0 = 0.01$, $\Psi_S(\rho_{max}/2) \approx 439$, while for $\beta_0 = 0.05$, $\Psi_S(\rho_{max}/2) \approx 430$. This is a hint that the shielding capability of these type of systems increase for increasing values of β_0.

Another important point to consider in analyzing the numerical results is the following. By invoking the boundary conditions for $\Psi_S(\rho)$ in Eq. (6.65), we may write:

$$\left.\frac{d\Psi_S(\rho)}{d\rho}\right|_{\rho=\rho_{max}} = \frac{2\psi_{ex}}{\rho_{max}}. \tag{6.113}$$

Because of the above relation, by the definition of the normalized local magnetic field ξ given in Eq. (6.59) and by considering Eq. (6.61), we have

$$\xi(\rho_{max}) = \frac{1}{2\kappa^2}\left.\frac{d\Psi_S(\rho)}{d\rho}\right|_{\rho=\rho_{max}} = \frac{\psi_{ex}}{\kappa^2 \rho_{max}^2}. \tag{6.114}$$

This value can indeed be detected in the curves for $\xi_S(\rho)$ in Fig. 6.23b and 6.24b, where we have taken $\kappa = 1$.

We now turn our attention to the problem of the magnetic history inside the system.

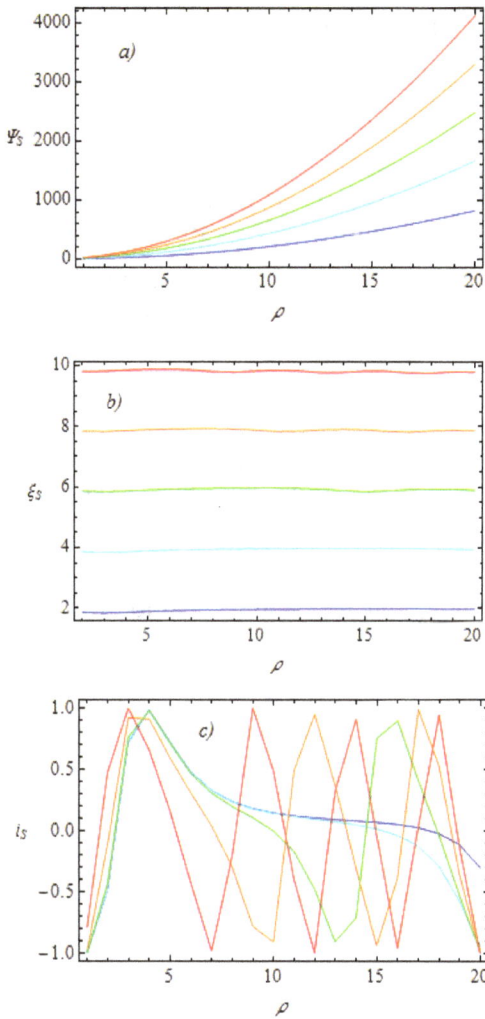

Fig. 6.24. a) Stationary solution Ψ_S of the continuous model of circular arrays of Josephson junctions; b) normalized field distribution ξ_S and c) current distribution i_S inside the model system for $\beta_0 = 0.05$. The normalized radial dimension of the system is $\rho_{max} = 20$. For all panels colored lines are used to distinguish the normalized quantities, calculated at increasing values of the forcing terms ψ_{ex}: 800 (blue); 1600 (cyan); 2400 (green); 3200 (orange); 4000 (red).

Let us consider the response of concentric circular arrays of Josephson junctions when the normalized externally applied flux ψ_{ex} varies over a complete cycle. Starting from Zero Field Cooling (ZFC) conditions, the initial conditions for an opportune auxiliary function $\Gamma(\rho, \tau)$, can be written as follows

$$\Gamma(\rho, 0) = -\frac{2\psi_{ex}}{\rho_{max}}\rho. \qquad (6.115)$$

This function had been already introduced in Section 6.5.1 to obtain homogenous boundary conditions and can be defined as in Eq. (6.71):

$$\Gamma(\rho, \tau) = \psi(\rho, \tau) - \frac{2\psi_{ex}}{\rho_{max}}\rho. \qquad (6.116)$$

Therefore, at $\psi_{ex} = 0$, we have $\Gamma(\rho, 0) = 0$. After having calculated the new field distribution at $\psi_{ex} = \Delta\psi_{ex} > 0$, we obtain a configuration $\Gamma(\rho, T)$ of the solution at a normalized time $T \gg 1$, so that $\Gamma(\rho, T)$ can be considered a stationary solution for the system. Therefore, for this incremented value of ψ_{ex} the new initial condition will be given by:

$$\Gamma(\rho, 0) \rightarrow \Gamma(\rho, T). \qquad (6.117)$$

By repeating this procedure for successive increments or decrements of ψ_{ex}, we may take account of the magnetic history of the system.

The boundary conditions of the system, on the other hand, can be obtained by Eq. (6.73b) and (6.73c):

$$\Gamma(0, \tau) = 0, \qquad (6118a)$$

$$\frac{\partial\Gamma(\rho, \tau)}{\partial\rho}\bigg|_{\rho=\rho_{max}} = 0. \qquad (6.118b)$$

The magnetic cycle is thus started by going from $\psi_{ex} = 0$ to $+\psi_{ex}^*$; it is reversed by going from $+\psi_{ex}^*$ to $-\psi_{ex}^*$; it is finally completed by going from $-\psi_{ex}^*$ to $+\psi_{ex}^*$. In all portions of the cycle the same step amplitude $\Delta\psi_{ex}$ is used. In order to represent the outcome of our analysis, we consider the normalized magnetization m of the system, defined in this case as follows:

$$m = \psi(\rho, T) - \psi_{ex} = \Psi_s(\rho_{max}) - \psi_{ex}. \qquad (6.119)$$

In Fig. 6.25 magnetization curves for concentric circular arrays of Josephson junctions with normalized radial dimension $\rho_{max} = 20$ and with parameters $\beta_0 = 0.10$ (red curve) and $\beta_0 = 0.20$ (black curve) are shown.

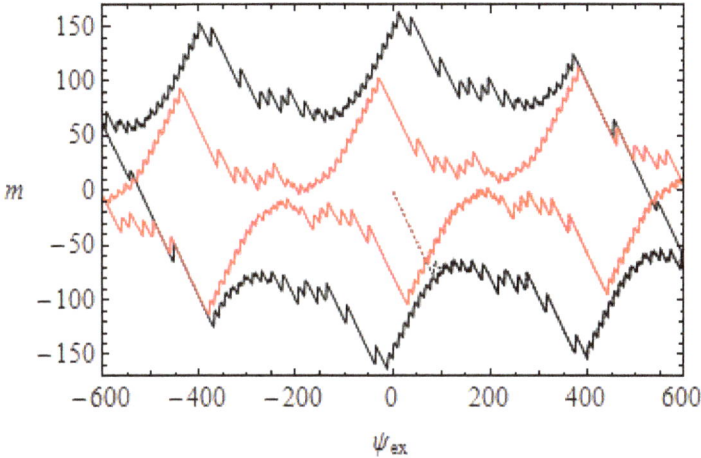

Fig. 6.25. Magnetization curves of concentric circular arrays of Josephson junctions with normalized radial dimension $\rho_{max} = 20$ and with parameter $\beta_0 = 0.15$ (red full line) and $\beta_0 = 0.20$ (black full line). The dotted curve starting from the origin represent the visible first magnetization portion of the cycle ($0 < \psi_{ex} < 600$). The upper parts of the curves represent the portion of the cycle for which ψ_{ex} decreases from 600 to -600. The portion of the magnetization cycle obtained by letting ψ_{ex} increase from -600 to 600 lies in the lower part of the two curves.

The characteristic diamond-like shape of these curves reproduces the hysteresis features already encountered in the discrete version of the model in Fig. 6.20. We may preliminary notice that the two curves differ in size: the higher the value of the parameter β_0 the bigger the size of the hysteresis loop of the magnetization cycle. This is due to the fact that lower Josephson couplings between adjacent junctions tend to favor flux penetration inside the system, as already remarked, for example, in Section 6.2.1. We also notice perfect diamagnetic behavior in the first magnetization portion of the magnetization cycles (dotted lines) up to a value of ψ_{ex} comprised between 50 and 100. After irreversible flux transition occurs, upon increasing the externally applied normalized magnetic flux, the first magnetization curve converges to the lower portion of the magnetization curve for both cycles.

The study of the magnetic response of Josephson junction arrays, as we have already remarked and as it has been seen from the subject of the present chapter, is interesting *per se*, given the possibility of applying analytic and numerical methods to different classes of networks of Josephson junctions, from the simplest to more artful ones. Nevertheless, in the following chapter we shall see that these types of networks are also useful in interpreting the low-field magnetic properties of granular superconductors. Therefore, the effort here made will not be wasted and will certainly be of help in considering important aspect of the vast class of granular superconductors.

Chapter 7

Superconducting Granular Systems

7.1. Granular superconductors

Granular superconductors can be seen as a collection of superconducting islands embedded in a non-superconducting medium (Ebner, 1985; Muller K. A., 1987). The superconducting grains are weakly coupled, through Josephson junctions, to their nearest neighbors (Deutscher, 1987; Clem, 1988). In this class of superconductors it is very important to distinguish between the physical properties arising from the granular nature of the material from those that are intrinsic to the material itself. A simple description of a two-dimensional granular superconductor is given in Fig. 7.1, where the grains are represented as thin identical discs.

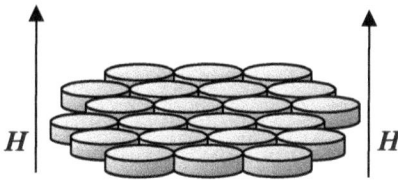

Fig. 7.1. A simple representation of a granular superconducting system in the presence of an externally applied magnetic field H.

It is therefore natural to argue that low-field magnetic response of granular superconductors can be qualitatively studied by means of Josephson junction network models (Wolf, 1993; Tinkham, 1989; Chen,

1994). In fact, soon after the discovery of high-T_c superconductivity (Bednorz & Muller, 1986), superconducting glass models (Muller K. A., 1987) were adopted by assuming that each superconducting island, labeled with an index k, could be described by a macroscopic superconducting wave function $\psi_k = \Delta_k e^{i\theta_k}$, where θ_k is the superconducting phase of the island itself. In this way, the interaction energy E_{kj} between adjacent k −th and j −th islands can be thought to be of Josephson type, so that the Hamiltonian of the system, in the presence of a magnetic field, could be written as follows (Ebner, 1985; Deutscher, 1987; Newrock, 2000):

$$H = \sum_{\langle jk \rangle} J_{jk} \cos \left(\theta_k - \theta_j - \frac{2\pi}{\Phi_0} \int_j^k \boldsymbol{A} \cdot d\boldsymbol{l} \right), \qquad (7.1)$$

where $\langle jk \rangle$ means that the sum is limited to the nearest neighbors, the exchange terms J_{jk} depend on the temperature, Φ_0 is the elementary flux quantum, $\boldsymbol{A}$ is the vector potential, and where the integral extends over the junction barrier, the indices j and k standing for the coordinates of two points on the junction electrodes. This spin-glass-like description of granular superconductors fits well samples which are in a superconducting magnetic state close to the superconducting–normal transition region in the H–T plane. Therefore, in this type of approach intra-granular flux motion can be taken into account (Tinkham, 1989). However, overall shielding current effects cannot be considered in these models. In fact, as we have already seen for Josephson junction networks, the Hamiltonian in Eq. (7.1) refers exclusively to the energy of the Josephson junctions. As a consequence, if we wish to describe the low-field magnetic properties of granular superconductors, these "bare" network models, where grains appear as dimensionless points, are to be extended to account for the diamagnetic response of the superconducting portion of the system. Were the magnetic energy of the circulating currents also considered in the superconducting networks studied in the previous chapter, the vanishing size of the grains in these models would still not allow to take account of the diamagnetic behavior of the superconducting matrix. As a consequence, in adopting Josephson junction networks as models of granular superconductors, the notion of "dressed" network needs to be introduced. In this way, the overall magnetic response, due to inter-granular and intra-granular flux motion, of superconducting granular systems can be described. In fact, both the low-field shielding properties of this class of superconductors and the diamagnetic response of the superconducting grains can be shown to be included in the "dressed" circuital model.

We therefore start by considering a simple four-grain two-dimensional system in the presence of an axial magnetic field. We shall show that the full response of this ideal sample can be obtained by means of a single equation for flux motion inside the system. The successive step in describing the low-field quantitative response of granular system is to consider a greater amount of grains and the geometrical and physical characteristics of real samples. This further generalization of granular superconducting systems will be made in the following section. Only in the next chapter a general approach to three-dimensional systems will be discussed.

7.2. A four-grain system

A simple granular system consisting of four identical cylindrical grains, whose centers are placed at the vertices of a square of radius $2R_g$, is shown in Fig. 7.2. Shielding currents, distributed along the grain side boundaries, are seen to flow when an axial magnetic field is applied to the system. These currents also flow through the four Josephson junctions located at the contact region between two adjacent grains. In Fig. 7.2 two effective currents are shown. These currents are denoted as I, if flowing in the outer part of the four-grain system, and i, if flowing in the inner part. In the same way, the fluxes threading the inner and outer upper loops are defined as Φ and Θ, respectively. Our aim is to carry out a quantitative analysis of this system, so that the extension of the grains can be considered. For axial low magnetic fields, i.e., field applied along the vertical direction (z −axis if the system lies in the $x - y$ plane), the simplified representation of the current distribution in Fig. 7.2 is sufficient. In fact, given the cylindrical symmetry of the problem, the only relevant currents in the system are those flowing in the azimuthal direction.

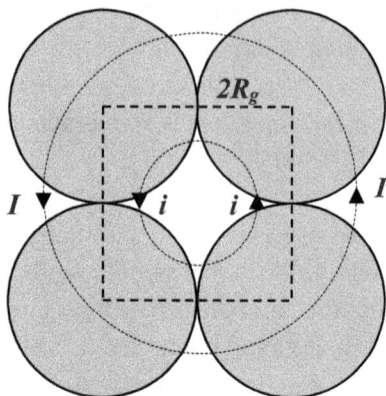

Fig. 7.2. Simple properties of a four-grain superconducting system. The effective current i is associated to the inner loop of effective area S_{in}. The effective current I is associated to the loop of effective area S_{out}. The effective current flowing in the Josephson junctions at the grain interfaces is $i + I$.

The fluxes threading the inner and outer loops can be expressed in terms of these currents, by taking into account both self- and mutual-inductance coefficients, in the following general way:

$$\Phi = li + mI + \mu_0 H S_{in}, \tag{7.2a}$$

$$\Theta = mi + LI + \mu_0 H S_{out}, \tag{7.2b}$$

where S_{in} and S_{out} are the inner and outer loop effective areas, respectively. If the outer current distribution now wraps the enclosed region in such a way that the flux inside the external loop is equal to the flux trapped in the inner inter-granular and intra-granular region, we can take the fluxes threading the inner and the outer loops as equal, so that $\Phi = \Theta$. We would like to point out that this assumption requires that, for increasing temperatures, for example, the outer and the inner current loops depicted in Fig. 7.2 move toward each other, in such a way to enclose only flux-free regions in the ring they both form. In addition, the magnetic response of the system is left unaltered by this assumption only if the magnetic permeability μ of the superconducting material is such that $\mu \approx \mu_0$. By solving for the effective current flowing in the Josephson junctions $I_B = i + I$, we get:

$$I_B = a\,\Phi - b\,\Phi_{ex}^{(out)} - c\,\Phi_{ex}^{(in)}, \qquad (7.3)$$

where

$$\Phi_{ex}^{(out)} = \mu_0 H S_{out}, \quad \Phi_{ex}^{(in)} = \mu_0 H S_{in}, \qquad (7.4)$$

and where:

$$a = \frac{l+L-2m}{lL-m^2}, \quad b = \frac{l-m}{lL-m^2}, \quad c = \frac{L-m}{lL-m^2}. \qquad (7.5)$$

By making the simplifying assumption $m \approx l$, we have:

$$a \approx \frac{1}{l}, \quad b \approx 0, \quad c \approx a. \qquad (7.6)$$

In this way, the current I_B is written as follows:

$$I_B = \frac{\Phi - \Phi_{ex}^{(in)}}{l}, \qquad (7.7)$$

Having obtained the expression for the current I_B flowing in the four identical Josephson junctions of the network, by the RSJ model (Barone, 1982) we may write:

$$\frac{d\phi}{d\tau} + \sin\phi = i_B, \qquad (7.8)$$

where ϕ is the superconducting phase difference across each junction, where $i_B = I_B/I_{J0}$ plays the role of the normalized bias current, I_{J0} being the maximum Josephson current of the four identical junctions at $T = 0$ K, and where τ is the usual normalized time. We may now write the fluxoid quantization equation as follows:

$$4\phi + 2\pi\frac{\Phi}{\Phi_0} = 2\pi n, \qquad (7.9)$$

where n is an integer. Under zero-field-cooling conditions, we may set $n = 0$. Therefore, Eq. (7. 9) can be written as follows:

$$\phi = -\frac{\pi}{2}\frac{\Phi}{\Phi_0} = -\frac{\pi}{2}\Psi, \qquad (7.10)$$

with the obvious definition of Ψ. By substituting Eq. (7.10) in Eq. (7.8) and by using Eq. (7.7), we may write:

$$\frac{\pi}{2}\frac{d\Psi}{d\tau} + \sin\left(\frac{\pi\Psi}{2}\right) + \frac{\Psi}{\beta} = \frac{\Psi_{ex}^{(in)}}{\beta}, \qquad (7.11)$$

where $\Psi_{ex}^{(in)} = \mu_0 H S_{in}/\Phi_0$, and where

$$\beta = \frac{l I_J}{\Phi_0}. \qquad (7.12)$$

Eq. (7.11) is formally equal to Eq. (6.11) derived for four junctions in a single plaquette in the previous chapter. However, from Eq. (7.11) we learn that, under the assumptions made, the effective loop area is to be considered S_{in}, which may depend on the temperature, as it is also true for the parameters $I_{J0} = I_J(T)$ and $l = l(T)$ contained in the definition of β of Eq. (7.12).

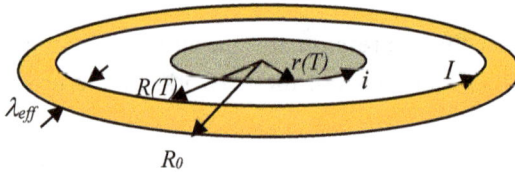

Fig. 7.3. Simple representation of the flux distribution inside a four-grain superconducting system. The effective currents i and I flow at the boundaries of the flux-free region (white ring of outer radius $R(T)$). The normalized flux Ψ threads the inner loop of radius $r(T)$ and effective area S_{in}. The flux-penetrated region (yellow ring) has a radial extension of $\lambda_{eff}(T) = R_0 - R(T)$.

We have thus replaced a bare Josephson network junction model (a single plaquette) by the more realistic dressed model, which allows for the actual extension of the grains and, as we shall see, for the intra-granular flux distribution. Moreover, this model makes it possible to keep track of the temperature dependence of the magnetic properties of this simple granular system; as an example, the temperature dependence of zero-field-cooling and field-cooling magnetic susceptibility will be simulated in what follows. Finally, the magnetic quantities of the model can be directly related to the physical properties of the system. In fact, we may assume that, at finite temperatures, the flux distribution in the sample can be represented by the simple scheme given in Fig. 7.3. In this figure, the effective radii $r(T)$ and $R(T)$ are given, in terms of the penetration length $\lambda_{eff}(T)$, by the following simple relation:

$$r(T) = r_0 + \lambda_{eff}(T), \tag{7.13a}$$

$$R(T) = R_0 - \lambda_{eff}(T), \tag{7.13b}$$

where the radii r_0 and R_0 define the zero-temperature values of the inner and outer loop areas and, so that: $S_{in}(0) = \pi r_0^2$ and $S_{out}(0) = \pi R_0^2$. In this way, by comparing these quantities with the geometric properties of the four-grain system in fig. 7.2, we may see that:

$$r_0 = \sqrt{\frac{4-\pi}{\pi}} R_g, \quad R_0 = \sqrt{\frac{4+3\pi}{\pi}} R_g. \tag{7.14}$$

From Eq. (7.13a–b) we notice the inner area S_{in} is seen to expand and the outer area S_{out} to contract with the temperature T, because the penetration length $\lambda_{eff}(T)$ is an increasing function of temperature, which can be expressed as follows (Schmidt, 1997):

$$\lambda_{eff}(T) = \lambda(T) \tanh \frac{R_0 - r_0}{2\lambda(T)}, \tag{7.15}$$

where

$$\lambda(T) = \lambda(0) \left[1 - \left(\frac{T}{T_c} \right)^4 \right]^{-\frac{1}{2}} \tag{7.16}$$

is the phenomenological expression of the penetration depth, T_c being the critical temperature of the grains. Notice that, for $T \to T_c$, $\lambda(T) \to \infty$, while the effective penetration length is equal to $(R_0 - r_0)/2$, in such a way that the system acquires the magnetic properties of a normal paramagnetic material at $T = T_c$. In order to define the temperature dependence of the parameter β in Eq. (7.12), we take the maximum Josephson current $I_J(T)$ to vary according to the Ambegaokar–Baratoff formula (Ambegaokar & Baratoff, 1963):

$$I_J(T) = \frac{\pi\Delta(T)}{eR}\tanh\frac{\Delta(T)}{2k_BT},\qquad(7.17)$$

where e is the absolute value of the electron charge, $\Delta(T)$ is the energy gap of the superconducting material, R is the normal resistance of the junction and k_B is the Boltzmann constant. In Eq. (7.17) the temperature dependence of the energy gap can be approximated as follows:

$$\Delta(T) \approx \Delta_0\sqrt{1 - \frac{T}{T_c}},\qquad(7.18)$$

where Δ_0 is the energy gap value at $T = 0$ K.

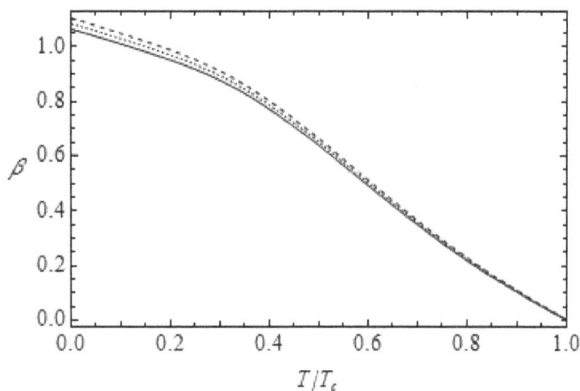

Fig. 7.4. Temperature dependence of the parameter β of the dressed Josephson junction model for a four-grain superconducting system. In the system the grain radius and height are both 10.0 μm, the critical temperature is set to 10.0 K, the zero-temperature value of the maximum Josephson current is 200 μA. The zero-temperature value $\lambda(0)$ of the penetration depth are as follows: 500 Å (full line), 1000 Å (dotted line), 1500 Å (dashed line).

We may finally express the inductance parameter l approximately as follows:

$$l \approx \mu_0 \frac{S_{in}}{d}, \tag{7.19}$$

where d is the temperature-independent height of each cylindrical grain and the expression on the right-hand side is strictly valid only for a thin cylindrical shell, and where

$$S_{in} = \pi r^2(T) = \pi \left[r_0 + \lambda_{eff}(T) \right]^2. \tag{7.20}$$

In this way, the temperature dependence of the parameter β in Eq. (7.11) can be finally found. A graphical representation of the function $\beta = \beta(T)$ is given in Fig. 7.4 for various values of the zero-temperature value $\lambda(0)$. One can notice that $\beta(T)$ is a monotonically decreasing function in the temperature. We also notice that this function is rather insensitive on the variation of $\lambda(0)$ from 500 Å to 1500 Å for the specific choice of parameters made in Fig. 7.2 ($R_g = h = 10.0$ μm, $T_c = 10.0$ K, $I_{J0} = 200$ μA).

7.2.1. Susceptibility of a four-grain system

Having derived the equation for magnetic flux motion within the system, namely, Eq. (7.11), and having specified the way the flux distributes inside the sample, we may simulate a susceptibility experiment for the four-grain system at a fixed temperature. Therefore, we can derive the χ vs. applied flux curves and compare these with those obtained for a single plaquette, in order to emphasize the difference between the two models.

Therefore, in order to derive these curves, assume that the system is at $T = 0$ K and that an externally applied flux is applied to it gradually, so that the value of $\Psi_{ex}^{(in)}$ goes from 0 to a given upper value in small steps. After each small increment $\Delta\Psi_{ex}^{(in)}$, the system is allowed to go to equilibrium and the susceptibility χ is recorded. In order to find χ, we first calculate the average magnetic field $\langle B \rangle$ as follows:

$$\langle B \rangle = \frac{\mu_0 H (S_{tot} - S_{out}) + \Phi}{S_{tot}}, \tag{7.21}$$

where S_{tot} is the area of the total extension of the system, which coincides with S_{out} at $T = 0$ K. The first addendum at the numerator in

Eq. (7.21) is absent at $T = 0$ K. However, we retain this term, in order to derive a general expression for χ. Therefore, we write:

$$\chi = \frac{\langle B \rangle - \mu_0 H}{\mu_0 H} = \frac{\Psi - \Psi_{ex}^{(out)}}{\Psi_{ex}^{(tot)}}, \qquad (7.22)$$

where $\Psi_{ex}^{(out)} = \mu_0 H S_{out}/\Phi_0$ and $\Psi_{ex}^{(tot)} = \mu_0 H S_{tot}/\Phi_0$. We remark once again that $\Psi_{ex}^{(tot)} = \Psi_{ex}^{(out)}$ at $T = 0$ K. By solving the nonlinear differential equation (7.11), we obtain the Ψ vs. $\Psi_{ex}^{(in)}$ curve in Fig. 7.5.

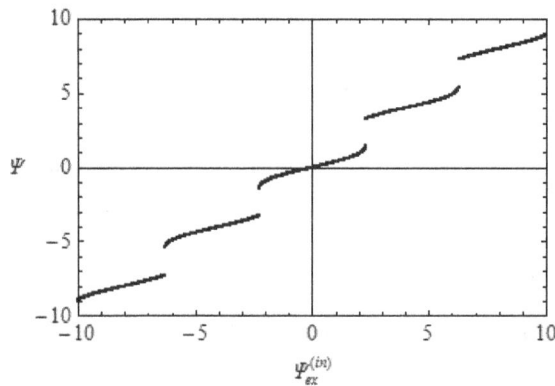

Fig. 7.5. Flux Ψ threading the inner loop in a four-grain system as a function of the forcing parameter $\Psi_{ex}^{(in)}$ at $T = 0$ K. In the dressed Josephson junction model the grain radius and height are both 10.0 μm, the critical temperature is set to 10.0 K, the zero-temperature value of the maximum Josephson current is 200 μA. The zero-temperature value of the penetration depth $\lambda(0)$ is 1000 Å, so that the parameter β is equal to 1.083 at $T = 0$ K.

Notice that Eq. (7.22) is formally equivalent to Eq. (6.16) in the previous chapter when $\beta = \beta_L$ and when the two forcing terms in Eq. (7.22), namely $\Psi_{ex}^{(out)}$ and $\Psi_{ex}^{(tot)}$, coincide, as they actually do at $T = 0$ K. Fig. 7.5 represents also the magnetic flux vs. ψ_{ex} curve of a single plaquette of area S_{in} for $\beta_L = \beta = 1.083$. However, we shall see that, because of the different normalization in the two cases and of the definition of the susceptibility, given by Eq. (7.22) for a dressed network model and by the first equality in Eq. (6.16) for a single plaquette of area S_{in}, the magnetic properties are quantitatively different for the two systems even at $T = 0$ K. In fact, the susceptibility vs. $\Psi_{ex}^{(in)}$ curves, even though qualitatively similar, are seen to be quantitatively different in Fig. 7.6a–b. In particular, in Fig. 7.6a, the magnetic response of a bare network of area S_{in} for $\beta = 1.083$ is shown. In Fig. 7.6b, on the other hand, the magnetic response of a dressed Josephson junction network model is reported for parameter values equal to those chosen in Fig. 7.5. In order to grasp more easily the quantitative difference between the two curves in Fig. 7.6a and Fig. 7.6b, in Fig. 7.7 we compare the results obtained in the latter figures.

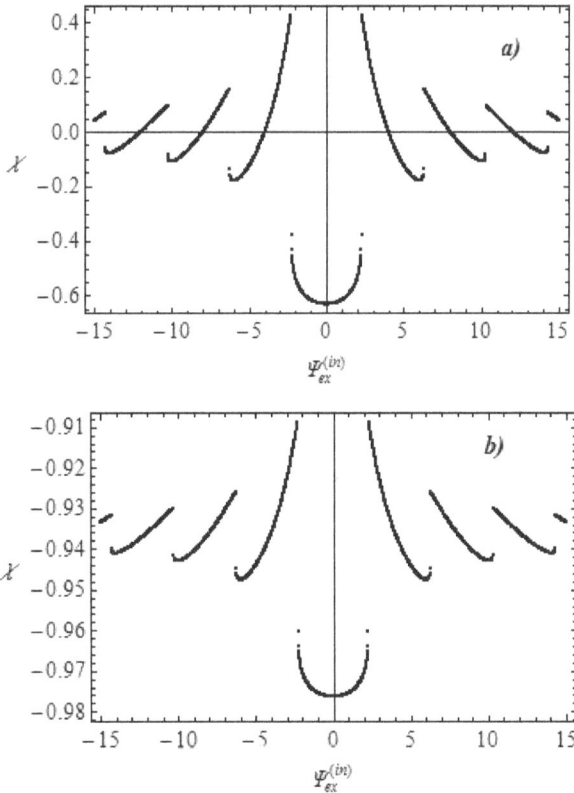

Fig. 7.6. Magnetic susceptibility of a four-grain system as a function of the forcing parameter $\Psi_{ex}^{(in)}$ at $T = 0$ K. In panel a the magnetic response of a bare network of area S_{in} is shown; in panel b the magnetic response of a dressed Josephson junction network

model is reported. The parameter β is equal to 1.083 for both models. For the dressed network model the same parameters as in Fig. 7.5 have been chosen.

By carefully looking at Fig. 7.6a and Fig. 7.6b, we could already notice the qualitative similarity between the two curves. This is due to the fact that flux penetration occurs in a similar way in both systems. However, the range of the two curves, as more opportunely shown in Fig. 7.7 on the same scale, is much different. In fact, while for a dressed model the χ vs. $\Psi_{ex}^{(in)}$ curves varies in the vicinity of the value $\chi = -0.93$, the susceptibility of the bare model in Fig. 7.6a goes from negative to positive values in the range shown. We conclude that much care must be taken when adopting a Josephson junction network model in quantitatively analyzing the magnetic properties of granular superconductors.

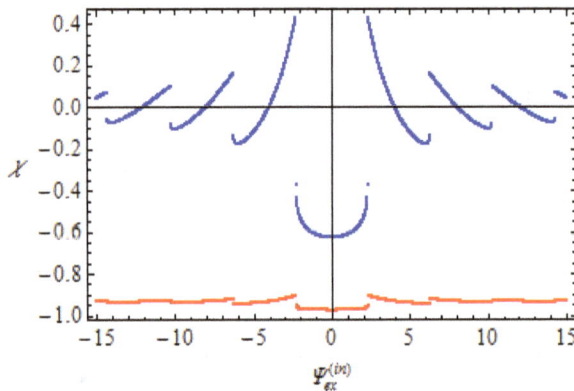

Fig. 7.7. Magnetic susceptibility of a four-grain system as a function of the forcing parameter $\Psi_{ex}^{(in)}$ at $T = 0$ K. The magnetic responses of a bare network of area S_{in} and of a dressed Josephson junction model are shown as a blue and a red curve, respectively. The parameter β is equal to 1.083 for both models. For the dressed network model the same parameters as in Fig. 7.5 have been chosen.

7.2.2. *Temperature dependence of susceptibility curves*

Having specified the way the temperature affects the physical properties of the system, we can now perform a second simulated experiment where the χ vs. T curves can be obtained either in zero-field-cooling (ZFC) either in field-cooling (FC) conditions. In order to correctly define the effective forcing quantity in terms of the magnetic field, we need to introduce an opportune scaling for the latter quantity. Therefore we set:

$$\Psi_{ex}^{(in)} = \frac{S_{in}}{S_{tot}} \frac{\mu_0 H S_{tot}}{\Phi_0} = \sigma_{in}(T) \Psi_{ex}^{(tot)}, \qquad (7.23)$$

with the obvious definition of the temperature-dependent ratio $\sigma_{in}(T)$, which can be expressed in terms of the radii $r(T)$ and R_0 as follows:

$$\sigma_{in}(T) = \left[\frac{r(T)}{R_0}\right]^2 = \left[\frac{r_0 + \lambda_{eff}(T)}{R_0}\right]^2. \qquad (7.24)$$

For future reference, we also define

$$\sigma_{out}(T) = \left[\frac{R(T)}{R_0}\right]^2 = \left[\frac{R_0 - \lambda_{eff}(T)}{R_0}\right]^2. \qquad (7.25)$$

With the new scaling option in Eq. (7.23) for the applied flux, Eq. (7.11) can be written as follows:

$$\frac{\pi}{2} \frac{d\Psi}{d\tau} + \sin\left(\frac{\pi\Psi}{2}\right) + \frac{\Psi}{\beta(T)} = \frac{\sigma_{in}(T)}{\beta(T)} \Psi_{ex}^{(out)}, \qquad (7.26)$$

where we have specified that the parameter β depends on the temperature. In order to simulate measurement of the susceptibility at various temperatures under ZFC and FC conditions, we proceed as follows. Starting from $T = 0$ K, a measuring field is gradually applied from $\Psi_{ex}^{(out)} = 0$. Once the final value of $\Psi_{ex}^{(out)}$ is reached by small enough increments in the magnetic field, the temperature is gradually raised to $T = T_c$, thus completing the ZFC part of the simulation. When the temperature $T = T_c$ is reached, the system is in the normal state and fully penetrated by magnetic flux. Therefore, when the temperature is gradually lowered below T_c, the FC part of the experiment is performed. Notice that, because of the new scaling option, we need to write the susceptibility in Eq. (7.22), in terms of the forcing term $\Psi_{ex}^{(tot)}$, as follows:

$$\chi = \frac{\Psi - \sigma_{out}(T)\Psi_{ex}^{(tot)}}{\Psi_{ex}^{(tot)}}. \qquad (7.27)$$

In order to reach an acceptable value of the measuring field, we iterate the first part of the calculation up to $\Psi_{ex}^{(tot)} = N$. By computing the measuring field, we write:

$$\mu_0 H = \frac{\Phi_0 \Psi_{ex}^{(tot)}}{S_{tot}} = \frac{N\Phi_0}{\pi R_0^2} = 0.246 \, N \, \mu T. \tag{7.28}$$

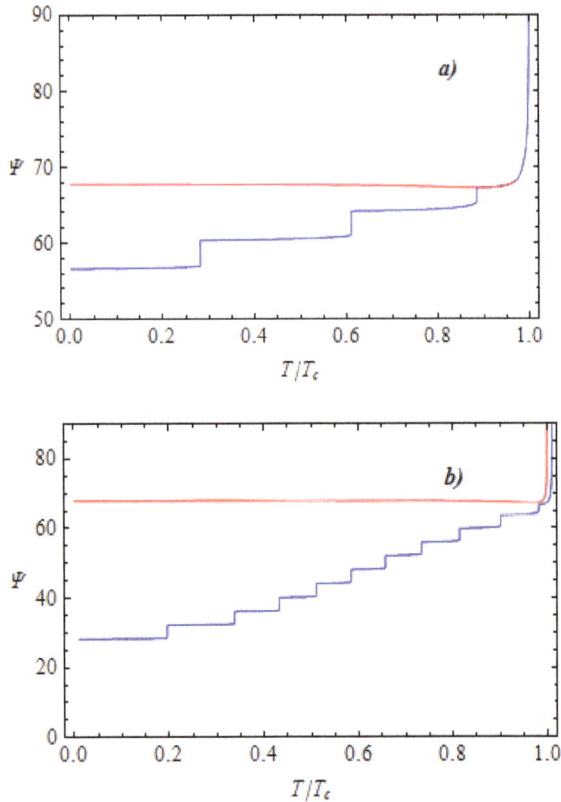

Fig. 7.8. Magnetic flux Ψ threading the inner part of area S_{in} of a four-grain system at various temperatures. The ZFC values of Ψ are represented in blue, the FC values in red. The grain radius is 25.0 μm (panel a) and 50.0 μm (panel b). The zero-temperature value of the maximum Josephson current and of the penetration depth are 300 μA and 1500 Å, respectively. The value of $\beta(0)$ is equal to 10.0 in panel a, and to 39.6 in panel b. The value of the measuring field is 0.246 mT for all curves.

Therefore, for $N = 1000$ we have $\mu_0 H = 0.246$ mT. By considering first $R_g = 25.0$ μm, $h = 10.0$ μm, $T_c = 10.0$ K, $I_{J0} = 300$ μA, and $\lambda(0) = 1500$ Å, we obtain $\beta(0) = 10.0$. Next, by considering $R_g = 50.0$ μm and by leaving unaltered the remaining parameters, we have $\beta(0) = 39.6$. We shall use these two values of $\beta(0)$ in showing two different set of curves representing the magnetic behavior of the system. In fact, taking $N = 1000$ in Eq. (7.28), so that the measuring field value results to be $0.246\,N$ mT, we compute the flux Ψ threading the inner loop of a four-grain system as a function of the temperature T. The results are shown in Fig. 7.8a–b. The blue curves in Fig. 7.8a–b represent the ZFC portion of the simulation. The red curves, on the other hand, represent the FC values of the normalized flux Ψ. From Fig. 7.8a–b we can argue that the shielding properties of the system are more effective at low temperatures, where the parameter β attains higher values. In fact, as the temperature increases after ZFC, more and more fluxons enter the inner loop of the four-grain system. In this way, as we had already noticed in the previous section, the shielding property of Josephson junction network increases for increasing values of the parameter β. This particular feature is also detectable by comparing the zero-temperature value of the normalized flux Ψ computed after ZFC (blue curve) in Fig. 7.8a, for which $\beta(0) = 10.0$, and the corresponding value of Ψ in Fig. 7.8b, for which $\beta(0) = 39.6$. In the first case, we have $\Psi_{ZFC}(0) \approx 57$; in the second case, $\Psi_{ZFC}(0) \approx 24$. This difference in normalized flux is greatly reduced after field-cooling. In fact, the zero-temperature values of the flux Ψ during FC (red curves) are the following: $\Psi_{FC}(0) \approx 68$ for $\beta(0) = 10.0$; $\Psi_{FC}(0) \approx 64$ for $\beta(0) = 39.6$.

One of the main features of the system is that irreversibility also appears under temperature cycling. In fact, the blue (ZFC) and red (FC) curves coincide only in a narrow temperature interval near T_c, in which reversible magnetic response is present.

It is instructive to determine the temperature T_i at which irreversibility sets in. Let us consider the stationary magnetic state at a given temperature T, described by Eq. (7.26), from which we write:

$$\Psi_{ex}^{(out)} = \frac{1}{\sigma_{in}(T)}\left[\Psi + \beta(T)\sin\left(\frac{\pi\Psi}{2}\right)\right]. \tag{7.29}$$

From Chapter 2 we have learned that magnetic reversible behavior of a single superconducting loop interrupted by a Josephson junction is obtained if the derivative of the externally applied field with respect to Ψ does not vanish. This condition is met, in this case, if $\beta(T) < 2/\pi$. Therefore, in order to find T_i, we set:

$$\beta(T_i) = \frac{2}{\pi}. \tag{7.30}$$

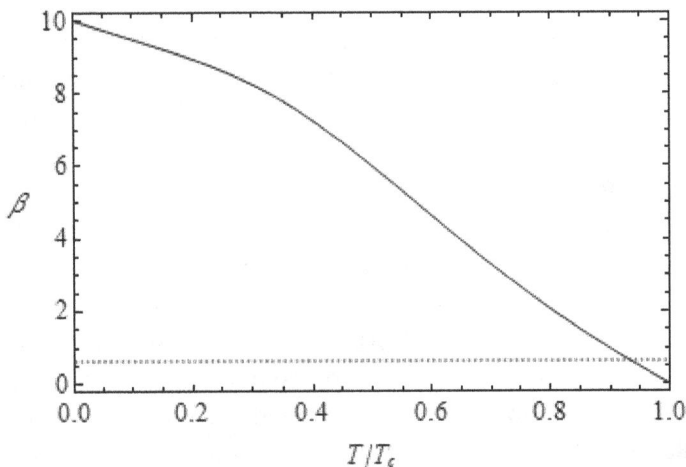

Fig. 7.9. Temperature dependence (full-line curve) of the parameter β of the dressed Josephson junction model for a four-grain superconducting system. In the system the grain radius is 25.0 μm, the height is 10.0 μm, the critical temperature is set to 10.0 K, the zero-temperature value of the maximum Josephson current is 300 μA. Finally, the zero-temperature value $\lambda(0)$ of the penetration depth is 1500 Å. The dashed horizontal line is at height $2/\pi$ and its intersection with the full-line curve determines the irreversibility temperature T_i.

In this way, when we plot the β vs. T graph and trace a dashed horizontal line at height $2/\pi$, the intersection between the two curves gives T_i. As an example, consider the case in which $\beta(0) = 10.0$. The temperature dependence of the parameter β of the dressed Josephson junction model in which the grain radius is 25.0 μm, the other parameters being those in Fig. 7.8a, is represented in Fig. 7.9 along with the horizontal dotted line of height $2/\pi$. We notice that the intersection between the two curves occurs at $T/T_c \approx 0.90$, in good agreement with the abscissa at which irreversibility in the Ψ vs. T curves is observed in Fig. 7.8a.

We next turn our attention to the susceptibility curves, shown in Fig. 7.10a–b and calculated by means of Eq. (7.27) and by the above shown Ψ vs. T data.

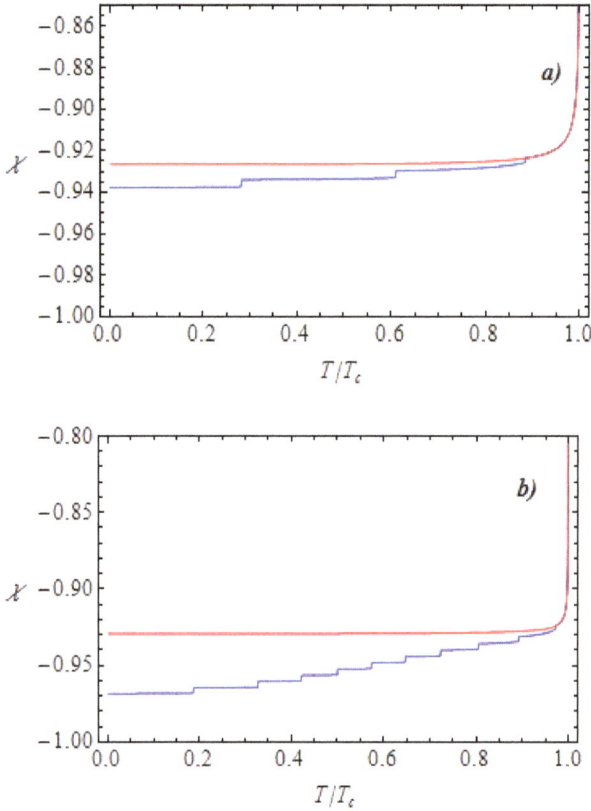

Fig. 7.10. Magnetic susceptibility χ of a four-grain system at various temperatures. The ZFC values of χ are represented in blue, the FC values in red. The grain radii are 25.0 μm (panel a) and 50.0 μm (panel b). The zero-temperature value of the maximum Josephson current and of the penetration depth are 300 μA and 1500 Å, respectively. The value of $\beta(0)$ is equal to 10.0 in panel a and to 39.6. in panel b. The value of the measuring field is 0.246 mT for all curves.

The parameters in Fig. 7.10a are the same as in Fig. 7.8a, while the parameters in Fig. 7.10b are the same as in Fig. 7.8b. The irreversibility character of these curves is rather evident. In fact, as in Fig. 7.9a–b, the blue curves represent susceptibility values for temperatures going from zero to T_c after ZFC, while the red curves represent susceptibility values for temperatures going from T_c to zero during FC. The reversibility temperature interval goes from T_i to T_c, as in the in the corresponding Ψ vs. T curves of Fig. 7.8a–b. The lower value of the susceptibility for $\beta(0) = 39.6$ (Fig. 7.11b), either in ZFC, or in FC, with respect to the susceptibility values for $\beta(0) = 10.0$ (Fig. 7.11a), indicate, once again, that shielding properties of the system increase with increasing value of the parameter β. Finally, to each discontinuity in the ZFC Ψ vs. T curves, a discontinuous variation in the corresponding ZFC χ vs. T curve is present. Discontinuities do not appear in the FC χ vs. T curves for this choice of parameters.

7.3. Concentric circular arrays: the discrete model

One of the possible ways of extending the simple four-grain model for a sintered superconducting system is to consider a two-dimensional cylindrical sample. As already mentioned in subsection 7.2.1, given the cylindrical symmetry of the problem, the only relevant currents in the system are those flowing in the azimuthal direction.

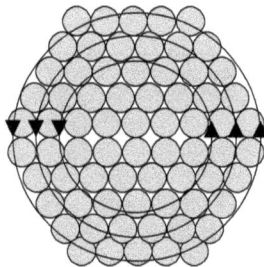

Fig. 7.11. Simplified current distribution in a two-dimensional quasi-cylindrical sample of granular superconductor under an axial magnetic field.

In granular systems, however, the currents are not strictly directed along the circular paths described in Fig. 7.11, since they are forced to flow along the grain borders and to go through the Josephson junctions at the contact points of adjacent grains, as shown in Fig. 7.12.

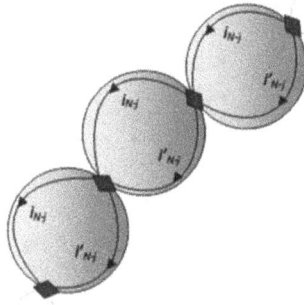

Fig. 7.12. Normalized currents i_{N-j} and i'_{N-j} flow in the ring labeled with the index $N - j$. These currents shield the grains from the external magnetic field. Being the grains considered as purely diamagnetic entities at $T = 0\,K$, the magnetic shielding is complete. Josephson junctions at the interfaces are represented as diamonds.

Therefore, a situation similar to the four-grain system reappears in this case. In order to develop a rather realistic model, let us consider N superconducting concentric rings intercalated by non-superconducting regions, as shown in Fig. 7.13. The N superconducting rings are taken to be perfectly diamagnetic and are labeled with indices $N - j$, with $j = 0, 1, 2, \ldots N - 1$, the outermost ring carrying index N, the innermost index 1.

If a uniform axial magnetic field $\vec{H}$ is applied to the system, shielding currents circulate as shown in Fig. 7.13, where the normalized currents i_{N-j} and i'_{N-j} are schematically represented. These currents are confined to flow on the outer and inner lateral surface of the ring and converge toward the Josephson junctions when the interface between grains is approached, as schematically represented in Fig. 7.12. The Josephson junctions in the rings are "embedded" in the superconducting matrix, since they separate adjacent superconducting grains, as shown in Fig. 7.13.

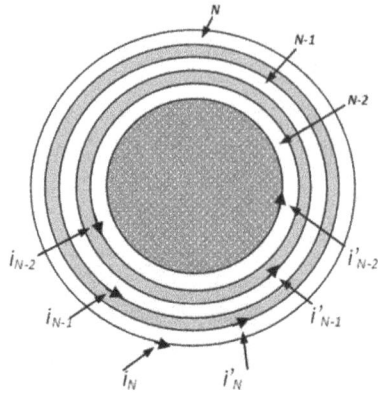

Fig. 7.13. External concentric diamagnetic rings (white regions), intercalated by non-superconducting regions (gray regions). The inner part of the system is not shown, being replaced by a dark circle. Normalized currents i_{N-j} and i'_{N-j} (shown only for $j = 0, 1, 2$) circulate in the azimuthal direction in the presence of an axial magnetic field.

The normalized effective current flowing through each Josephson junction is given by:

$$i^{(B)}_{N-j} = i_{N-j} + i'_{N-j}. \tag{7.31}$$

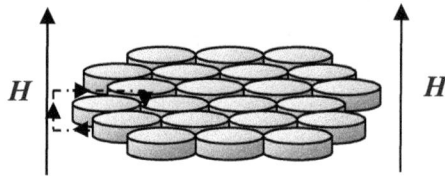

Fig. 7.14. Schematic representation of the rectangular path over which the line integral in Ampere's law is calculated.

By Ampere's law we can find an approximate expression for the local magnetic field h_{N-j} in the inner adjacent non-superconducting shell, located at the perimeter of the circle of area S_{N-j}. We may remark that, in order for the result to be exact, the sample should be infinitely long in the axial dimension. However, if we integrate on a rectangular path sufficiently close to the sample, we can obtain acceptable values of the quantities h_{N-j}, for $j = 0, 1, 2, \ldots N - 1$. Therefore, by following the path shown in Fig. 7.14, we may write:

$$\mu_0\left(H - h_{N-j}\right)d = -\mu_0 I_{j0} \sum_{k=0}^{j} i_{N-k}^{(B)}, \tag{7.32}$$

where, d is the height of the cylindrical sample and I_{j0} is the maximum Josephson current of all identical junctions in the sample. We would now like to find h_{N-j}. We may proceed as in the previous chapter, being careful in making an important distinction: flux penetrated areas are only portions of the sample depicted as gray regions in Fig. 7.13. Nevertheless, the local magnetic field can be safely defined in terms of the magnetic fluxes Φ_{N-j} threading the non-superconducting shells of total area S_{N-j} as done in the previous chapter, so that, by setting $k = N - j$ in Eq. (6.40), we have:

$$\mu_0 h_{N-j} = \frac{\Phi_{N-j} - \Phi_{N-j-1}}{S_{N-j} - S_{N-j-1}}, \tag{7.33}$$

provided that the areas S_{N-j}, as previously stated, take account of only non-superconducting regions. As in the previous chapter, let us introduce the normalized quantity ξ_{N-j}, defined as follows:

$$\xi_{N-j} = \frac{\mu_0 h_{N-j} S_0}{\Phi_0} = \frac{\Psi_{N-j} - \Psi_{N-j-1}}{\frac{S_{N-j} - S_{N-j-1}}{S_0}}, \tag{7.34}$$

where $\Psi_{N-j} = \Phi_{N-j}/\Phi_0$, and S_0 is the area of a single inter-granular site. The term in the denominator thus represents the number n_{N-j} of inter-granular sites, so that:

$$S_{N-j} = n_{N-j}^{(i)} S_0. \tag{7.35}$$

Therefore, Eq. (7.34) can be written as follows:

$$\xi_{N-j} = \frac{\Psi_{N-j} - \Psi_{N-j-1}}{n_{N-j}^{(i)} - n_{N-j-1}^{(i)}}. \tag{7.36}$$

The difference in the denominator of the right hand side of Eq. (7.36) represents the number of inter-granular sites and thus the number n_{N-j} of junctions along the shell of radius R_{N-j}, as measured from the center of the sample. Here we shall only consider simple geometrical arrangements of the grains, in such a way that the difference $n_{N-j}^{(i)} - n_{N-j-1}^{(i)}$ can be readily calculated.

As a first example, let us consider a square lattice of grains as in Fig. 7.2. By calculating the normal fraction α_S as the ratio between the inter-granular and the total area of a single elementary cell formed by a square of radius $2R_g$ (see Figure 7.2), we have:

$$\alpha_S = \frac{S_0}{4R_g^2} = 1 - \frac{\pi}{4}. \tag{7.37}$$

The average number of normal inter-granular sites within a region of radius $R_k = 2kR_g$, which comprises the inner k superconducting shells, can be obtained by dividing the total normal area inside this region $(\alpha_S \pi R_k^2)$ by the area S_0 of a single inter-granular site, so that, considering Eq. (7.37), we have:

$$n_k^{(i)} = \alpha_S \frac{\pi R_k^2}{S_0} = \pi k^2. \tag{7.38}$$

In this way, the number n_k of inter-granular sites within a single shell of radius R_k, is given by:

$$n_k = n_k^{(i)} - n_{k-1}^{(i)} = \gamma_S(2k - 1), \tag{7.39}$$

where $\gamma_S = \pi$ here represents the form factor for the square array and where $k = 1, 2, \ldots, N$.

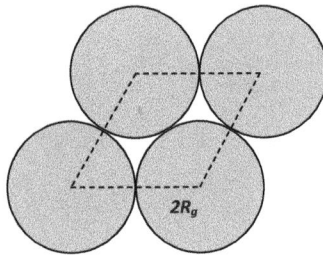

Fig. 7.15. Geometry of a two-dimensional close-packed lattice of cylindrical grains.

Had we chosen a two-dimensional close-packed lattice of grains, whose elementary cell is shown in Fig. 7.15, we would have written a different expression for the normal ratio. In fact, by referring to Fig. 7.15, we notice that the normal to total are a ratio is given by:

$$\alpha_{CP} = \frac{S_0}{\sqrt{3}R_g^2} = 1 - \frac{\pi}{2\sqrt{3}}. \tag{7.40}$$

Therefore, considering that a single shell has radius $\sqrt{3}R_g/2$, so that $R_k = \sqrt{3}R_g k$, we write the equation defining $n_k^{(i)}$, similar to Eq. (7.39), as follows:

$$n_k^{(i)} = \alpha_{CP}\frac{\pi R_k^2}{S_0} = \sqrt{3}\pi k^2. \tag{7.41}$$

In this case, the number n_k of inter-granular sites within a single shell is given, for $k = 1, 2, ..., N$, by the following expression:

$$n_k = n_k^{(i)} - n_{k-1}^{(i)} = \sqrt{3}\pi(2k - 1), \tag{7.42}$$

so that the form factor of the lattice in which grains are closed packed is $\gamma_{CP} = \sqrt{3}\pi$, which is greater than the corresponding value γ_S for a square lattice. Therefore, for a generic lattice L, with form factor γ_L, we may write:

$$n_{N-j} = \gamma_L[2(N - j) - 1]. \tag{7.43}$$

This analysis allows us to write Eq. (7.36) in the following way:

$$\xi_{N-j} = \frac{\Psi_{N-j} - \Psi_{N-j-1}}{n_{N-j}} = \frac{1}{\gamma_L}\frac{\Psi_{N-j} - \Psi_{N-j-1}}{2(N-j)-1}. \tag{7.44}$$

Moreover, Eq. (7.32), when written in terms of normalized quantities, reads:

$$\psi_{ex} - \xi_{N-j} = -\beta \sum_{k=0}^{j} i_{N-k}^{(B)}, \tag{7.45}$$

where $\psi_{ex} = \mu_0 H S_0/\Phi_0$, and $\beta = L_0 I_{j0}/\Phi_0$, with $L_0 = \mu_0 S_0/d$. Notice that ψ_{ex} refers to the applied flux to a single inter-granular site. This quantity relates to the external flux $\Psi_{ex} = \mu_0 H S_N/\Phi_0$ applied to the entire sample as follows:

$$\Psi_{ex} = \frac{S_N}{S_0}\psi_{ex} = \frac{\pi R_N^2}{S_0}\psi_{ex}. \qquad (7.46)$$

From Eq. (7.38) and (7.41) we may notice that the ratio $\pi R_N^2/S_0$ can be expressed as $\gamma_L N^2/\alpha_L$, where γ_L and α_L are the form factor and the normal ratio, respectively, for a lattice denoted by the index L. In fact, from Eq. (7.38) and (7.41) we may set $n_N^{(i)} = (\alpha_L \pi R_N^2)/S_0 = \gamma_L N^2$. In this way Eq. (7.46) can be written as follows:

$$\Psi_{ex} = \frac{\gamma_L}{\alpha_L}N^2\psi_{ex}. \qquad (7.47)$$

This result will be useful in comparing this analysis with the one previously carried out in Section 6.4.2 for bare Josephson junction networks. Recall, also, that ψ_{ex} has been used as a forcing term in Chapter 6 in Eq. (6.7b).

By now considering Eq. (7.44), we may argue that Eq. (7.45) represents the flux-current relation for the system of concentric circular superconducting shells. In order to obtain the effective currents flowing in the junctions, we may subtract two successive terms in Eq. (7.45), so that we get:

$$i_{N-j}^{(B)} = \frac{1}{\beta}\left(\xi_{N-j} - \xi_{N-j+1}\right), \qquad (7.48)$$

for $j = 0, 1, 2, \ldots N - 1$, where, for $j = 0$, $\xi_{N+1} = \psi_{ex}$. To obtain the latter relation, we may compare Eq. (7.45), and Eq. (7.48), both specified for $j = 0$.

Let us now turn our attention to other constitutive equations for the model: the fluxoid quantization relations. The reader is by now already familiar with this subject. We thus write these relations as follows:

$$2\pi\Psi_{N-j} + n_{N-j}\phi_{N-j} = 2\pi m_{N-j}, \qquad (7.49)$$

for $j = 0, 1, 2, \ldots N - 1$, where the quantities ϕ_{N-j} are the superconducting phase difference across any of the n_{N-j} junctions in the superconducting ring with index $(N - j)$, and where the integers m_{N-j} are null if we consider zero-field-cooling conditions. Therefore, we may set $m_{N-j} = 0$ for simplicity, assuming that the initial flux trapped in the sample is zero. Finally, the dynamic equation for a single Josephson junction in the ring with index $(N - j)$ can be written by means of the RSJ model (Barone, 1982) as follows:

$$\frac{\Phi_0}{2\pi R} \frac{d\phi_{N-j}}{dt} + \sin \phi_{N-j} = i_{N-j}^{(B)}, \tag{7.50}$$

where $j = 0, 1, 2, ... N - 1$ and R is the resistive parameter of the junctions. In order to compare the model obtained in the previous chapter and the dynamic equations we are going to write in the present section, let us write Eq. (7.50) in terms of the normalized flux variables Ψ_k. Therefore, by first making use of Eq. (7.48) and (7.49), we write:

$$\frac{d\Psi_{N-j}}{d\tau} + n_{N-j}\beta \sin\left(\frac{2\pi\Psi_{N-j}}{n_{N-j}}\right) = n_{N-j}(\xi_{N-j+1} - \xi_{N-j}), \tag{7.51}$$

where $j = 0, 1, 2, ... N - 1$ and $\tau = Rt/L_0$. We now express the right hand side of Eq. (7.51) in terms of the normalized flux variables by means of Eq. (7.44), so that, we have:

$$\frac{d\Psi_{N-j}}{d\tau} + n_{N-j}\beta \sin\left(\frac{2\pi\Psi_{N-j}}{n_{N-j}}\right) = \frac{n_{N-j}\Psi_{N-j+1}}{n_{N-j+1}} - \left(\frac{n_{N-j}}{n_{N-j+1}} + 1\right)\Psi_{N-j} + \Psi_{N-j-1}. \tag{7.52}$$

In the above expression, among all equations for $j = 0, 1, 2, ... N - 1$, we write down explicitly, in the order, those for $j = 0$ and for $j = N - 1$ as follows:

$$\frac{d\Psi_N}{d\tau} + n_N \beta \sin\left(\frac{2\pi\Psi_N}{n_N}\right) + \Psi_N - \Psi_{N-1} = n_N\psi_{ex}, \tag{7.53a}$$

$$\frac{d\Psi_1}{d\tau} + \beta \gamma_L \sin\left(\frac{2\pi\Psi_1}{\gamma_L}\right) + \frac{4\Psi_1 - \Psi_2}{3} = 0. \tag{7.53b}$$

In Eq. (7.53a) we have made use of Eq. (7.44) to find, for $j = -1$, what follows:

$$\xi_{N+1} = \psi_{ex} = \frac{\Psi_{N+1} - \Psi_N}{n_{N+1}}. \tag{7.54}$$

In Eq. (7.53b), on the other hand, we made use of Eq. (7.43) to express the number of junctions in terms of the parameter γ_L, by setting $n_1 = \gamma_L$ and $n_2 = 3\gamma_L$.

The above dynamic equations, Eq. (7.52) and Eq. (7.53a–b), can be numerically integrated in time for the normalized flux variables Ψ_{N-j} with $j = 0, 1, 2, \ldots N - 1$. The initial conditions for the superconducting phases are given by the ZFC conditions $\Psi_{N-j} = 0$. As for the boundary conditions on the sample, these have been taken care of in Eq. (7.53a–b), where we have set: $\Psi_0 = 0$, $\Psi_{N+1} = n_N \psi_{ex} + \Psi_N$. These equations are different, in the coefficients, with respect to Eq. (6.43) for bare Josephson networks, since different numbering of the junctions have been adopted, due to a different configuration of the system. We may also remark that the dimensions of the cylindrical grains are comparable with the dimensions of the sample itself, so that N is not very large. In this way, it would not be appropriate, a priori, to take the continuous limit of Eq. (7.52). In the following section, as done in the previous chapter, we shall consider the situations in which the continuous limit may be taken. Numerical results in the discrete case are found and discussed in the following subsection.

7.3.1. *Numerical results for a square lattice*

As it can be argued from Eq. (7.44) and (7.48), knowledge of the solution of Eq. (7.52) and Eq. (7.53a–b) for the normalized flux variables allows the calculation of the current and field distribution inside a sample in which the junctions inside are arranged according to a given ordered scheme. We start by considering a system in which grains are arranged in a square lattice.

Therefore, after having numerically solved Eq. (7.52) and Eq. (7.53a–b) for $N = 20$, the results for the normalized magnetic field ξ_k and the total flux Ψ_k are reported in Fig. 7.17a–c and in Fig. 7.18a–c, respectively, for $\beta = 0.05$. In solving for the normalized magnetic flux variables Ψ_k, at each integration step the values of these quantities at the previous step are used as initial conditions. In this way, the irreversible magnetic behavior of the system may be accounted for. When starting the ψ_{ex} cycle, we notice that, except for a small amount of magnetic flux inside the most external shells, the inner part of the system does not allow flux penetration. As ψ_{ex} is increased, the magnetic field starts to penetrate inside the sample.

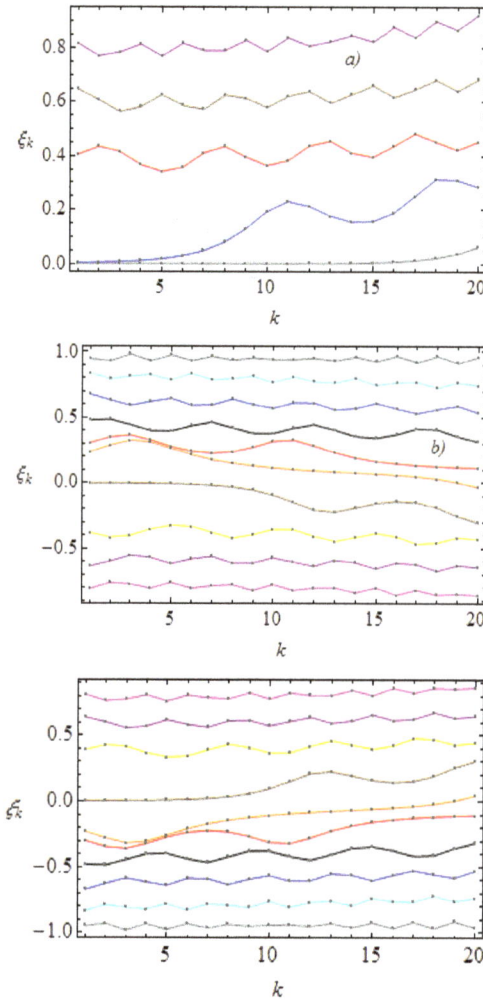

Fig. 7.16. Normalized magnetic field ξ_k in the shell model of a granular sample at $T = 0\,K$. The Josephson junctions are arranged in a square lattice and $\beta = 0.05$. The index k ranges from 1 (inner shell) to 20 (external shell). The lines are used to distinguish the values of the normalized field values ξ_k (grey points) at various forcing terms ψ_{ex}. In panel a, the increasing ψ_{ex} values are: 0.10 (gray); 0.30 (blue); 0.50 (red); 0.90 (purple). In panel b, the decreasing ψ_{ex} values are: 0.90 (gray); 0.70 (cyan); 0.50 (blue); 0.30 (black); 0.10 (red); -0.10 (orange); -0.30 (brown); -0.50 (yellow); -0.70 (purple); -0.90 (magenta). In panel c, the increasing ψ_{ex} values are: -0.90 (gray); -0.70 (cyan); -0.50 (blue); -0.30 (black); -0.10 (red); 0.10 (orange); 0.30 (brown); 0.50 (yellow); 0.70 (purple); 0.90 (magenta).

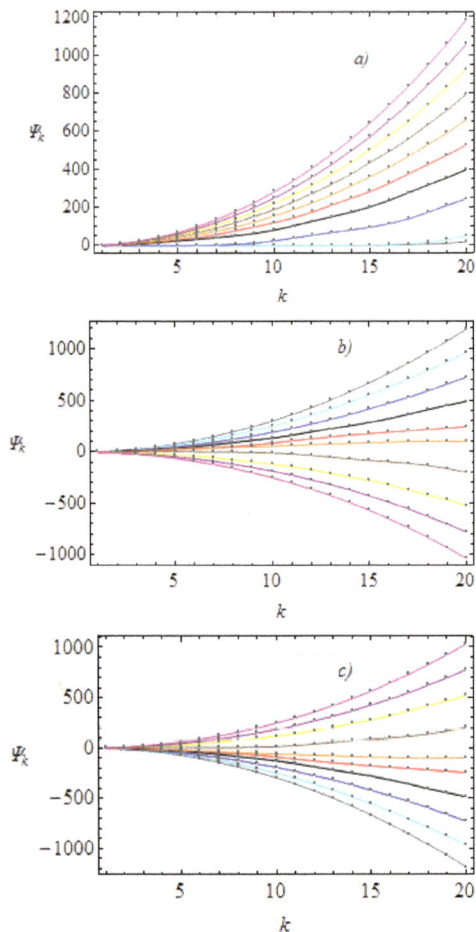

Fig. 7.17. Total flux within the k-th shell in the discrete model of a granular sample at $T = 0\,K$. The Josephson junctions are arranged in a square lattice and $\beta = 0.05$. The index k ranges from 1 (inner shell) to 20 (external shell). The lines are used to distinguish the values of the normalized flux values Ψ_k (grey points) at various forcing terms ψ_{ex}. In panel a, the increasing ψ_{ex} values are: 0.10 (gray); 0.20 (cyan); 0.30 (blue); 0.40 (black); 0.50 (red); 0.60 (orange); 0.70 (brown); 0.80 (yellow); 0.90 (purple); 1.00 (magenta). In panel b, the decreasing ψ_{ex} values are: 0.90 (gray); 0.70 (cyan); 0.50 (blue); 0.30 (black); 0.10 (red); -0.10 (orange); -0.30 (brown); -0.50 (yellow); -0.70 (purple); -0.90 (magenta). In panel c, the increasing ψ_{ex} values are: -0.90 (gray); -0.70 (cyan); -0.50 (blue); -0.30 (black); -0.10 (red); 0.10 (orange); 0.30 (brown); 0.50 (yellow); 0.70 (purple); 0.90 (magenta).

The mechanisms for flux penetration are similar to those described for simpler bare networks.

In Fig. 7.16a the normalized magnetic field distributions at various normalized applied fluxes, increasing from $\psi_{ex} = 0.0$ to $\psi_{ex} = +1.00$, or $\Psi_{ex} = +5856$, are shown for ψ_{ex} in the interval $[0.1, +0.90]$. For the distributions in Fig. 7.16b the normalized applied flux is varied from $\psi_{ex} = +1.00$ to $\psi_{ex} = -1.00$, while it is shown in the range going from $\psi_{ex} = +0.90$ to $\psi_{ex} = -0.90$. Finally, in Fig. 7.16c, the normalized applied flux is brought from $\psi_{ex} = -1.00$ back to $\psi_{ex} = +1.00$ and is shown for $\psi_{ex} = -0.90$ back to $\psi_{ex} = +0.90$. In Fig. 7.17a the Ψ_k vs. k curves at various normalized applied fluxes, increasing from $\psi_{ex} = 0.0$ to $\psi_{ex} = +1.00$, are shown. The curves in Fig. 7.17b, on the other hand, represent the normalized magnetic flux Ψ_k for decreasing values of the normalized applied flux, from $\psi_{ex} = +0.90$ to $\psi_{ex} = -0.90$. Finally, in the Ψ_k vs. k curves in Fig. 7.17c, the normalized applied flux goes from $\psi_{ex} = -0.90$ back to $\psi_{ex} = +0.90$. The latter curves represent the direct solutions of the problem. From these quantities the $i_k^{(B)}$ vs. k distributions can be derived by means of Eq. (7.44) and (7.48). From these curves it can be argued once more that the inter-granular flux penetration mechanisms are regulated by the Josephson junctions, which bridge flux quanta in the radial direction from one inter-granular site to next-neighboring ones. As we have already seen in simpler systems, flux motion occurs when the maximum Josephson current of the junctions in the k-th shell (let us say) is reached. In this case, in fact, the resistive state of all junctions in the same k-th shell is attained and a non-null value of the time derivative of the quantity Ψ_k is indicative of flux motion inside the system. We may further remark that, since the shell we are considering is sandwiched in between an external and an internal normal region, flux can penetrate from the former to the latter because of phase slip in the junctions embedded in the superconducting ring. This flux penetration mechanism is similar to the critical state models in type-II superconductors (Schmidt, 1997). However, magnetic field and current distribution in the latter class of superconductors are obtained by different flux dynamics and for higher values of the applied magnetic field. Nonetheless, the similarities between the two penetration mechanisms, those occurring in granular superconductor at very low fields and those present in type-II superconductors above the first critical magnetic field H_{c1} (Schmidt, 1997), have led researchers to propose critical state models for describing low-field magnetic properties of granular superconductors (Ginzburg, 1991; Muller, 1992).

Let us now take a closer look at Fig. 7.16a–c, in order to further corroborate the above reasoning.

In Fig. 7.16a we see that for small values of the increasing externally applied flux ψ_{ex} (gray curve) the magnetic field distribution is confined close to the external boundary of the sample, leaving inner portion of the system flux-free. As ψ_{ex} increases, shielding currents in the proximity of the external boundary increase, so that flux penetration can occur when the Josephson current of the junctions is locally exceeded. Therefore, the blue curve in Fig. 7.16a shows that flux penetrates gradually inside the sample, from the outer border to the innermost part. When the external magnetic field is further increased (see red curve of Fig. 7.16a), all inter-granular sites are penetrated by flux quanta, the latter concentrating more in regularly spaced regions. For further increase of the forcing term ψ_{ex}, the spacing between these regions decreases. For decreasing externally applied forcing terms, the distributions are shown in Fig. 7.16b. We notice how these curves start decreasing from the outer part of the sample, since the system needs to satisfy the boundary condition $\xi_{N+1} = \psi_{ex}$ and ξ_N tends to approach this value. When the externally applied flux is close to zero, the fluxons migrate out of the system, except for residual resilient inner fluxons (see red and orange curves in Fig. 17.6b). For negative values of ψ_{ex}, anti-fluxons start penetrating the sample, locating first in the outer part of the system and then distributing, as in the case of fluxons, inside the sample. By letting ψ_{ex} increase again from the lowest value $\psi_{ex} = -1$ in the cycle toward $\psi_{ex} = +1$, in Fig. 7.16c the same scenario described above reappears, with distributions qualitatively similar, although quantitatively different, for correspondingly equal values of ψ_{ex}.

The total flux Ψ_k within the shell of radius R_k shown in Fig. 7.17a-c follow a trend very similar to corresponding curves shown for a bare network in the previous chapter (see, for example, Fig. 6.22a).

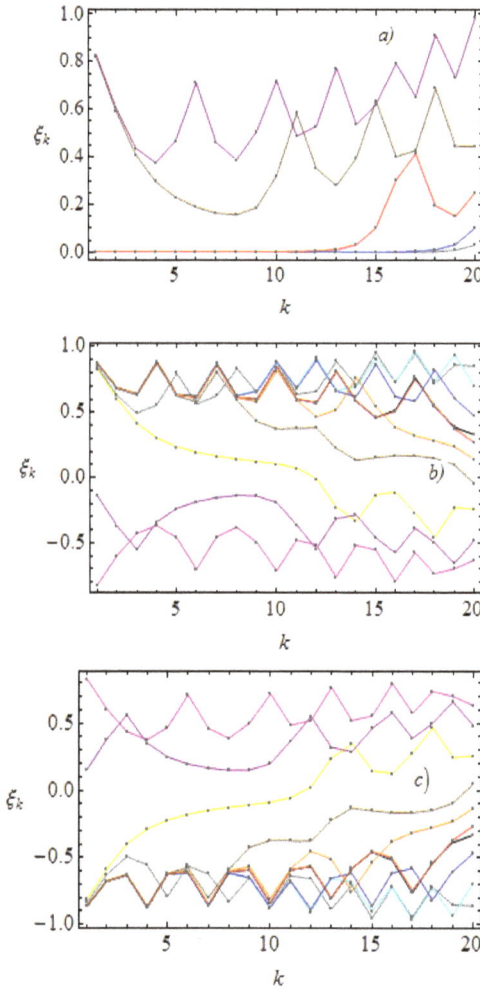

Fig. 7.18. Normalized magnetic field ξ_k in the shell model of a granular sample at $T = 0\ K$. The Josephson junctions are arranged in a square lattice and $\beta = 0.25$. The index k ranges from 1 (inner shell) to 20 (external shell). The lines are used to distinguish the values of the normalized field values ξ_k (grey points) at various forcing terms ψ_{ex}. In panel a, the increasing ψ_{ex} values are: 0.10 (gray); 0.30 (blue); 0.50 (red); 0.90 (purple). In panel b, the decreasing ψ_{ex} values are: 0.90 (gray); 0.70 (cyan); 0.50 (blue); 0.30 (black); 0.10 (red); -0.10 (orange);-0.30 (brown); -0.50 (yellow); -0.70 (purple); -0.90 (magenta). In panel c, the increasing ψ_{ex} values are: -0.90 (gray); -0.70 (cyan); -0.50 (blue); -0.30 (black); -0.10 (red); 0.10 (orange); 0.30 (brown); 0.50 (yellow); 0.70 (purple); 0.90 (magenta).

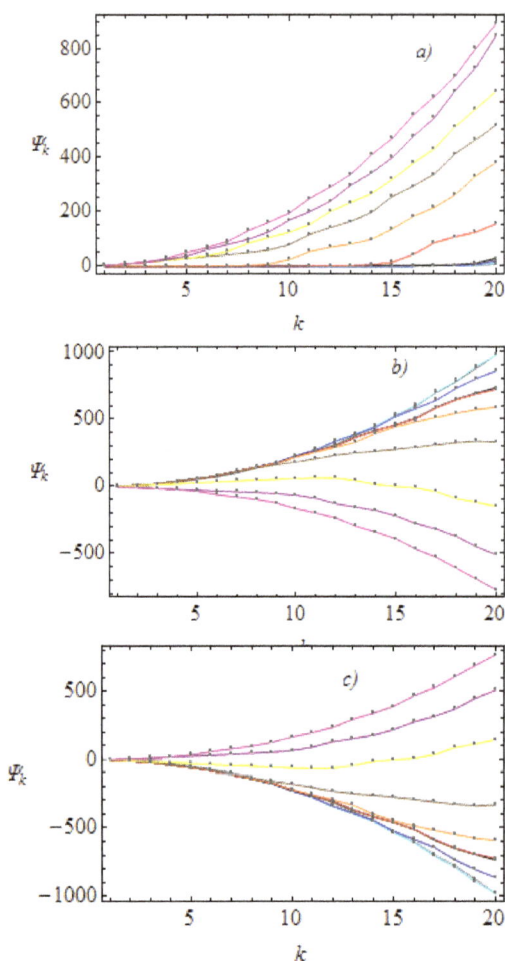

Fig. 7.19. Total flux Ψ_k within the k-th shell in the discrete model of a granular sample at $T = 0\,K$. The Josephson junctions are arranged in a square lattice and $\beta = 0.25$. The index k ranges from 1 (inner shell) to 20 (external shell). The lines are used to distinguish the values of the normalized flux values Ψ_k (grey points) at various forcing terms ψ_{ex}. In panel a, the increasing ψ_{ex} values are: 0.10 (gray); 0.20 (cyan); 0.30 (blue); 0.40 (black); 0.50 (red); 0.60 (orange); 0.70 (brown); 0.80 (yellow); 0.90 (purple); 1.00 (magenta). The gray, cyan, and blue curves are very close to each other. In panel b, the decreasing ψ_{ex} values are: 0.90 (gray); 0.70 (cyan); 0.50 (blue); 0.30 (black); 0.10 (red); -0.10 (orange); -0.30 (brown); -0.50 (yellow); -0.70 (purple); -0.90 (magenta). In panel c, the increasing ψ_{ex} values are: -0.90 (gray); -0.70 (cyan); -0.50 (blue); -0.30 (black); -0.10 (red); 0.10 (orange); 0.30 (brown); 0.50 (yellow); 0.70 (purple); 0.90 (magenta).

Therefore, even differing from the quantitative point of view, these curves can be commented as done in Chapter 6.

We now turn our attention on the effects of an increasing value of the parameter β. The results of a numerical computational of the ξ_k vs. k and of the Ψ_k vs. k curves are thus shown in Fig. 7.18a–c and in Fig. 7.19a–c, respectively, for $\beta = 0.25$.

By comparing Fig. 7.18a with Fig. 7.16a obtained for $\beta = 0.05$, we see that first irreversible penetration now occurs at a normalized applied flux value close to $\psi_{ex} = 0.50$ (see red curve, where a first distribution of fluxons is visible as a hump in the plot), whereas the corresponding value of the normalized lower threshold field $\psi_{ex}^{(1)}$ for $\beta = 0.05$ occurs at $\psi_{ex} = 0.10$. Therefore, a steady increase of $\psi_{ex}^{(1)}$ with β can be hypothesized. This hypothesis will be verified later by a detailed analytic and numerical analysis. When ψ_{ex} exceeds the lower threshold, magnetic flux lines penetrates gradually inside the sample, from the outer border to the innermost part, following the same pattern already seen in Fig. 7.16a. On the other hand, for decreasing values of ψ_{ex}, the distributions in Fig. 7.18b show that fluxons appear to be more resilient with respect to those embedded in a system with a lower value of β. In other words, the pinning strength, to which fluxons in inter-granular sites are subject, increases for increasing values of β. In fact, by looking at the yellow curve in Fig. 7.18b, we notice that a congruous number of fluxons still resist inside the sample even though the externally applied field is well below zero and even though anti-fluxons have already started to penetrate from the outer border inside the system. The same type of behavior, *mutatis mutandis*, can be seen in Fig. 7.18c when ψ_{ex} is made to increase from the lowest value $\psi_{ex} = -1$ toward $\psi_{ex} = +1$.

As for the normalized total flux Ψ_k within the shell of radius R_k shown in Fig. 7.19a–c, the characteristic behavior already encountered in Fig. 7.16a–c is here also present. In Fig. 7.19a, in particular, we may notice how the three curves obtained for $\psi_{ex} = 0.10$ (gray), $\psi_{ex} = 0.20$ (cyan), $\psi_{ex} = 0.30$ (blue), $\psi_{ex} = 0.40$ (black), coalesce into a single one, because all four values of ψ_{ex} are less than the normalized lower threshold field value $\psi_{ex}^{(1)}$, which is about equal to 0.455, as numerically determined. In this case, then, the fact these curves attain low values of Ψ_k signifies that the externally applied magnetic flux is shielded by the sample through currents not exceeding the maximum Josephson current value in all shells. As soon as $\psi_{ex} > \psi_{ex}^{(1)}$ the usual shape of these curves appear. Notice, however, how all curves in Fig. 7.19a–c are less smooth than the corresponding curves in Fig. 7.17a–c. This is another aspect of the increase of the parameter β which can be connected to a feature already detected in the normalized magnetic field distributions, namely, the increase of the pinning strength exerted on fluxons residing in the inter-granular regions for increasing values of β.

Another important characteristic quantity of these systems is the normalized lower threshold field value $\psi_{ex}^{(1)}$. In order to estimate $\psi_{ex}^{(1)}$ in this case, a procedure similar to one adopted in Section 6.4.2 is followed. By starting, as done in Eq. (6.49), from the stationary portion of the dynamic equation defining flux motion inside the system, we write:

$$n_{N-j}\,\beta\,\sin\left(\frac{2\pi\Psi_{N-j}}{n_{N-j}}\right) = \frac{n_{N-j}\,\Psi_{N-j+1}}{n_{N-j+1}} - \left(\frac{n_{N-j}}{n_{N-j+1}}+1\right)\Psi_{N-j} + \Psi_{N-j-1}, \quad (7.55)$$

for $(j = 1, 2, ..., N-1)$. By assuming that the sine term can be approximated by its argument and, for simplicity, that the ratios n_{N-j}/n_{N-j+1} can approximated by the constant quantity r $(0 < r < 1)$ to be evaluated by a best fit procedure later, we rewrite Eq. (7.55) as follows:

$$r\Psi_{N-j+1} - (1+r+2\pi\beta)\Psi_{N-j} + \Psi_{N-j-1} = 0. \qquad (7.56)$$

By now setting

$$\Psi_{N-j} = S^{N-j}\Psi_N, \qquad (7.57)$$

Eq. (7.56) reads:

$$rS^2 - (1+r+2\pi\beta)S + 1 = 0. \qquad (7.58)$$

We thus have two positive roots of the quantity S:

$$S_{\pm}(\beta, r) = \frac{(1+r+2\pi\beta)\pm\sqrt{(1+r+2\pi\beta)^2-4r}}{2r}, \qquad (7.59)$$

where $S_+(\beta, r) > 1$ and $0 < S_-(\beta, r) < 1$. Since Ψ_{N-j} needs to decrease as the index j increases, we need to choose $S = S_-(\beta, r)$ for our purposes. Let us extend the validity of Eq. (7.57) also for $j = -1$, and let us impose the boundary condition in Eq. (7.54), so that:

$$\psi_{ex} = \frac{\Psi_{N+1} - \Psi_N}{n_{N+1}} = \frac{1 - S_-(\beta, r)}{n_{N+1} S_-(\beta, r)} \Psi_N. \tag{7.60}$$

When first irreversible flux transition takes place, the current in the outermost shell approaches the maximum Josephson current, so that Ψ_N attains a value which makes the argument of the sine term, defining the current in the junction, close to $\pi/2$, i. e., $\Psi_N \approx n_N/4$. Therefore, according to this picture, the normalized lower threshold field value is

$$\psi_{ex}^{(1)}(\beta, r) \approx \frac{n_N}{n_{N+1}} \frac{1 - S_-(\beta, r)}{4 S_-(\beta, r)}, \tag{7.61}$$

where we retain n_N/n_{N+1} unaltered. By a patient numerical determination of $\psi_{ex}^{(1)}$ for various values of the parameter β, we get Table 7.1.

β	0.01	0.05	0.10	0,15	0.20	0.25	0.30
$\psi_{ex}^{(1)}$	0.104	0.195	0.274	0.336	0.394	0.452	0.510

Table. 7.1. Numerically determined values of the lower threshold field $\psi_{ex}^{(1)}$ for various value of the parameter β.

A best-fit procedure on the quantity r, utilizing the expression in Eq. (7.61), we find the best-fit curve is given by $r = 0.717$. A graph of the best-fit expression for $\psi_{ex}^{(1)}$ as a function of the parameter β is given in Fig. 7.20 in the range $[0, 0.3]$ as a full line along with the numerically determined points taken from Table 7.1. Considering the simplifying assumptions made in assuming the ratio r to be a constant, we may affirm that the predictions of Eq. (7.61) are in fairly good agreement with what were found by numerical analysis. Naturally, a solution of the difference equation (7.55) with ratios n_{N-j}/n_{N-j+1} accounted for in exact way, would give a better fit of the function $\psi_{ex}^{(1)}(\beta, r)$. However, we notice that the monotonically increasing character of the normalized lower threshold field, seen as a function of β, can be captured by this rather simplified analysis.

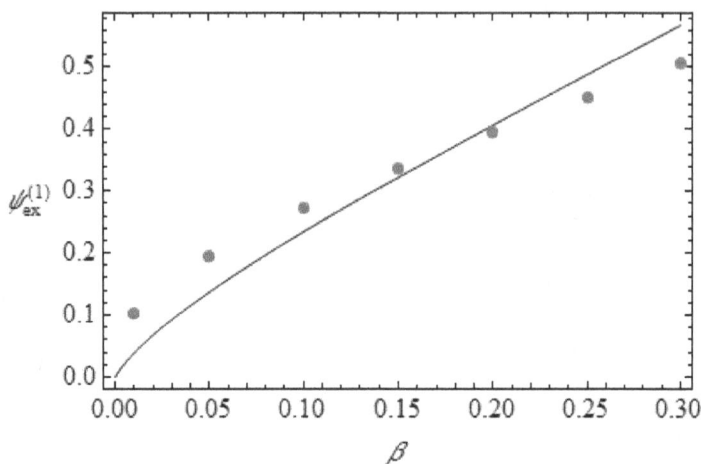

Fig. 7.20. Numerical evaluation (dots) of the normalized lower threshold field of a granular superconducting system modeled as a collection of cylindrical grains in a planar square-lattice arrangement. The full-line curve represents a best fit, according to the analytic procedure adopted leading to Eq. (7.61), where the function $\psi_{ex}^{(1)}(\beta, r)$ is specified. The optimum value of the fitting parameter r is found to be $r = 0.717$.

7.3.2. *Magnetization curves for a square array*

Let us now consider the magnetization curves of the concentric shell model. We first calculate the average magnetic induction $\langle B \rangle$ in the system, by keeping in mind that we have hypothesized that all superconducting rings are perfectly diamagnetic entities. Therefore, considering that only the normal portions of the sample can be penetrated by magnetic flux, we may set:

$$\langle B \rangle = \frac{\Phi_N}{S_{TOT}} = \frac{\Phi_0}{S_{TOT}} \Psi_N, \qquad (7.62)$$

where S_{TOT} is the total cross-section of the sample.

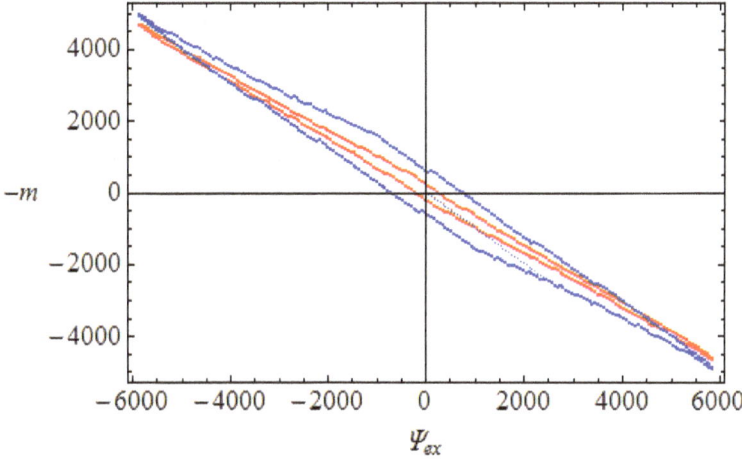

Fig. 7.21. Magnetization curves, at $T = 0\,K$, of granular systems in which grains are in a square-lattice arrangement. The red curve is for $\beta = 0.05$, the blue curve for $\beta = 0.25$. First magnetization portions are shown as dotted lines.

Therefore, by definition, the magnetization is $M = \langle B \rangle/\mu_0 - H$, we have:

$$m = \Psi_N - \frac{\mu_0 S_{TOT}}{\Phi_0} H = \Psi_N - \Psi_{ex}, \qquad (7.63)$$

where $m = \mu_0 S_{TOT} M/\Phi_0$. By making use of the numerical analysis made in the previous sub-section, we may get, by Eq. (7.63), the magnetization curves for $\beta = 0.05$ and $\beta = 0.25$. Therefore, in Fig. 7.21 the m vs. Ψ_{ex} curves are shown for the above two values of the parameter β. Here we prefer to show these curves in terms of $\Psi_{ex} = \mu_0 S_{TOT} H/\Phi_0 = \gamma_S N^2 \psi_{ex}/\alpha_S$ and not simply ψ_{ex}, in order to later compare the properties of granular systems with square-lattice arrangement with the analogous properties of closed-packed superconducting grain systems.

In Fig. 7.21 we may soon notice two main features. The first is an enlargement of the cycle for $\beta = 0.25$ with respect to the one obtained for $\beta = 0.05$. This particular aspect depends on the number of flux quanta and on the way the latter are distributed within the system for a given cycle, since the area of the magnetization cycle represents the energy dissipated by the system in one cycle.

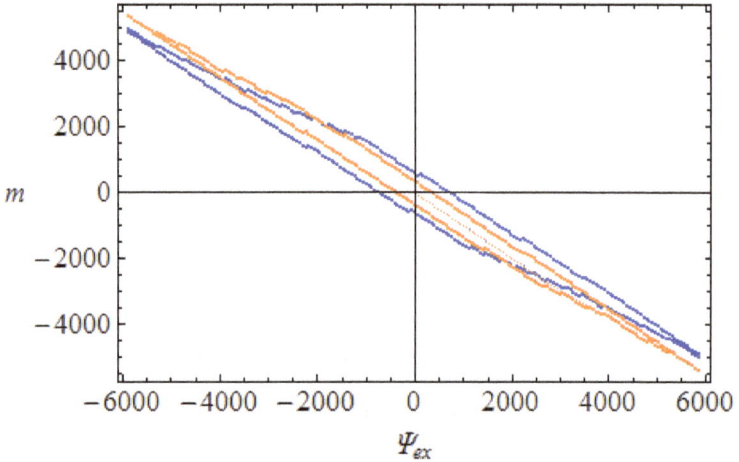

Fig. 7.22. Magnetization curves, at $T = 0\ K$, of granular systems in which grains are in a square-lattice arrangement. The blue curve is for $\beta = 0.25$, as in Fig. 7.21, the orange curve is for $\beta = 0.35$. First magnetization portions are shown as dotted lines.

Therefore, the enlargement of the magnetization cycle is not a mere consequence of the increase of the parameter β. As a matter of fact, by looking at Fig. 7.22, where the magnetization cycles for $\beta = 0.25$ and $\beta = 0.35$ are shown, we may notice that, by increasing the value of this parameter from $\beta = 0.25$ to $\beta = 0.35$, the magnetization cycle (in orange) reduces in size, when compared to the blue curve obtained for $\beta = 0.25$. In order to understand this type of behavior, we need to consider the field distribution in the range $-1 < \psi_{ex} < +1$, as shown in Fig. 7.23a–c.

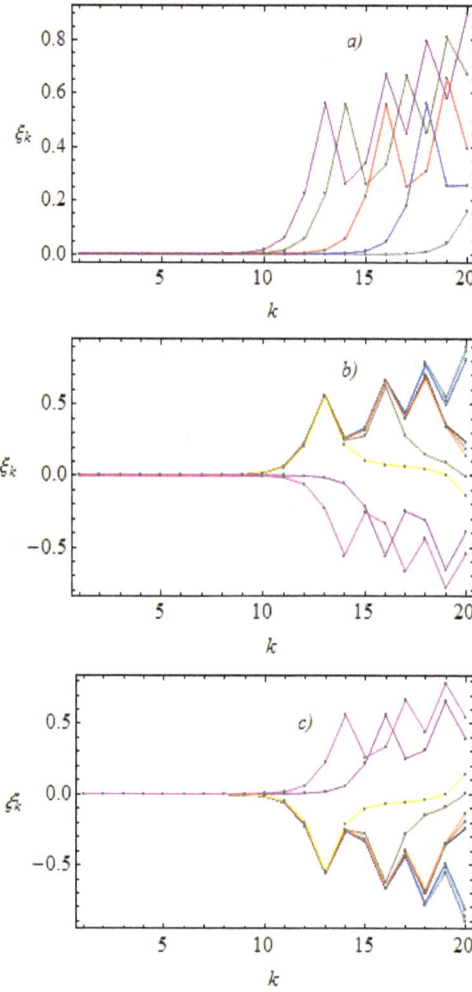

Fig. 7.23. Normalized magnetic field ξ_k in the shell model of a granular sample at $T = 0\,K$. The Josephson junctions are arranged in a square lattice and $\beta = 0.35$. The index k ranges from 1 (inner shell) to 20 (external shell). The lines are used to distinguish the values of the normalized field values ξ_k (grey points) at various forcing terms ψ_{ex}. In panel a, the increasing ψ_{ex} values are: 0.50 (gray); 0.60 (blue); 0.70 (red); 0.90 (brown), 1.00 (purple). In panel b, the decreasing ψ_{ex} values are: 0.90 (gray); 0.70 (cyan); 0.50 (blue); 0.30 (black); 0.10 (red); −0.10 (orange);−0.30 (brown); −0.50 (yellow); −0.70 (purple); −0.90 (magenta). In panel c, the increasing ψ_{ex} values are: −0.90 (gray); −0.70 (cyan); −0.50 (blue); −0.30 (black); −0.10 (red); 0.10 (orange); 0.30 (brown); 0.50 (yellow); 0.70 (purple); 0.90 (magenta).

In these distributions it is clear that, in the range of the magnetization cycle, the flux quanta penetrate only the outer half part of the sample. Being the peaks of the normalized field values comparable to those in Fig. 7.18a–c, where a full penetration of the sample is present, we conclude that the number of fluxons which are forced in and out the sample during the cycle is less for $\beta = 0.35$ than for $\beta = 0.25$. Furthermore, the pinning strength is not significantly different in the two cases, since the values of β are comparable. We therefore argue that the energy dissipated during a magnetization cycle to force fluxons in and out the granular system is less for $\beta = 0.35$ than for $\beta = 0.25$, thus explaining the more enlarged shape of the magnetization curve in the latter case. Similar arguments can be used to explain the difference in size of the magnetization cycles for $\beta = 0.05$ and for $\beta = 0.25$, noticing that this time the lower value of the parameter β determines a less marked hysteresis. In this way, we may finally state that the size of the magnetization cycle does not depend, in a monotonically way, on the parameter β.

The second feature, depending monotonically on β this time, is the slope of the asymptote of the magnetization curve, corresponding to the symmetry line of the m vs. Ψ_{ex} cycle. This aspect is clear both from Fig. 7.21 and Fig. 7.22, where, for increasing values of the parameter β, the magnetization curves are inclined more and more with respect to the horizontal axis. This feature can be explained by the following semi-quantitative argument.

Let us consider the slope $a(\beta)$ of the asymptote m_A of the magnetization curves, whose equation can be written as follows:

$$m_A = - |a(\beta)| \, \Psi_{ex}. \tag{7.64}$$

By denoting with m_A^* and Ψ_{ex}^* the values obtained at the lowest part of the magnetization curve, we have:

$$|a(\beta)| = - \frac{m_A^*}{\Psi_{ex}^*} = \frac{\Psi_{ex}^* - \Psi_N^*}{\Psi_{ex}^*}, \tag{7.65}$$

where Ψ_N^* is the total flux in the granular system at $\Psi_{ex} = \Psi_{ex}^*$. The quantity Ψ_N^* can be obtained, in terms of $\Psi_{ex}^* = \gamma_S N^2 \psi_{ex}^*/\alpha_S$, with $\psi_{ex}^* = 1$, from Eq. (7.60), so that:

$$\Psi_N^* = \frac{n_{N+1}S_-(\beta)}{1-S_-(\beta)} \psi_{ex}^* = \alpha_S \frac{2N+1}{N^2} \frac{S_-(\beta)}{1-S_-(\beta)} \Psi_{ex}^*, \tag{7.66}$$

where $S_-(\beta)$ is the function $S_-(\beta, r)$, defined in Eq. (7.59), calculated at $r = 0.717$. In this way by Eq. (7.65), we have:

$$|a(\beta)| = 1 - \alpha_S \frac{2N+1}{N^2} \frac{S_-(\beta)}{1-S_-(\beta)}. \tag{7.67}$$

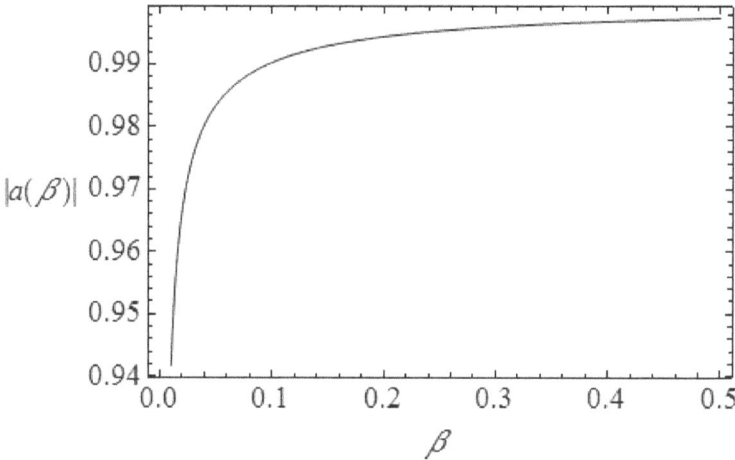

Fig. 7.24. Approximate absolute value of the slope of the asymptote of the magnetization curves obtained for granular systems at $T = 0$ K.

Therefore, confiding in the qualitative validity of the analysis made above in determining the lower threshold field, we may determine the monotonic behavior of $|a(\beta)|$ from Eq. (7.67). Therefore, in Fig. 7.24 we show the qualitative dependence on β of the absolute value of the slope of the asymptote of the magnetization curve. The evident monotonic increasing behavior of the curve in Fig. 7.24 supports the thesis that the slope of the asymptote of the magnetization curves, obtained for granular systems at $T = 0$ K, depends monotonically on β.

7.3.3. *Close-packed granular systems*

In the present sub-section we consider the magnetic properties of close-packed granular superconductors in order to detect similarities and differences with systems with square-lattice symmetry.

Therefore, we proceed to numerically solve Eq. (7.52) and Eq. (7.53a–b) for $N = 20$, with $\gamma_{CP} = \sqrt{3}\pi$ and α_{CP} given by Eq. (7.40). Being the intrinsic geometric parameters γ_{CP} and α_{CP} different from the corresponding parameters in the case of a square-lattice granular system, assuming samples of equal dimensions, we need to consider field ranges which are comparable with the one adopted in the latter case. Therefore, by Eq. (7.47), we may set:

$$\frac{\gamma_S}{\alpha_S}\psi_{ex}^{(S)} = \frac{\gamma_{CP}}{\alpha_{CP}}\psi_{ex}^{(CP)}, \tag{7.68}$$

where $\psi_{ex}^{(S)}$ and $\psi_{ex}^{(CP)}$ are the maximum excursions of the normalized magnetic flux in the two cases. In this way, being $\psi_{ex}^{(S)} = 1$, we have:

$$\psi_{ex}^{(CP)} = \frac{\gamma_S}{\gamma_{CP}}\frac{\alpha_{CP}}{\alpha_S} = \frac{2\sqrt{3}-\pi}{6\left(1-\frac{\pi}{4}\right)} \approx 0.25. \tag{7.69}$$

While we had previously shown field distributions for value of the normalized external applied flux ψ_{ex} in the interval $-1 < \psi_{ex} < +1$, we now need to choose the following range: $-0.25 < \psi_{ex} < +0.25$. This is because we would like to let the externally applied flux, considered over the whole sample (Ψ_{ex}) and not just on a single inter-granular site (ψ_{ex}), to attain values comparable to the homologous forcing term in the case of granular systems with square-lattice arrangement already analyzed in Section 7.3.1 and 7.3.2. The reason why we chose to deal with the quantity ψ_{ex} from the beginning will be clear soon. A comparison between the low-field magnetic properties of the system with different intrinsic geometric properties, in fact, will be easier when we deal with both types of forcing terms, namely, ψ_{ex} and Ψ_{ex}. In fact, before showing the field distributions at various values of ψ_{ex}, let us substitute the values of n_{N-j} as given in Eq. (7.43) into the right-hand side of the stationary equation (7.55), so that:

$$n_{N-j}\,\beta\,\sin\left(\frac{2\pi\Psi_{N-j}}{n_{N-j}}\right) = \frac{2(N-j)-1}{2(N-j)+1}\Psi_{N-j+1}$$

$$-\frac{4(N-j)}{2(N-j)+1}\Psi_{N-j} + \Psi_{N-j-1}. \tag{7.70}$$

While the left-hand side of Eq. (7.70) depends on the intrinsic geometric parameter γ_L, the right-hand side does not. However, when we approximate the sine term with its argument in Eq. (7.70), the dependence from γ_L completely disappears. In this way, we may conclude that the analysis made to determine the lower threshold value $\psi_{ex}^{(1)}$ can be taken, in good approximation, not to depend on the parameter γ_L. Moreover, the stationary equations (7.70) do not depend on α_L either, so that the value of $\psi_{ex}^{(1)}$ could be univocally defined for this class of networks, in the approximation set forth in the previous subsection. Inspection of Eq. (7.61) and a patient numerical analysis in determining a table similar to Table 7.1 in the case of close-packed arrangement of grains shows that this conclusion is indeed true. This fact makes the normalization of the externally applied flux in terms of ψ_{ex} rather convenient, since we may analyze the magnetic properties of networks with different intrinsic geometric properties, taking Table 7.1 as a reference point to distinguish between reversible and irreversible magnetic regimes of these rather simple models of granular superconducting systems. In fact, by determining the quantity $\psi_{ex}^{(1)}$, we may compute the lower threshold field for flux penetration inside the superconducting granular system.

We now turn to consider the outcome of numerical integration of Eq. (7.52) and Eq. (7.53a–b) for the normalized flux variables Ψ_k in the case of close-packed arrangement of the superconducting grains.

The ξ_k vs. k and the Ψ_k vs. k curves are reported in Fig. 7.25a–c and in Fig. 7.26a–c, respectively, for $\beta = 0.015$ and for values of the forcing term ψ_{ex} in the interval $[-0.25, +0.25]$. By referring to Table 7.1, we see that $\psi_{ex}^{(CP)} > \psi_{ex}^{(1)}$, being $0.107 < \psi_{ex}^{(1)} < 0.195$. In this way, we expect the normalized local magnetic field to be spread over the entire sample for $\psi_{ex} > \psi_{ex}^{(1)} \approx 0.11$. In fact, as we see in Fig. 7.25a, as soon as ψ_{ex} gets larger than $\psi_{ex}^{(1)}$ (see, for example, the red curve for which $\psi_{ex} = 0.15$) non-null values of ξ_k are detected over the entire sample ($1 \leq k \leq 20$).

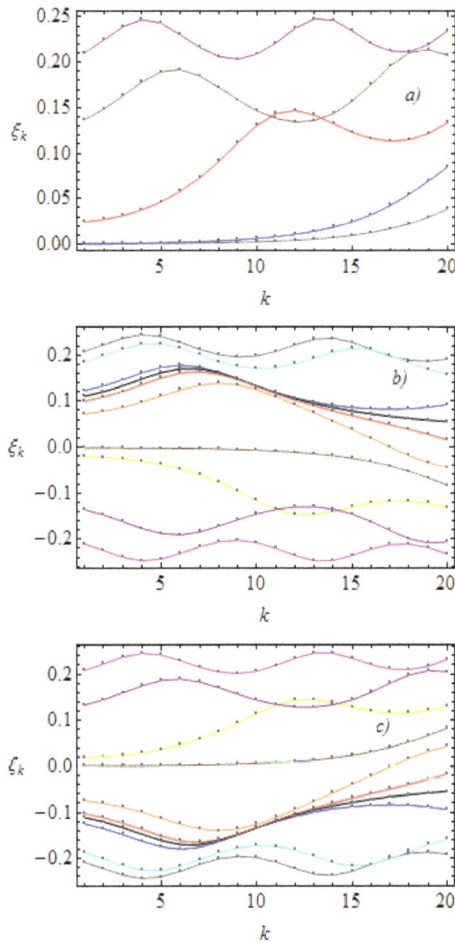

Fig. 7.25. Normalized magnetic field ξ_k in the shell model of a granular sample at $T = 0\ K$. The grains in the system are close-packed and $\beta = 0.015$. The index k ranges from 1 (inner shell) to 20 (external shell). The lines are used to distinguish the values of the normalized field values ξ_k (grey points) at various forcing terms ψ_{ex}. In panel a, the increasing ψ_{ex} values are: 0.05 (gray); 0.10 (blue); 0.15 (red); 0.20 (brown), 0.25 (purple). In panel b, the decreasing ψ_{ex} values are: 0.20 (gray); 0.15 (cyan); 0.10 (blue); 0.05 (black); 0.00 (red); -0.05 (orange); -0.10 (brown); -0.15 (yellow); -0.20 (purple); -0.25 (magenta). In panel c, the increasing ψ_{ex} values are: -0.20 (gray); -0.15 (cyan); -0.10 (blue); -0.05 (black); 0.00 (red); 0.05 (orange); 0.10 (brown); 0.15 (yellow); 0.20 (purple); 0.25 (magenta).

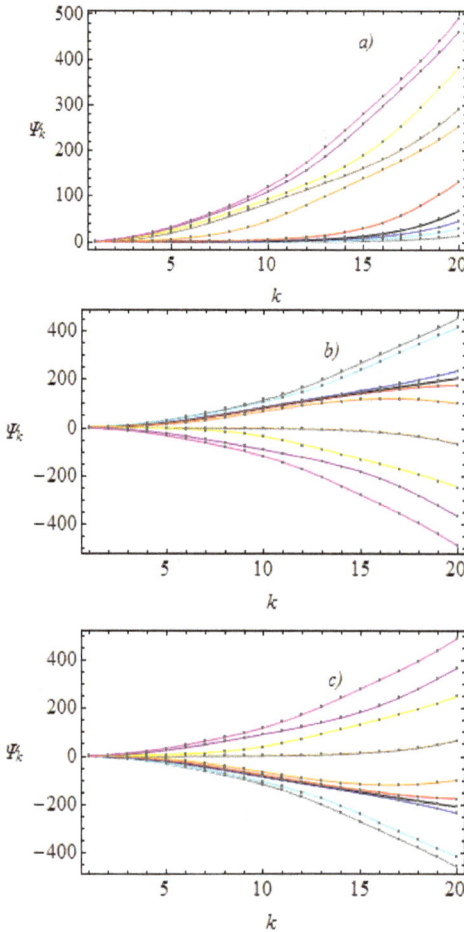

Fig. 7.26. Total flux Ψ_k within the k-th shell in the discrete model of a granular sample at $T = 0\,K$. The grains in the system are close-packed and $\beta = 0.015$. The index k ranges from 1 (inner shell) to 20 (external shell). The lines are used to distinguish the values of the normalized flux values Ψ_k (grey points) at various forcing terms ψ_{ex}. In panel a, the increasing ψ_{ex} values are: 0.025 (gray); 0.050 (cyan); 0.075 (blue); 0.100 (black); 0.125 (red); 0.150 (orange); 0.175 (brown); 0.200 (yellow); 0.225 (purple); 0.250 (magenta). In panel b, the decreasing ψ_{ex} values are: 0.20 (gray); 0.15 (cyan); 0.10 (blue); 0.05 (black); 0.00 (red); -0.05 (orange); -0.10 (brown); -0.15 (yellow); -0.20 (purple); -0.25 (magenta). In panel c, the increasing ψ_{ex} values are: -0.20 (gray); -0.15 (cyan); -0.10 (blue); -0.05 (black); 0.00 (red); 0.05 (orange); 0.10 (brown); 0.15 (yellow); 0.20 (purple); 0.25 (magenta).

As for Fig. 7.25b and 7.25c, we notice that the magnetic behavior is similar to what was already seen in Fig. 7.16b and 7.16c, respectively. The typical magnetic behavior already detected in Fig. 7.17a, 7.17b, and 7.17c, can also be observed in Fig. 7.26a, 7.26b, and 7.26c, respectively. Here we only notice how the curves tend to group together below and above the lower threshold value in Fig. 7.26a.

This type of behavior is still more pronounced if we consider the ξ_k vs. k and the Ψ_k vs. k curves reported in Fig. 7.27a–c and in Fig. 7.28a–c, respectively, for $\beta = 0.075$. In these figures the same range of the forcing term ψ_{ex} as in Fig. 7.25a–c and 7.26a–c is used. Let us start by considering the behavior of the system for ψ_{ex} values increasing from zero to $\psi_{ex}^{(CP)} = 0.25$. By interpolating the values of the lower threshold field reported in Fig. 7.20, for $\beta = 0.075$ we obtain the following estimate: $\psi_{ex}^{(1)} \approx 0.22$. In this way, as also arguable from Fig. 7.27a, only the purple curve, obtained for $\psi_{ex} = 0.25$, shows a flux penetrated region well inside the sample. When the applied field is made to decrease, the bundle of fluxons close to the outer boundary translates toward the center of the sample, reaching a rather stable configuration, even when ψ_{ex} reaches negative values. In fact, for $\psi_{ex} = -0.20$, a region close to the outer border, in which negative values of ξ_k occur, and the inner region of the sample with the same bundle of fluxons are clearly shown by the purple curve in Fig. 7.27b. Because of the stability of the inner fluxons, the outer part of the distribution does not vary appreciably when the normalized externally applied flux values are lowered from $\psi_{ex} = 0.20$ to $\psi_{ex} = -0.15$, giving rise to the folding- fan feature already detected in Fig. 7.27a. The bunch of fluxons close to the center of the sample disappears at $\psi_{ex} = -0.25$, as shown by the magenta curve in Fig. 7.27b. For this value of ψ_{ex}, only a bunch of anti-fluxons, close to the outer borders, are detectable. Finally, when the applied field is again increasing, from Fig. 7.27c we may see that the bundle of anti-fluxons close to the outer boundary translates toward the center and a second stable configuration, symmetric to the one in Fig. 7.27b, is reached. Because of the stability of this configuration, a second "folding fan", symmetric to the first, can be noticed in the curves obtained for values of the normalized applied flux going from $\psi_{ex} = -0.20$ to $\psi_{ex} = +0.15$.

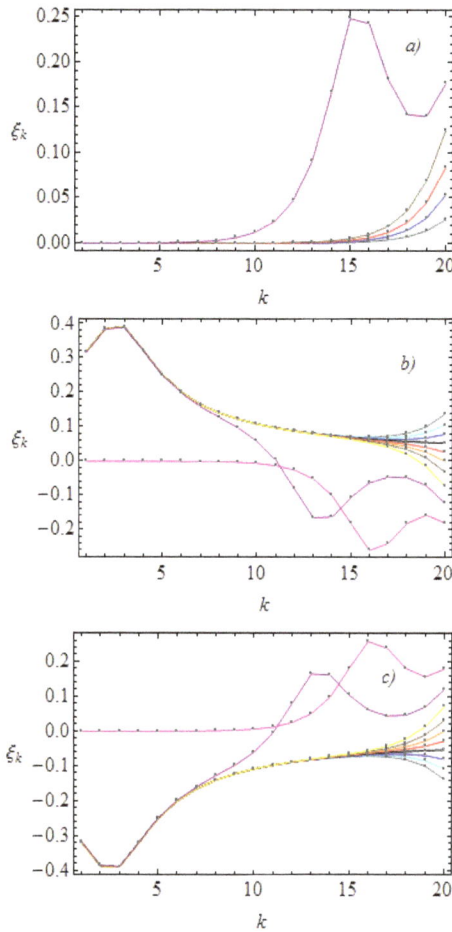

Fig. 7.27. Normalized magnetic field ξ_k in the shell model of a granular sample at $T = 0\ K$. The grains in the system are close-packed and $\beta = 0.075$. The index k ranges from 1 (inner shell) to 20 (external shell). The lines are used to distinguish the values of the normalized field values ξ_k (grey points) at various forcing terms ψ_{ex}. In panel a, the increasing ψ_{ex} values are: 0.05 (gray); 0.10 (blue); 0.15 (red); 0.20 (brown), 0.25 (purple). In panel b, the decreasing ψ_{ex} values are: 0.20 (gray); 0.15 (cyan); 0.10 (blue); 0.05 (black); 0.00 (red); -0.05 (orange); -0.10 (brown); -0.15 (yellow); -0.20 (purple); -0.25 (magenta). In panel c, the increasing ψ_{ex} values are: -0.20 (gray); -0.15 (cyan); -0.10 (blue); -0.05 (black); 0.00 (red); 0.05 (orange); 0.10 (brown); 0.15 (yellow); 0.20 (purple); 0.25 (magenta).

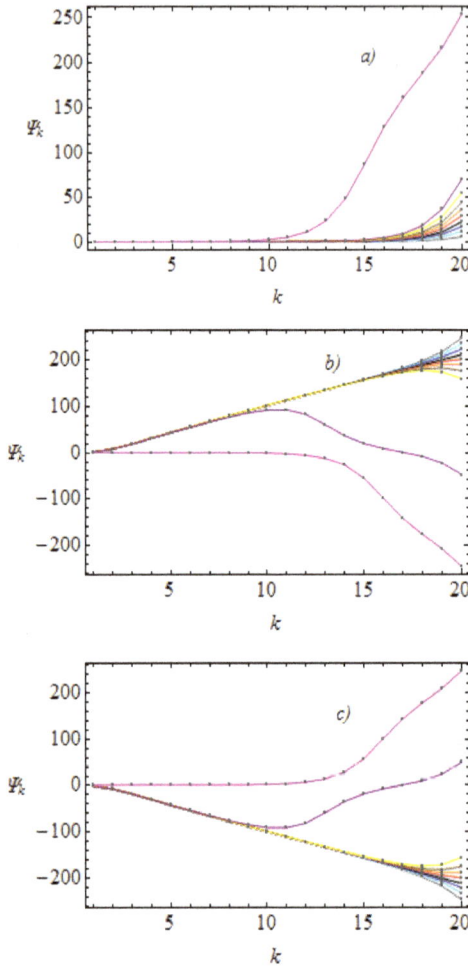

Fig. 7.28. Total flux Ψ_k within the k-th shell in the discrete model of a granular sample at $T = 0\,K$. The grains in the system are close-packed and $\beta = 0.075$. The index k ranges from 1 (inner shell) to 20 (external shell). The lines are used to distinguish the values of the normalized flux values Ψ_k (grey points) at various forcing terms ψ_{ex}. In panel a, the increasing ψ_{ex} values are: 0.025 (gray); 0.050 (cyan); 0.075 (blue); 0.100 (black); 0.125 (red); 0.150 (orange); 0.175 (brown); 0.200 (yellow); 0.225 (purple); 0.250 (magenta). In panel b, the decreasing ψ_{ex} values are: 0.20 (gray); 0.15 (cyan); 0.10 (blue); 0.05 (black); 0.00 (red); -0.05 (orange); -0.10 (brown); -0.15 (yellow); -0.20 (purple); -0.25 (magenta). In panel c, the increasing ψ_{ex} values are: -0.20 (gray); -0.15 (cyan); -0.10 (blue); -0.05 (black); 0.00 (red); 0.05 (orange); 0.10 (brown); 0.15 (yellow); 0.20 (purple); 0.25 (magenta).

The folding-fan feature disappears for $\psi_{ex} = +0.20$ (purple curve of Fig. 7.27c). At the latter value of ψ_{ex} a bunch of fluxons appears close to the outer border, while the inner conglomerate of anti-fluxons is still visible in the purple curve. The latter bundle disappears at $\psi_{ex} = +0.25$ (magenta curve of Fig. 7.27c), leaving only a conglomerate of fluxons close to outer border of the system, similar to that in the magenta curve of Fig. 7.27a.

Because of this characteristic behavior, the Ψ_k vs. k curves in Fig. 7.28a show a more pronounced tendency to grouping together for value of ψ_{ex} below and above $\psi_{ex}^{(1)} \approx 0.22$ than the corresponding curves obtained for $\beta = 0.015$ in Fig. 7.26a. The folding-fan feature is also detectable in the Ψ_k vs. k curves in Fig. 7.28b and 7.28c. This type of grouping, however, is not an exclusive character of close-packed superconducting granular systems, since we can detect a similar, although less evident, "folding fan", made by only three curves, in Fig. 7.23b–c. In fact, when we choose to show distributions for values of ψ_{ex} in the vicinity of the threshold $\psi_{ex}^{(1)}$, folding-fan features naturally arise, irrespectively of the particular inner geometry of the granular system under exam.

We conclude our analysis on discrete models of Josephson junction networks of granular superconductors by analyzing the magnetization curves for systems consisting of cylindrical grains in a close-packed arrangement for the same range of values of ψ_{ex} as in Fig. 7.27a–c and 7.28a–c. Therefore, in Fig. 7.29a–c magnetization cycles obtained for a system made of $N = 20$ concentric superconducting shells at $T = 0\,K$ are shown. We can interpret the difference between the cycles in Fig. 7.29a, for which $\beta = 0.015$, and the one in Fig. 7.29b, for which $\beta = 0.075$, by referring to the normalized magnetic field distributions in Fig. 7.25a–c and 7.27ac. In fact, in Fig. 7.25a–c we notice that, apart from an interval of ψ_{ex} close to the origin, in which flux exclusion or expulsion from the sample is more effective, non-null values of ξ_k are present inside the whole system. In this way, for values of ψ_{ex} far from the origin, the magnetization curve approaches an asymptotic behavior, whose slope is approximately given by Eq. (7.67), when α_S is replaced with α_{CP}.

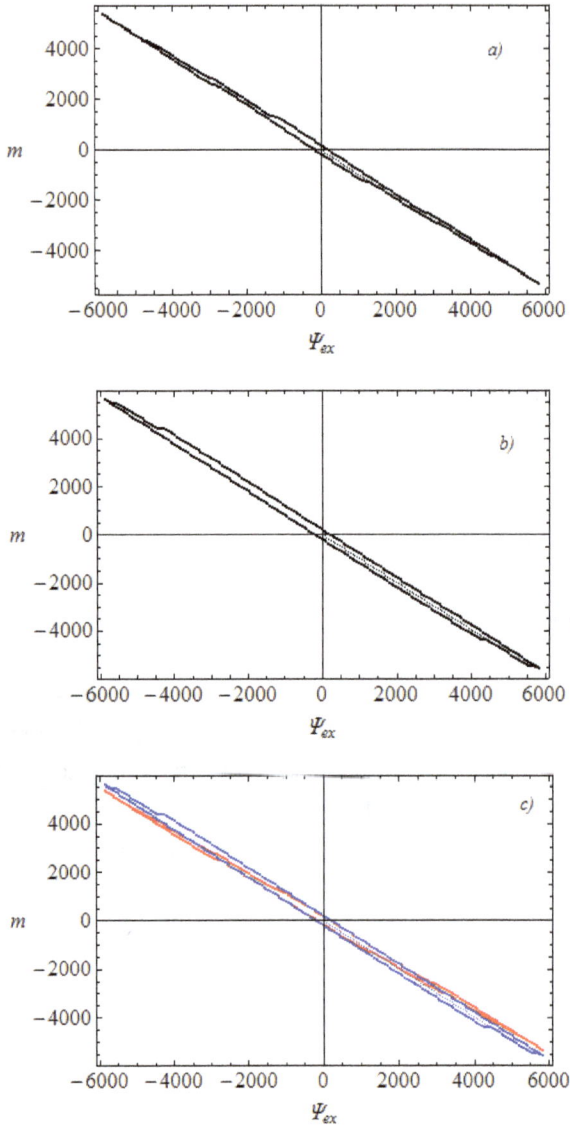

Fig. 7.29. Magnetization curves, at $T = 0\ K$, of granular systems in which grains are in a close-packed arrangement. The magnetization curve in panel a is obtained for $\beta = 0.015$, while the curve in panel b is for $\beta = 0.075$. First magnetization portions are shown as dotted lines in both curves. In panel c the two curves in a and b are reported on the same frame, the first in red, the second in blue, in order to detect their relative size and their different slope with respect to the ψ_{ex}-axis.

Therefore, for large values of ψ_{ex}, the grains, being considered strict diamagnetic entities, are mainly responsible for the response of the whole system, whose magnetization approaches the simple expression $m \approx -[1 - \alpha_{eff}(\beta)]\Psi_{ex}$, where $\alpha_{eff}(\beta)$ is an effective normal ratio, which depends on the parameter β. In fact, in order to qualitatively explain the dependence on β of the slope $a(\beta) = -[1 - \alpha_{eff}(\beta)]$ of the symmetry curve of the magnetization cycle, we had derived Eq. (7.67) in the case of cylindrical grains in a square-lattice arrangement.

By replacing α_S with α_{CP} in Eq. (7.67), we obtain the same type of behavior also for a system of weakly coupled superconducting cylindrical grains in a close-packed arrangement. Therefore, in Fig. 7.29b the magnetization curve for this class of superconductors is shown for $\beta = 0.075$. When the latter curve is compared with the magnetization cycle shown in Fig. 7.29a, as it is done in Fig. 7.29c, we notice that Eq. (7.67), predicting that the slope $|a(\beta)|$ increases for increasing values of β, can be used to make useful qualitative estimates also in the case of granular superconducting systems consisting of cylindrical grains in a close-packed arrangement.

The study of the temperature dependence of the magnetization curves and of the zero-field-cooled and field-cooled susceptibility for these systems goes beyond the scope of the present introductory text. However, the patient reader can well try to apply the ideas in Section 7.2 to derive these magnetic properties, not without a great deal of attention.

7.4. Concentric shell model in the continuous limit

In the present section we derive, as done for a bare network of concentric circular arrays of Josephson junction in Section 6.5, the dynamic equation for a very large number of concentric shells in the dressed discrete model described in the previous sections of the present chapter. Let us then start by taking the continuous limit of Eq. (7.51), obtaining:

$$\frac{\partial \Psi(\rho,\tau)}{\partial \tau} + n(\rho)\beta \sin\left(\frac{2\pi\Psi(\rho,\tau)}{n(\rho)}\right) = n(\rho)\frac{\partial \xi(\rho,\tau)}{\partial \rho}, \qquad (7.71)$$

where the variable ρ is a normalized distance, replacing the variable k, as already seen in Section 6.5. Looking at Eq. (7.43), we argue that the quantity $n(\rho)$ is given by

$$n(\rho) = \gamma_L(2\rho - 1). \tag{7.72}$$

Notice how Eq. (7.72) is similar to Eq. (6.61), written for a bare network. In addition, according to Eq. (7.54), the time-dependent normalized local field $\xi(\rho, \tau)$ is given by:

$$\xi(\rho, \tau) = \frac{1}{n(\rho)} \frac{\partial \Psi(\rho, \tau)}{\partial \rho}. \tag{7.73}$$

Finally, according to Eq. (7.48), the normalized effective current flowing in the junctions at distance ρ from the center of the cylindrical sample, is

$$i_{N-j}^{(B)} = -\frac{1}{\beta} \frac{\partial \xi(\rho, \tau)}{\partial \rho}. \tag{7.74}$$

Having defined all relevant quantities in the continuum limit, we may rewrite the dynamic equations as follows:

$$\frac{\partial \Psi(\rho, \tau)}{\partial \tau} + (2\rho - 1)\gamma_L \beta \sin\left(\frac{2\pi\Psi(\rho, \tau)}{\gamma_L(2\rho - 1)}\right) = (2\rho - 1)\frac{\partial}{\partial \rho}\left[\frac{1}{2\rho-1}\frac{\partial\Psi(\rho, \tau)}{\partial \rho}\right], \tag{7.75}$$

where ρ ranges in the interval $[1, \rho_{max}]$, ρ_{max} being the normalized radius of the sample. In order to simplify the notation of the above expression, let us set:

$$x = \gamma_L(2\rho - 1). \tag{7.76}$$

In this way, Eq. (7.75) will finally read:

$$\frac{\partial \Psi(x, \tau)}{\partial \tau} + \beta x \sin\left(\frac{2\pi\Psi(x, \tau)}{x}\right) = 4\gamma_L^2 x \frac{\partial}{\partial x}\left[\frac{1}{x}\frac{\partial\Psi(x, \tau)}{\partial x}\right], \tag{7.77}$$

where x ranges in the interval $[\gamma_L, x_{max}]$, with $x_{max} = \gamma_L(2\rho_{max} - 1)$. The boundary conditions can be determined by setting the normalized total magnetic flux equal to zero at the center of the sample ($\rho = 1$ in this case) and by expressing Eq. (7.54) in the continuous limit, so that we write:

$$\Psi(\gamma_L, \tau) = 0, \tag{7.78a}$$

$$\frac{\partial\Psi(x,\tau)}{\partial x}\bigg|_{x=x_{max}} = \frac{x_{max}}{2\gamma_L}\psi_{ex}. \tag{7.78b}$$

The initial condition is the following zero-field-cooling conditions:

$$\Psi(x,0) = 0. \tag{7.79}$$

In order to verify Eq. (7.78b), we recall Eq. (7.54) and write:

$$\frac{1}{n(\rho_{max})}\frac{\partial\Psi(\rho,\tau)}{\partial\rho}\bigg|_{\rho=\rho_{max}} = \psi_{ex}. \tag{7.80}$$

By Eq. (7.72), we may set $n(\rho_{max}) = \gamma_L(2\rho_{max} - 1)$, so that, we rewrite Eq. (7.80) as follows:

$$\frac{\partial\Psi(\rho,\tau)}{\partial\rho}\bigg|_{\rho=\rho_{max}} = \gamma_L(2\rho_{max} - 1)\psi_{ex}. \tag{7.81}$$

Finally, by definition of x in Eq. (7.76) and by the chain rule, we finally find Eq. (7.78b).

Let us now notice that, apart from the multiplicative constant factor on the right-hand side $4\gamma_L^2$ and on the boundary conditions 7.77a–b, Eq. (7.78) is equivalent to Eq. (6.64) in Chapter 6. The diligent reader can thus carry out the analysis for small β values, similarly to what already done in Chapter 6 for the continuous concentric circular array model. Here we skip the analytic part, already treated in detail in Section 6.5.1, and go directly to approach the problem numerically in order to obtain the magnetic field distributions and the magnetization curves for various values of the parameter β.

In Fig. 7.30a–c and in Fig. 7.31a–c, the normalized magnetic field distributions and the Ψ vs. x curves, respectively, are shown for systems with square-lattice arrangement of grains for $\beta = 0.10$. These curves are obtained by numerical integration of Eq. (7.77) and are shown for normalized applied fluxes first increasing from $\psi_{ex} = 0.0$ to $\psi_{ex} = +0.30$, then decreasing to $\psi_{ex} = -0.30$, finally increasing again to $\psi_{ex} = +0.30$.

In particular, in Fig. 7.30a we notice the characteristic slowly-varying behavior of the functions $\xi(x)$ at various normalized applied fluxes.

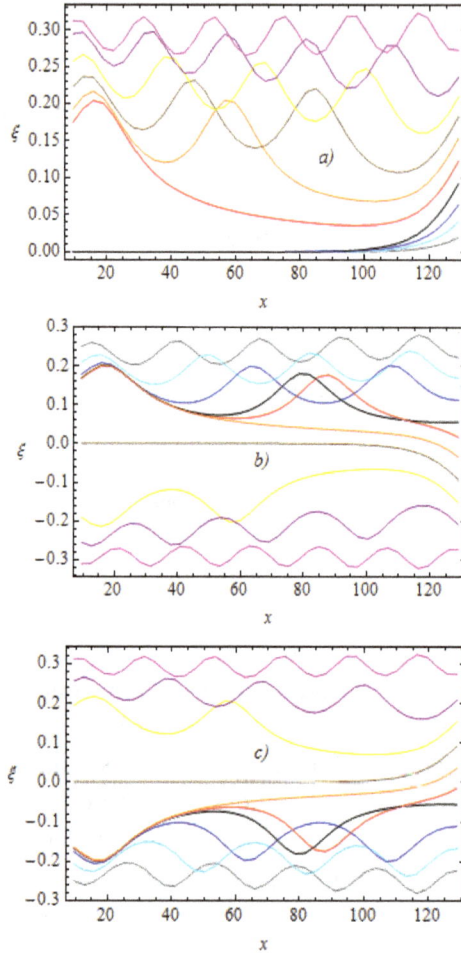

Fig. 7.30. Normalized magnetic field ξ in the shell model of a granular sample at $T = 0\,K$. The grains in the system are in a square-lattice arrangement and $\beta = 0.10$. The variable x ranges from γ_S (inner shell) to $41\gamma_S$ (external shell). In panel a, the increasing ψ_{ex} values are: 0.03 (gray); 0.06 (cyan); 0.09 (blue); 0.12 (black); 0.15 (red); 0.18 (orange); 0.21 (brown); 0.24 (yellow); 0.27 (purple); 0.30 (magenta). In panel b, the decreasing ψ_{ex} values are: 0.24 (gray); 0.18 (cyan); 0.12 (blue); 0.06 (black); 0.00 (red); -0.06 (orange); -0.12 (brown); -0.18 (yellow); -0.24 (purple); -0.30 (magenta). In panel c, the increasing ψ_{ex} values are: -0.24 (gray); -0.18 (cyan); -0.12 (blue); -0.06 (black); -0.00 (red); $+0.06$ (orange); $+0.12$ (brown); $+0.18$ (yellow); $+0.24$ (purple); $+0.30$ (magenta).

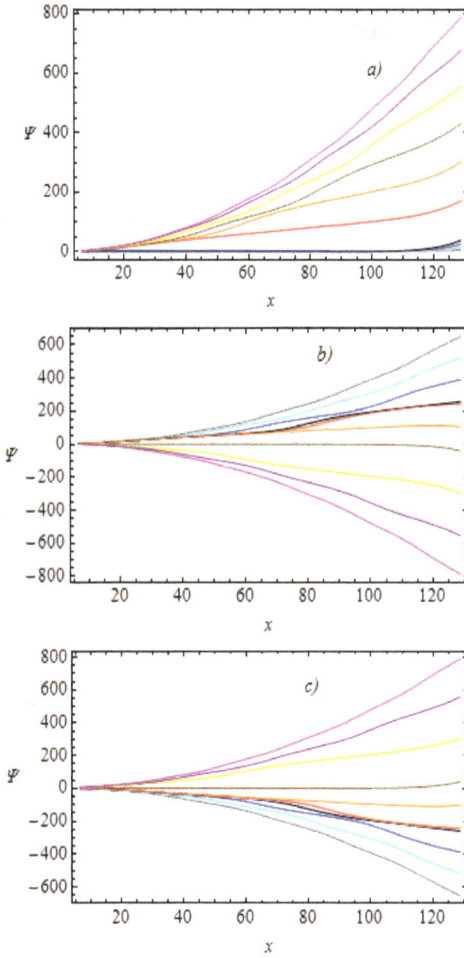

Fig. 7.31. Total flux Ψ in the discrete model of a granular sample at $T = 0\ K$. The grains in the system are in a square-lattice arrangement and $\beta = 0.10$. The variable x ranges from γ_S (inner shell) to $41\ \gamma_S$ (external shell). The values of the normalized flux values Ψ (grey points) at various forcing terms ψ_{ex}. In panel a, the increasing ψ_{ex} values are: 0.03 (gray); 0.06 (cyan); 0.09 (blue); 0.12 (black); 0.15 (red); 0.18 (orange); 0.21 (brown); 0.24 (yellow); 0.27 (purple); 0.30 (magenta). In panel b, the decreasing ψ_{ex} values are: 0.24 (gray); 0.18 (cyan); 0.12 (blue); 0.06 (black); 0.00 (red); -0.06 (orange); -0.12 (brown); -0.18 (yellow); -0.24 (purple); -0.30 (magenta). In panel c, the increasing ψ_{ex} values are: -0.24 (gray); -0.18 (cyan); -0.12 (blue); -0.06 (black); -0.00 (red); $+0.06$ (orange); $+0.12$ (brown); $+0.18$ (yellow); $+0.24$ (purple); $+0.30$ (magenta).

The grouping of these curves is indicative, as in discrete models, of the crossing of ψ_{ex} through the lower threshold field value. Notice also that, after first irreversible penetration, the number of peaks and their height in the $\xi(x)$ distributions grow for increasing values of ψ_{ex}. Furthermore, we might also notice how, for $\psi_{ex} > \psi_{ex}^{(1)}$ and for this particular value of β, fluxons in the system start occupying the innermost portion of the sample. Let us here briefly justify this apparently unusual behavior, different with respect to what seen in the discrete model. In fact, in Fig. 7.16a and 7.18a for discrete systems with a square-lattice arrangement and with $\beta = 0.05$ and $\beta = 0.25$, respectively, we have noticed that fluxons entering the system allocate close to the outer boundary. In this case, the inter-granular sites act as effective pinning centers, blocking flux quanta to propagate far inside the sample. With a continuous distribution of inter-granular sites inside the system, however, the energy configuration of the system of concentric circular shells for small values of β can be modified by small variations of ψ_{ex}. In this way, flux motion takes place inside the system at relatively low values of ψ_{ex} and the repulsive energy between the inner fluxons and the externally applied field can be minimized by maximizing the distance between the mobile flux bundle and the outer boundary. Nevertheless, the lower threshold field $\psi_{ex}^{(1)}$ maintains the characteristic monotonic increasing dependence on β.

Going back to Fig. 7.30a, we notice that further increase of the externally applied field will only introduce more and more fluxons inside the system, which dispose inside the sample, following the regular scheme indicated in Fig. 7.30a. Similarly smooth and slowly-varying curves are obtained in Fig. 7.30b–c.

In Fig. 7.31a the characteristic monotonically increasing behavior of the solutions $\Psi(x)$ is again detected. The grouping of the Ψ vs. x curves is also evident in Fig. 7.31a, giving rise to the "folding fan" structure for $\psi_{ex} < \psi_{ex}^{(1)}$ also detected in the discrete case. In Fig. 7.31a–c the solutions $\Psi(x)$ are seen to strictly obey the boundary conditions (7.77a–b), thus going to zero at $x = \gamma_S$ and showing an increasing absolute value of the derivative at $x = x_{max}$.

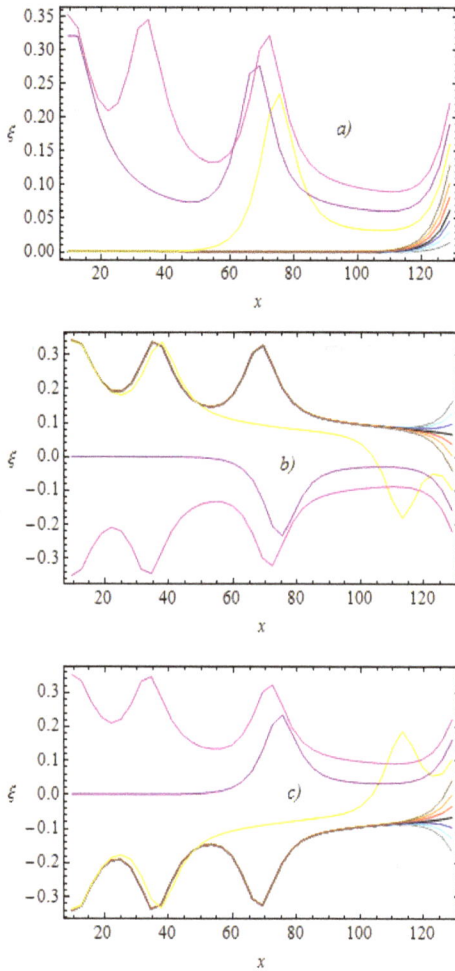

Fig. 7.32. Normalized magnetic field ξ in the shell model of a granular sample at $T = 0\,K$. The grains in the system are in a square-lattice arrangement and $\beta = 0.35$. The variable x ranges from γ_S (inner shell) to $41\gamma_S$ (external shell). In panel a, the increasing ψ_{ex} values are: 0.03 (gray); 0.06 (cyan); 0.09 (blue); 0.12 (black); 0.15 (red); 0.18 (orange); 0.21 (brown); 0.24 (yellow); 0.27 (purple); 0.30 (magenta). In panel b, the decreasing ψ_{ex} values are: 0.24 (gray); 0.18 (cyan); 0.12 (blue); 0.06 (black); 0.00 (red); -0.06 (orange); -0.12 (brown); -0.18 (yellow); -0.24 (purple); -0.30 (magenta). In panel c, the increasing ψ_{ex} values are: -0.24 (gray); -0.18 (cyan); -0.12 (blue); -0.06 (black); -0.00 (red); $+0.06$ (orange); $+0.12$ (brown); $+0.18$ (yellow); $+0.24$ (purple); $+0.30$ (magenta).

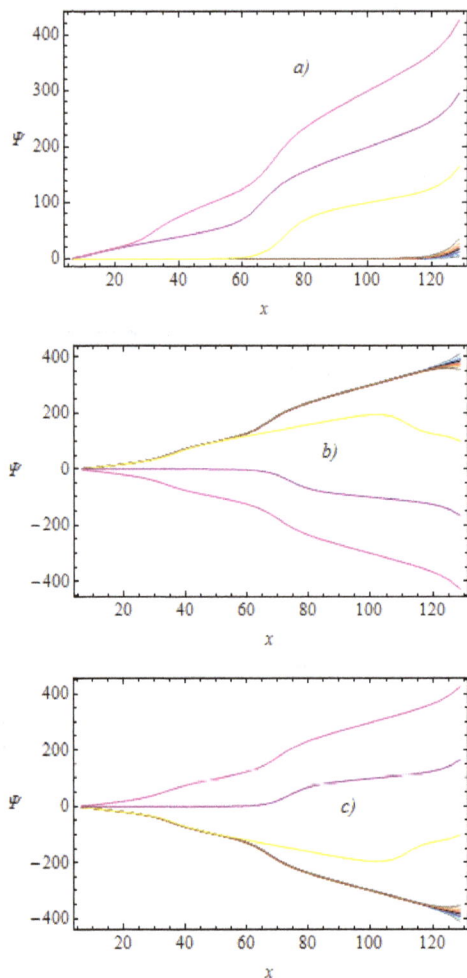

Fig. 7.33. Total flux Ψ in the discrete model of a granular sample at $T = 0\ K$. The grains in the system are in a square-lattice arrangement and $\beta = 0.35$. The variable x ranges from γ_S (inner shell) to $41\,\gamma_S$ (external shell). The values of the normalized flux values Ψ (grey points) at various forcing terms ψ_{ex}. In panel a, the increasing ψ_{ex} values are: 0.03 (gray); 0.06 (cyan); 0.09 (blue); 0.12 (black); 0.15 (red); 0.18 (orange); 0.21 (brown); 0.24 (yellow); 0.27 (purple); 0.30 (magenta). In panel b, the decreasing ψ_{ex} values are: 0.24 (gray); 0.18 (cyan); 0.12 (blue); 0.06 (black); 0.00 (red); -0.06 (orange); -0.12 (brown); -0.18 (yellow); -0.24 (purple); -0.30 (magenta). In panel c, the increasing ψ_{ex} values are: -0.24 (gray); -0.18 (cyan); -0.12 (blue); -0.06 (black); -0.00 (red); $+0.06$ (orange); $+0.12$ (brown); $+0.18$ (yellow); $+0.24$ (purple); $+0.30$ (magenta).

By increasing the value of the parameter β, in Fig. 7.32a-c and in Fig. 7.33a–c the normalized magnetic field distributions and the Ψ vs. x curves, respectively, are shown for systems with square-lattice arrangement of grains for $\beta = 0.30$. These curves are again shown for normalized applied fluxes first increasing from $\psi_{ex} = 0.0$ to $\psi_{ex} = +0.30$, then decreasing to $\psi_{ex} = -0.30$, finally increasing again to $\psi_{ex} = +0.30$.

In particular, in Fig. 7.32a we notice a noteworthy difference with respect to Fig. 7.30a. In fact, in this case, with a higher value of the parameter β, the first allocation of the bunch of fluxons entering the system after first irreversible penetration is visible in the middle of the sample. Further increase of the field push fluxons in the inner part of the sample. The penetration mechanisms are thus different from those described in Fig. 7.30a for $\beta = 0.10$. When we let the normalized externally applied magnetic flux values go from $+30$ to -30 in Fig. 7.32b, we again notice the characteristic "folding fan" feature. The same feature can be detected also in Fig. 7.32c. It is also important to notice the concomitant presence of fluxons and anti-fluxons, inside the system, as shown in the yellow curves of Fig. 7.32b and 7.32c. Similarly to the "folding fan" feature, this additional characteristic magnetic behavior is indicative of the capacity of the system as a whole to retain, in its inner part, magnetic flux lines. These features are all recognizable also in Fig. 7.33a–c. In particular, the "folding fan" is visible in the upper part of Fig. 7.33b and in the lower portion of Fig. 7.33c. The yellow curves in Fig. 7.33b–c do present a stiffness in the inner part. In fact, even though the most external portion follows the increase of the externally applied flux, the inner part shows an opposite slope, coherently with the opposite value of the normalized local magnetic field.

For the same values of the parameter β and for the same field range utilized above, in Fig. 7.34a–c we show the normalized magnetization, as defined in Eq. (7.63). In particular, in Fig. 7.34a the magnetization cycle for $\beta = 0.10$ is shown. The first magnetization curve is represented by a dashed line. In Fig. 7.34b the magnetization cycle for $\beta = 0.30$ is shown. In Fig. 7.34c the two curves reported in panel a and panel b are shown on the same plot, in order to detect similarity and differences between the two magnetization cycles.

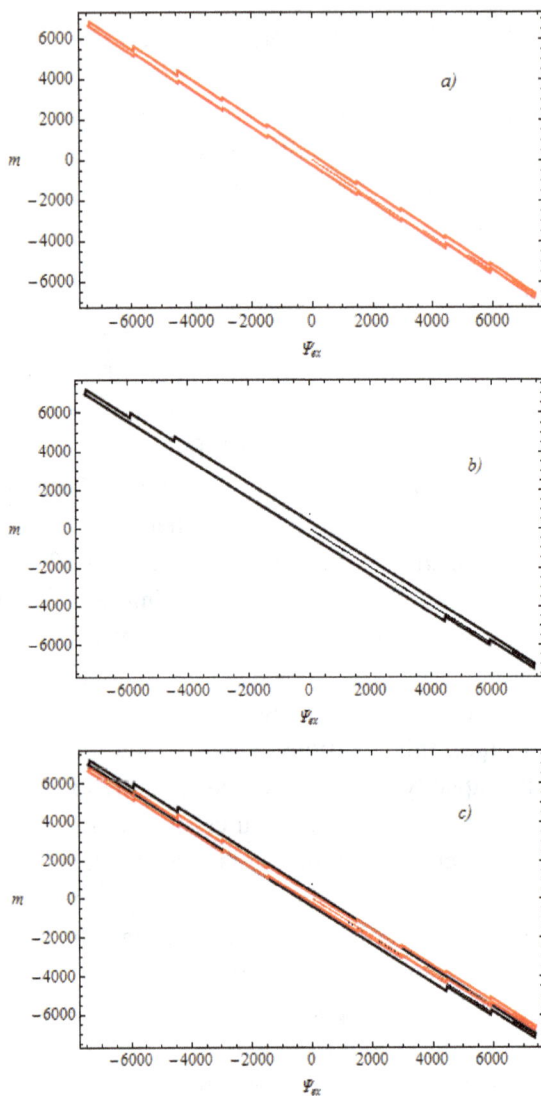

Fig. 7.34. Magnetization curves, at $T = 0\ K$, of granular systems in which grains are in a square-lattice arrangement. The magnetization curve in panel a is obtained for $\beta = 0.10$, while the curve in panel b is for $\beta = 0.35$. First magnetization portions are shown as dotted lines in both curves. In panel c the two curves in a and b are reported on the same frame, the first in red, the second in black, in order to detect their relative size and their different slope with respect to the Ψ_{ex}-axis.

Both magnetization curves present a sharper shape, rather than the rounded appearance of the magnetization cycles shown in Fig. 7. 21 and 7.22, obtained for the discrete version of the concentric circular shells model. On the basis of what previously noticed, while in the discrete circular concentric shells model, at least for not very small value of the parameter β, the inter-granular regions show a rather marked tendency to pinning flux lines, the same is not true for the continuous version of the model, where the properties of the system as a whole come into play.

Conversely, when the curves of the concentric circular shells model are compared with the corresponding ones in Chapter 6, for similar values of β as in Fig. 6.25 ($\beta = 0.15$ and $\beta = 0.20$) for the continuous version of the concentric circular array model, we notice that the diamond-like shape is completely lost. This is due to the preponderant magnetic role of the diamagnetic grains, which are completely absent in the bare models described in Chapter 6.

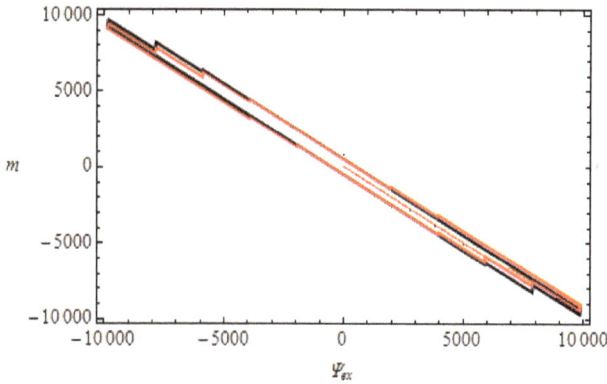

Fig. 7.35. Magnetization curves, at $T = 0\ K$, of granular systems in which grains are in a square-lattice arrangement. The magnetization cycle in red is obtained for $\beta = 0.35$, while the cycle in black is obtained for $\beta = 0.55$. First magnetization portions are shown as dotted lines in both cycles.

We also remark that the characteristic feature of the increasing slope of the symmetry line of the magnetization cycles with the increasing value of β is clearly visible in Fig. 7.34c, in which the curve for $\beta = 0.10$ is reported in red and the curve for $\beta = 0.30$ in black. The symmetry line of the black curve is seen to possess a higher slope when compared to the red curve, as predicted by Eq. (7.67) and represented in Fig. 7.34c. As already noted in Section 7.3.1, in fact, the slope of the symmetry lines increases monotonically with the parameter β.

Finally, in Fig. 7.35 magnetization cycles for $\beta = 0.35$ (in red) and for $\beta = 0.55$ (in black) are shown. Once again the symmetry line of the black curve, with higher value of β, has a higher slope when compared to the red curve, with lower value of β, confirming once more the validity of the analysis leading to Eq. (7.67).

Chapter 8

Three-dimensional Networks of Josephson Junctions

8.1. Cubic networks

In Chapter 6, the dynamic equations of Josephson junctions in two-dimensional (2D) networks were written in a rather compact form. It is therefore the object of the present Chapter: the definition and study of the equations of motion of the phase differences across over-damped Josephson junctions in a three-dimensional (3D) network. As in Chapter 6, we do not consider 3D networks generated by 2D elementary cells different from squares.

From the analysis of the Josephson junction dynamics, the magnetic properties of these systems will be investigated. Therefore we proceed as in previous chapters, first defining the constitutive equations for the network and next studying some specific significant cases. Naturally, the difficulty in analyzing these systems will increase for obvious reasons.

Let us then start by considering a two-dimensional planar element of a 3D Josephson junction network (3DJJN), as shown in Fig. 8.1 and as already seen in Fig. 6.1. This planar portion of the network is taken to lie in the $x - y$ plane, so that we label the flux and current in Fig. 8.1 as $\Phi_z(\vec{r})$ and $I_z(\vec{r})$, respectively. The currents $I_\xi(\vec{r})$ are taken to be positive according to the right-hand rule, as applied by an observer whose right-hand thumb points along the positive ξ-axis ($\xi = x, y, z$). By neglecting mutual inductance coefficients and utilizing the constitutive equations (6.7a-c) for a 2DJJN, we may write in the order, the fluxoid quantization condition, the flux-current relation, and the dynamic equations for the superconducting phase difference in the planar element of Fig. 8.1 as follows:

$$2\pi\psi_z(\vec{r}, t) + \Delta_x\phi_y(\vec{r}, t) - \Delta_y\phi_x(\vec{r}, t) = 2\pi n_z(\vec{r}), \qquad (8.1a)$$

$$\psi_z(\vec{r}, t) = \hat{z} \cdot \vec{\psi}_{ex}(\vec{r}) + \beta i_z(\vec{r}, t), \qquad (8.1b)$$

$$\frac{\Phi_0}{2\pi R I_{J0}} \frac{d\phi_\xi(\vec{r}, t)}{dt} + \sin\phi_\xi(\vec{r}, t) = \varepsilon_{\xi\eta z}\, \Delta_{-\eta} i_z(\vec{r}, t), \qquad (8.1c)$$

where ξ and η can be either x or y, where the forward and backward difference operators Δ_η and $\Delta_{-\eta}$ are defined in Eq. (6.6a) and (6.6b), respectively, and where the external magnetic field $\vec{H}$ is considered to be applied in an arbitrary direction in space so that $\vec{\psi}_{ex}(\vec{r}) = \mu_0\vec{H}A_0/\Phi_0$, A_0 being the area of the planar element. In the definition of $\vec{\psi}_{ex}$ we keep the dependence on $\vec{r}$, despite the uniformity of $\vec{H}$, for reasons which will be clear later. In Eq. (8.1b), the characteristic parameter of the loop is $\beta = 4L_0 I_{J0}/\Phi_0$. Finally, in all the above equations, the currents and fluxes are normalized, so that: $\psi_\eta(\vec{r}) = \Phi_\eta(\vec{r}, t)/\Phi_0$ and $i_\eta(\vec{r}, t) = I_\eta(\vec{r}, t)/I_{J0}$.

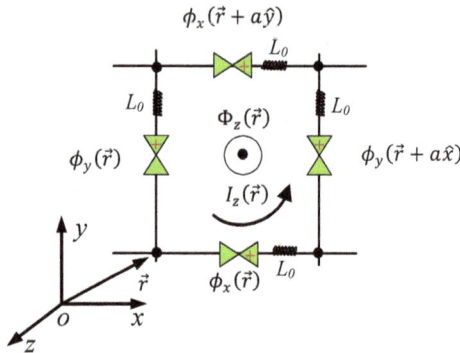

Fig. 8.1. Schematic representation of the stationary (time-independent) physical quantities associated to a planar element of a three-dimensional Josephson network. The lower left vertex of the cell is at a position $\vec{r} = a(i\hat{x} + j\hat{y} + k\hat{z})$, a being the length of the cell side and i, j, k integers. An inductance $4L_0$ is associated to each planar loop of the network.

As it is clear from Eq. (8.1b), only self-inductance coefficients are considered, while mutual inductance coefficients are neglected, because of the hypothesis set forth in this analysis. Studies of two-dimensional or three-dimensional Josephson junction networks which take account of magnetic interaction between loop currents are present in the literature (De Leo, 2001) (De Luca R. D., 1998). A brief account of the consequences of the introduction of mutual inductance coefficients will be considered in the last part of the present Chapter.

In order to generalize Eq. (8.1a–c) to three dimensions, let us consider the labeling of the currents in Fig. 8.2. Having already defined the positive direction for the circulating currents, we may specify, by referring to Fig. 8.2, where stationary currents are represented for simplicity, the way the labeling works.

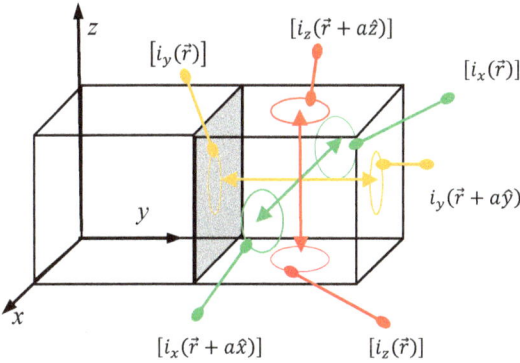

Fig. 8.2. Labeling of stationary currents in the three-dimensional Josephson junction network. The currents are indicated as colored circles: in green $i_x(\vec{r})$ and $i_x(\vec{r} + a\hat{x})$; in yellow $i_y(\vec{r})$ and $i_y(\vec{r} + a\hat{y})$; in red $i_z(\vec{r})$ and $i_z(\vec{r} + a\hat{z})$. A similar type of labeling is adopted for fluxes threading each face of the cubic structure of the network.

First of all, the lower index indicates the direction orthogonal to the plane on which the current circulates. Therefore, the stationary currents $i_x(\vec{r})$ and $i_x(\vec{r} + a\hat{x})$ (green circles in Fig. 8.2) circulate on the $y - z$ plane, the stationary currents $i_y(\vec{r})$ and $i_y(\vec{r} + a\hat{y})$ (yellow circles in Fig. 8.2) circulate on the $x - z$ plane, and the stationary currents $i_z(\vec{r})$ and $i_z(\vec{r} + a\hat{z})$ (red circles in Fig. 8.2) circulate on the $y - z$ plane. A similar labeling can be adopted for the stationary magnetic flux threading each loop in the network. Time-dependent quantities are simply denoted by showing the explicit time dependence, as, for example, in $i_\xi(\vec{r}, t)$.

Before generalizing Eq. (8.1a–c) for planar elements lying in the $x - z$ and in the $y - z$ planes, let us rewrite Eq. (8.1a) as follows:

$$2\pi\psi_z(\vec{r}) + \left[\vec{\Delta}^{(+)} \times \vec{\phi}(\vec{r})\right]_z = 2\pi n_z(\vec{r}), \qquad (8.2)$$

where:

$$\vec{\Delta}^{(\pm)} = \Delta_{\pm x}\,\hat{x} + \Delta_{\pm y}\,\hat{y} + \Delta_{\pm z}\hat{z}, \qquad (8.3a)$$

$$\vec{\phi}(\vec{r}) = \phi_x(\vec{r}, t)\hat{x} + \phi_y(\vec{r}, t)\hat{y} + \phi_z(\vec{r}, t)\hat{z}. \qquad (8.3b)$$

We now need to define both the forward and the backward difference vector operators in Eq. (8.3a). Similarly to what was already done in Chapter 6, therefore, the single components act of $\vec{\Delta}^{(\pm)}$ act on a scalar function $f(\vec{r})$ as follows:

$$\Delta_\xi^{(+)} f(\vec{r}) = f(\vec{r} + a\hat{\xi}) - f(\vec{r}), \qquad (8.4a)$$

$$\Delta_\xi^{(-)} f(\vec{r}) = f(\vec{r}) - f(\vec{r} - a\hat{\xi}), \qquad (8.4b)$$

where ξ can be x, y, or z. Even though the above definitions do not have any profound physical implications, their efficacy can be readily verified. In fact, we may notice that Eq. (8.2) calls for the following additional definitions:

$$\vec{\psi}(\vec{r}, t) = \psi_x(\vec{r}, t)\hat{x} + \psi_y(\vec{r}, t)\hat{y} + \psi_z(\vec{r}, t)\hat{z}, \qquad (8.5a)$$

$$\vec{n}(\vec{r}) = n_x(\vec{r})\hat{x} + n_y(\vec{r})\hat{y} + n_z(\vec{r})\hat{z}. \qquad (8.5b)$$

We are now able to generalize Eq. (8.1a–c) by writing:

$$2\pi\vec{\psi}(\vec{r}, t) + \vec{\Delta}^{(+)} \times \vec{\phi}(\vec{r}, t) = 2\pi\vec{n}(\vec{r}), \qquad (8.6a)$$

$$\vec{\psi}(\vec{r}, t) = \vec{\psi}_{ex}(\vec{r}) + \beta\,\vec{\imath}(\vec{r}, t), \qquad (8.6b)$$

$$\frac{\Phi_0}{2\pi R I_{J0}} \frac{d\phi_\xi(\vec{r},t)}{dt} + \sin\phi_\xi(\vec{r}) = \left[\vec{\Delta}^{(-)} \times \vec{\iota}(\vec{r},t)\right]_\xi, \tag{8.6c}$$

where ξ can be x, y, or z in Eq. (8.6c), and where $\vec{\psi}_{ex}(\vec{r})$ is taken to depend on the planar element of the 3D network, in order to take account of the boundary conditions. Starting from the three sets of equations above, the dynamics for all junctions in the network can be obtained by applying the translator operator $\hat{T}_\eta$ ($\eta = x, y, z$) to the left- and right-hand side terms of Eq. (8.6a–c). In applying this operator, we need to remark that, as seen in Chapter 6, the currents, fluxes, or phase differences along eventually fictitiously added branches are to be taken equal to zero. In fact, some positions $\vec{r}$ that do not correspond to any branch of the network, can be introduced to take account of boundary conditions, or to write the constitutive equations (8.6a–c) at the borders. Notice also that the translation operator $\hat{T}_\eta$ ($\eta = x, y, z$) commutes with the difference operators. In this way, $\hat{T}_\eta$ ($\eta = x, y, z$) can be directly applied to quantities which depend on the position $\vec{r}$ in Eq. (8.6a–c). In order to prove that the translation operators commute with the difference operators, it suffices to say that the three components of the difference operator $\vec{\Delta}^{(\pm)}$ can be expressed in terms of the corresponding translator operators $\hat{T}_\eta$ ($\eta = x, y, z$) as follows:

$$\Delta_\eta^{(+)} = \Delta_{+\eta} = \left(\hat{T}_\eta - \mathbb{I}\right), \tag{8.7a}$$

$$\Delta_\eta^{(-)} = \Delta_{-\eta} = \left(\mathbb{I} - \hat{T}_{-\eta}\right), \tag{8.7b}$$

where $\mathbb{I}$ is the identity operator.

In what follows, we shall deal with systems starting from Zero Field Cooling (ZFC) conditions, so that we may set $\vec{n}(\vec{r}) = 0$ in Eq. (8.6a). Furthermore, by this same assumption, the initial conditions of our system will be

$$\vec{\phi}(\vec{r}, 0) = 0, \tag{8.8}$$

for all positions $\vec{r}$ in the network. By therefore setting $\vec{n}(\vec{r}) = 0$ in Eq. (8.6a), we can express the normalized fluxes $\vec{\psi}$ in terms of the phase differences as follows:

$$\vec{\psi}(\vec{r},t) = -\frac{1}{2\pi}\vec{\Delta}^{(+)} \times \vec{\phi}(\vec{r},t). \tag{8.9}$$

Furthermore, we can express the currents $\vec{\imath}(\vec{r},t)$ in terms of the superconducting phase differences $\vec{\phi}(\vec{r},t)$ by means of Eq. (8.6b) and (8.9), so that:

$$\vec{\imath}(\vec{r},t) = -\frac{1}{2\pi\beta}\left[\vec{\Delta}^{(+)} \times \vec{\phi}(\vec{r},t) + 2\pi\vec{\psi}_{ex}(\vec{r})\right]. \tag{8.10}$$

Eq. (8.10) can now be substituted into Eq. (8.6c). In this way, the following dynamic equation for the superconducting phase differences $\vec{\phi}(\vec{r},t)$ are obtained:

$$\frac{4L_0}{R}\frac{d\phi_\xi(\vec{r},t)}{dt} + 2\pi\beta \sin\phi_\xi(\vec{r},t)$$
$$= -\left[\vec{\Delta}^{(-)} \times \vec{\Delta}^{(+)} \times \vec{\phi}(\vec{r},t) + 2\pi\vec{\Delta}^{(-)} \times \vec{\psi}_{ex}(\vec{r})\right]_\xi, \tag{8.11}$$

In Eq. (8.11), despite the uniformity of $\vec{\psi}_{ex}(\vec{r})$, we have still retained the term $\vec{\Delta}^{(-)} \times \vec{\psi}_{ex}(\vec{r})$, which is non-null only for peripheral elements. We thus need to be very careful in determining this term when applying Eq. (8.11) to 3DJJNs. In fact, it is possible that a non-null contribution may arise from peripheral cells for which, when the position $\vec{r}$ is varied by a distance a in the direction opposite to one or two axes, no other network element is encountered. This delicate issue will be explicitly addressed when simple 3DJJNs will be analyzed.

By using the time normalization

$$\tau = \frac{R}{4L_0}t, \tag{8.12}$$

where t is the laboratory time, we may write Eq. (8.11) in a more convenient form:

$$\frac{d\phi_\xi(\vec{r},\tau)}{d\tau} + 2\pi\beta \sin\phi_\xi(\vec{r},\tau) =$$
$$-\left[\vec{\Delta}^{(-)} \times \vec{\Delta}^{(+)} \times \vec{\phi}(\vec{r},\tau) + 2\pi\,\vec{\Delta}^{(-)} \times \vec{\psi}_{ex}(\vec{r})\right]_\xi. \tag{8.13}$$

By finally introducing the non-linear operator, defined as:

$$O_J\left[\phi_\xi(\vec{r},\tau)\right] = \frac{d\phi_\xi(\vec{r},\tau)}{d\tau} + 2\pi\beta \sin\phi_\xi(\vec{r},\tau). \tag{8.14}$$

Eq. (8.13) can be completely cast in vector form as follows:

$$O_J[\vec{\phi}(\vec{r},\tau)] = -\vec{\Delta}^{(-)} \times \vec{\Delta}^{(+)} \times \vec{\phi}(\vec{r},\tau) - 2\pi\vec{\Delta}^{(-)} \times \vec{\psi}_{ex}. \quad (8.15)$$

Eq. (8.15) is the most compact form of the dynamic equations. In fact, the cross-products could still be expanded. We shall do this only at a later stage. The form in which the dynamic equations for the superconducting phase differences are written, and the compact definition of the flux and currents given, respectively, in Eq. (8.9) and (8.10), provide a first starting point for the analysis of three-dimensional Josephson junction networks.

8.2. Simple networks

As we have done for two-dimensional Josephson junction networks, we start by considering simple structures before analyzing more complex cases. In this section, we first analyze in more details the general properties of Eq. (8.15). We then consider the specific form these equations take for three orthogonal loops for negligible mutual inductance coefficients. Finally, we develop a detailed analysis of a closed structure consisting of a single cube without current bias.

Let us then start by considering Eq. (8.15) to obtain a more detailed form of these non-linear ordinary differential equations. We had previously mentioned that vector calculus algebra is rather similar to the algebra of difference operators. In fact, we may prove that the triple cross product obeys the following relation:

$$\vec{\Delta}^{(-)} \times \vec{\Delta}^{(+)} \times \vec{\phi} = -[\vec{\Delta}^{(-)} \cdot \vec{\Delta}^{(+)}]\vec{\phi} + \vec{\Delta}^{(+)}[\vec{\Delta}^{(-)} \cdot \vec{\phi}], \quad (8.16)$$

where the dot represents the scalar product. Eq. (8.16) closely recalls the analogous property (Gradshteyn, 1979) of the nabla ($\vec{\nabla}$) operator. Briefly, the proof goes as follows. First write the i-th component (here we abandon, for a moment, the Greek symbols and express the indices in terms of the natural numbers 1, 2, 3) of the cross products as follows:

$$\{\vec{\Delta}^{(-)} \times \vec{\Delta}^{(+)} \times \vec{\phi}\}_i = \varepsilon_{ijk}\Delta_{-j}(\varepsilon_{klm}\Delta_l \phi_m), \quad (8.17)$$

where the sum over repeated indices convention is adopted and the plus sign in the index of Δ_l is omitted. Recalling the property linking Levi-Civita's symbols to the Kronecker delta (Arfken, 2012), by which:

$$\varepsilon_{ijk}\varepsilon_{klm} = \delta_{il}\delta_{jm} - \delta_{im}\delta_{jl}, \tag{8.18}$$

Eq. (8.17) can be written as follows:

$$\left\{\vec{\Delta}^{(-)} \times \vec{\Delta}^{(+)} \times \vec{\phi}\right\}_i = \left(\delta_{il}\delta_{jm} - \delta_{im}\delta_{jl}\right)\Delta_{-j}\Delta_l\phi_m \tag{8.19}$$

or

$$\left\{\vec{\Delta}^{(-)} \times \vec{\Delta}^{(+)} \times \vec{\phi}\right\}_i = \Delta_{-j}\Delta_i\phi_j - \Delta_{-j}\Delta_j\phi_i, \tag{8.20}$$

which completes the proof. Because of Eq. (8.16), Eq. (8.15) can be written as follows:

$$O_J[\phi_\xi(\vec{r},\tau)] = \Delta_{-\lambda}\Delta_\lambda\phi_\xi(\vec{r},\tau) - \Delta_{-\lambda}\Delta_\xi\phi_\lambda(\vec{r},\tau) + B_\xi(\vec{r}), \tag{8.21}$$

Where $\vec{B}(\vec{r}) = -2\pi\vec{\Delta}^{(-)} \times \vec{\psi}_{ex}(\vec{r})$ is a boundary term and where Greek symbols are again used (ξ and λ take on the values x,y,z) and the convention of summing over repeated indices is still adopted. Eq. (8.21) can be explicitly written in terms of $\phi_x(\vec{r},\tau)$, $\phi_y(\vec{r},\tau)$, $\phi_z(\vec{r},\tau)$ as follows:

$$O_J[\phi_x(\vec{r},\tau)] = \left(\Delta_{-y}\Delta_y + \Delta_{-z}\Delta_z\right)\phi_x(\vec{r},\tau) - \Delta_{-y}\Delta_x\phi_y(\vec{r},\tau) -$$
$$\Delta_{-z}\Delta_x\phi_z(\vec{r},\tau) + B_x(\vec{r}), \tag{8.22a}$$

$$O_J[\phi_y(\vec{r},\tau)] = \left(\Delta_{-x}\Delta_x + \Delta_{-z}\Delta_z\right)\phi_y(\vec{r},\tau) - \Delta_{-x}\Delta_y\phi_x(\vec{r},\tau) -$$
$$\Delta_{-z}\Delta_y\phi_z(\vec{r},\tau) + B_y(\vec{r}), \tag{8.22b}$$

$$O_J[\phi_z(\vec{r},\tau)] = \left(\Delta_{-x}\Delta_x + \Delta_{-y}\Delta_y\right)\phi_z(\vec{r},\tau) - \Delta_{-x}\Delta_z\phi_x(\vec{r},\tau) -$$
$$\Delta_{-y}\Delta_z\phi_y(\vec{r},\tau) + B_z(\vec{r}). \tag{8.22c}$$

Notice that the boundary terms can be written, according to the definition given above, as follows:

$$B_x(\vec{r}) = -2\pi\left(\Delta_{-y}\psi_{ex}^{(z)}(\vec{r}) - \Delta_{-z}\psi_{ex}^{(y)}(\vec{r})\right), \tag{8.23a}$$

$$B_y(\vec{r}) = +2\pi \left(\Delta_{-x}\psi_{ex}^{(z)}(\vec{r}) - \Delta_{-z}\psi_{ex}^{(x)}(\vec{r}) \right), \qquad (8.23b)$$

$$B_z(\vec{r}) = -2\pi \left(\Delta_{-x}\psi_{ex}^{(y)}(\vec{r}) - \Delta_{-y}\psi_{ex}^{(x)}(\vec{r}) \right), \qquad (8.23c)$$

where $\psi_{ex}^{(\xi)}(\vec{r}) = \hat{\xi} \cdot \vec{\psi}_{ex}(\vec{r})$ is the ξ-component of the vector $\vec{\psi}_{ex}(\vec{r})$ ($\xi = x, y, z$).

Starting from Eq. (8.22a–c), in the following subsection we shall consider the case of a network consisting of three orthogonal square elements.

8.2.1. *Three orthogonal square elements*

As a first example of a simple 3DJJN, let us consider a system consisting of three orthogonal square elements as shown in Fig. 8.3. Each orthogonal face of the structure is similar to the square element represented in Fig. 8.1.

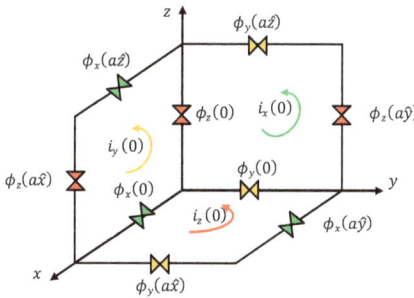

Fig. 8.3. Labeling in the three-dimensional Josephson junction network consisting of three orthogonal square elements. All stationary superconducting phase differences and currents in the three-dimensional Josephson junction network are shown with the same convention of colors as in Fig. 8.2. All currents have positive sign if they circulate in the same direction of the shown arrows. The time dependence of all quantities has been omitted for simplicity.

In Fig. 8.3 all superconducting phase differences are shown in correspondence with Josephson junctions (JJs) of different colors (green for x-oriented JJs, yellow for y-oriented JJs, red for z-oriented JJs). The convention on the sign of the superconducting phase differences across these JJs is the same adopted in Chapter 6, also shown in Fig. 8.1: the minus-sign electrode precedes the plus-sign electrode as one moves along the positive direction of the axis on which the JJ lies.

Let us now write down the dynamic equations of the system. These structures are not only interesting from an academic point of view, but have been proposed, in a more simplified version, to describe the behavior of an ultra-sensitive vector sensor (Oppenländer J. H., 1999). In Section 8.3 we shall analyze in more details the response of so called 3D SQUIDs (De Luca R. R., 2002), to which this class of networks can be ascribed. For the time being, let us get familiar with the dynamic equations and with the algebra of difference operators. Let us then explicitly write down the dynamic equations, starting form Eq. (8.22a–c), for the system under exam. In order to do so, we proceed as follows.

Let us start by writing Eq. (8.22a–c) for $\vec{r} = 0$. By making use of the definition of difference operators and of the relation $\Delta_{-\xi} = \mathbb{I}$ for $\xi = x, y, z$, valid only for $\vec{r} = 0$, we have:

$$O_J[\phi_x(0,\tau)] = (\Delta_y + \Delta_z)\phi_x(0,\tau) - \Delta_x\phi_y(0,\tau) -$$

$$\Delta_x\phi_z(0,\tau) + B_x(0), \qquad (8.24a)$$

$$O_J[\phi_y(0,\tau)] = (\Delta_x + \Delta_z)\phi_y(0,\tau) - \Delta_y\phi_x(0,\tau) -$$

$$\Delta_y\phi_z(0,\tau) + B_y(0), \qquad (8.24b)$$

$$O_J[\phi_z(0,\tau)] = (\Delta_x + \Delta_y)\phi_z(0,\tau) - \Delta_z\phi_x(0,\tau) -$$

$$\Delta_z\phi_y(0,\tau) + B_z(0). \qquad (8.24c)$$

Here the boundary terms $B_x(0)$, $B_y(0)$, and $B_z(0)$ are expressed, according to Eq. (8.23a–c) and the prescription relation $\Delta_{-\xi} = \mathbb{I}$ for $\xi = x, y, z$, as follows:

$$B_x(0) = -2\pi(\hat{z} - \hat{y}) \cdot \vec{\psi}_{ex} \qquad (8.25a)$$

$$B_y(0) = +2\pi(\hat{z} - \hat{x}) \cdot \vec{\psi}_{ex} \qquad (8.25b)$$

$$B_z(0) = -2\pi(\hat{y} - \hat{x}) \cdot \vec{\psi}_{ex}. \qquad (8.25c)$$

Therefore, by explicitly writing the difference operators in terms of the translation operators and by taking into account Eq. (8.25a–c), we may recast Eq. (8.24a–c) in the following way:

$$O_J[\phi_x(0,\tau)] = (\hat{T}_y + \hat{T}_z - 2\mathbb{I})\phi_x(0,\tau) - (\hat{T}_x - \mathbb{I})\phi_y(0,\tau) - $$
$$(\hat{T}_x - \mathbb{I})\phi_z(0,\tau) - 2\pi(\hat{z} - \hat{y}) \cdot \vec{\psi}_{ex}, \qquad (8.26a)$$

$$O_J[\phi_y(0,\tau)] = (\hat{T}_x + \hat{T}_z - 2\mathbb{I})\phi_y(0,\tau) - (\hat{T}_y - \mathbb{I})\phi_x(0,\tau) - $$
$$(\hat{T}_y - \mathbb{I})\phi_z(0,\tau) + 2\pi(\hat{z} - \hat{x}) \cdot \vec{\psi}_{ex}, \qquad (8.26b)$$

$$O_J[\phi_z(0,\tau)] = (\hat{T}_x + \hat{T}_y - 2\mathbb{I})\phi_z(0,\tau) - (\hat{T}_z - \mathbb{I})\phi_x(0,\tau) - $$
$$(\hat{T}_z - \mathbb{I})\phi_y(0,\tau) - 2\pi(\hat{y} - \hat{x}) \cdot \vec{\psi}_{ex}. \qquad (8.26c)$$

Finally, by letting the translator operators act on the superconducting phase differences $\phi_\xi(0,\tau)$, for $\xi = x, y, z$, we have:

$$O_J[\phi_x(0,\tau)] = \phi_x(a\hat{y},\tau) - 2\phi_x(0,\tau) + \phi_x(a\hat{z},\tau) + \phi_y(0,\tau) + $$
$$\phi_z(0,\tau) - \phi_y(a\hat{x},\tau) - \phi_z(a\hat{x},\tau) - 2\pi(\hat{z} - \hat{y}) \cdot \vec{\psi}_{ex}, \quad (8.27a)$$

$$O_J[\phi_y(0,\tau)] = \phi_y(a\hat{x},\tau) - 2\phi_y(0,\tau) + \phi_y(a\hat{z},\tau) + \phi_x(0,\tau) + $$
$$\phi_z(0,\tau) - \phi_x(a\hat{y},\tau) - \phi_z(a\hat{y},\tau) + 2\pi(\hat{z} - \hat{x}) \cdot \vec{\psi}_{ex}, \quad (8.27b)$$

$$O_J[\phi_z(0,\tau)] = \phi_z(a\hat{x},\tau) - 2\phi_z(0,\tau) + \phi_z(a\hat{y},\tau) + \phi_x(0,\tau) + $$
$$\phi_y(0,\tau) - \phi_x(a\hat{z},\tau) - \phi_y(a\hat{z},\tau) - 2\pi(\hat{y} - \hat{x}) \cdot \vec{\psi}_{ex}. \quad (8.27c)$$

In order to obtain the remaining six dynamic equations we proceed rather slowly as follows.

First, consider Eq. (8.22a–c) written in terms of the difference operators for $\vec{r} \neq 0$. In this case, we notice that these boundary junctions, different from those at position $\vec{r} = 0$, do not connect two orthogonal sides of the structure. In this way, we expect that the dynamic equations for these junctions will only contain terms belonging to the plane in which they lie. Let us then first consider Eq. (8.22a) specifically for $\vec{r} = a\hat{y}$. By deleting the terms containing the operators Δ_z and Δ_{-z} in this case, by noticing that $\Delta_y\phi_x(a\hat{y},\tau) = -\phi_x(a\hat{y},\tau)$, we may write:

$$O_J[\phi_x(a\hat{y},\tau)] = -\Delta_{-y}\phi_x(a\hat{y},\tau) - \Delta_x \Delta_{-y}\phi_y(a\hat{y},\tau) + B_x(a\hat{y}). \quad (8.28)$$

The first addendum on the right-hand side gives $-\phi_x(a\hat{y}, \tau) + \phi_x(0, \tau)$, by the definition of the operator Δ_{-y}. For the second addendum, we first write $\Delta_x \phi_y(0, \tau)$, by recognizing that $\Delta_{-y} \phi_y(a\hat{y}, \tau) = -\phi_y(0, \tau)$, then we obtain $\phi_y(a\hat{x}, \tau) - \phi_y(0, \tau)$ by definition of the forward difference operator Δ_x. In this way, Eq. (8.28) can be written as follows:

$$O_J[\phi_x(a\hat{y}, \tau)] = +\phi_x(0, \tau) - \phi_x(a\hat{y}, \tau) - \phi_y(0, \tau) + \phi_y(a\hat{x}, \tau) + B_x(a\hat{y}). \tag{8.29}$$

We now proceed in the same way in determining the dynamic equation for the phase difference $\phi_x(a\hat{z}, \tau)$. We will still proceed step by step in this case, in order to make sure that we can master and safely generalize the procedure. By considering Eq. (8.22a) specifically for $\vec{r} = a\hat{z}$, we delete the terms containing the operators Δ_y and Δ_{-y} and set $\Delta_z \phi_x(a\hat{z}, \tau) = -\phi_x(a\hat{z}, \tau)$, we may write:

$$O_J[\phi_x(a\hat{z}, \tau)] = -\Delta_{-z} \phi_x(a\hat{z}, \tau) - \Delta_x \Delta_{-z} \phi_z(a\hat{z}, \tau) + B_x(a\hat{z}). \tag{8.30}$$

The first addendum on the right-hand side gives $-\phi_x(a\hat{z}, \tau) + \phi_x(0, \tau)$, by directly applying the operator Δ_{-z}. The second addendum first gives $\Delta_x \phi_z(0, \tau)$, by recognizing that $\Delta_{-z} \phi_z(a\hat{z}, \tau) = -\phi_z(0, \tau)$. Finally, by definition of Δ_x, the second addendum becomes $\phi_z(a\hat{x}, \tau) - \phi_z(0, \tau)$. In this way, Eq. (8.30) can be written as follows:

$$O_J[\phi_x(a\hat{z}, \tau)] = -\phi_x(a\hat{z}, \tau) + \phi_x(0, \tau) + \phi_z(a\hat{x}, \tau) - \phi_z(0, \tau) + B_x(a\hat{z}). \tag{8.31}$$

Notice that we could have simply obtained Eq. (8.31) by letting $y \to z$ in Eq. (8.29). We did not proceed in doing so, only because we wanted to make sure that we were doing things correctly.

In order to now determine the boundary forcing terms $B_x(a\hat{y})$ and $B_x(a\hat{z})$, we may write Eq. (8.23a) as follows:

$$B_x(a\hat{y}) = -2\pi \left(\Delta_{-y} \psi_{ex}^{(z)}(a\hat{y}) - \Delta_{-z} \psi_{ex}^{(y)}(a\hat{y}) \right), \tag{8.32a}$$

$$B_x(a\hat{z}) = -2\pi \left(\Delta_{-y} \psi_{ex}^{(z)}(a\hat{z}) - \Delta_{-z} \psi_{ex}^{(y)}(a\hat{z}) \right), \tag{8.32b}$$

where only $\psi_{ex}^{(\mu)}(0)$, for $\mu = x, y, z$, has to be taken into account. In this way, we get:

$$B_x(a\hat{y}) = -2\pi\Delta_{-y}\psi_{ex}^{(z)}(a\hat{y}) = 2\pi\psi_{ex}^{(z)}(0), \qquad (8.33a)$$

$$B_x(a\hat{z}) = -2\pi\Delta_{-z}\psi_{ex}^{(y)}(a\hat{z}) = -2\pi\psi_{ex}^{(y)}(0), \qquad (8.33b)$$

We may thus finally write Eq. (8.29) and Eq. (8.31) as follows:

$$O_J[\phi_x(a\hat{y}, \tau)] = +\phi_x(0, \tau) - \phi_x(a\hat{y}, \tau) - \phi_y(0, \tau) + \phi_y(a\hat{x}, \tau)$$
$$+2\pi\hat{z} \cdot \vec{\psi}_{ex}, \qquad (8.34a)$$

$$O_J[\phi_x(a\hat{z}, \tau)] = +\phi_x(0, \tau) - \phi_x(a\hat{z}, \tau) - \phi_z(0, \tau) + \phi_z(a\hat{x}, \tau) - 2\pi\hat{y} \cdot$$
$$\vec{\psi}_{ex}. \qquad (8.34b)$$

Of course, these relations can be more easily derived by means of direct application of the constitutive equations. The reader may indeed prove the validity of these equations by direct calculations on the network.

Analogously, by considering Eq. (8.22b) and by proceeding as above, we may derive the following relations:

$$O_J[\phi_y(a\hat{x}, \tau)] = -\phi_x(0, \tau) + \phi_x(a\hat{y}, \tau) + \phi_y(0, \tau) - \phi_y(a\hat{x}, \tau) - 2\pi\hat{z} \cdot$$
$$\vec{\psi}_{ex}, \qquad (8.35a)$$

$$O_J[\phi_y(a\hat{z}, \tau)] = +\phi_y(0, \tau) - \phi_y(a\hat{z}, \tau) - \phi_z(0, \tau) + \phi_z(a\hat{y}, \tau) + 2\pi\hat{x} \cdot$$
$$\vec{\psi}_{ex}, \qquad (8.35b)$$

Finally, by considering Eq. (8.22c), we may write:

$$O_J[\phi_z(a\hat{x}, \tau)] = -\phi_x(0, \tau) + \phi_x(a\hat{z}, \tau) + \phi_z(0, \tau) - \phi_z(a\hat{x}, \tau) + 2\pi\hat{y} \cdot$$
$$\vec{\psi}_{ex}, \qquad (8.36a)$$

$$O_J[\phi_z(a\hat{y}, \tau)] = -\phi_y(0, \tau) + \phi_y(a\hat{z}, \tau) + \phi_z(0, \tau) - \phi_z(a\hat{y}, \tau) - 2\pi\hat{x} \cdot$$
$$\vec{\psi}_{ex}. \qquad (8.36b)$$

Eq. (8.34a–b), (8.35a–b), and (8.36a–b), together with Eq. (8.27a–c) are the nine dynamic equations for the junctions network consisting of three orthogonal square loops, as depicted in Fig. 8.3. These expressions form a system of nonlinear coupled ordinary differential equations. By inspecting, in the order, Eq. (8.34a) and Eq. (8.35a), Eq. (8.34b) and Eq. (8.36a), and Eq. (8.35b) and Eq. (8.36b), we may notice the validity of the following relations:

$$O_J[\phi_x(a\hat{y},\tau)] = -O_J[\phi_y(a\hat{x},\tau)], \qquad (8.37a)$$

$$O_J[\phi_x(a\hat{z},\tau)] = -O_J[\phi_z(a\hat{x},\tau)], \qquad (8.37b)$$

$$O_J[\phi_y(a\hat{z},\tau)] = -O_J[\phi_z(a\hat{y},\tau)]. \qquad (8.37c)$$

This means that the same current flows, in opposite directions, in the pair of JJs characterized by the phase differences reported in Eq. (8.37a–c). This can be seen from Fig. 8.3, where the currents are shown, recalling that the sign of the phase differences is determined by the convention set forth in Chapter 6: the minus-sign electrode precedes the plus-sign electrode in a junction as one moves along the positive direction of the axis on which the JJ lies. From Eq. (8.37a–c) we may thus obtain:

$$\phi_x(a\hat{y},\tau) = -\phi_y(a\hat{x},\tau), \qquad (8.38a)$$

$$\phi_x(a\hat{z},\tau) = -\phi_z(a\hat{x},\tau), \qquad (8.38b)$$

$$\phi_y(a\hat{z},\tau) = -\phi_z(a\hat{y},\tau), \qquad (8.38c)$$

These relations lead to the following reduced system of six non-linear differential equations for the network consisting of four orthogonal square elements:

$$O_J[\phi_x(0,\tau)] = -2\phi_x(0,\tau) + 2\phi_x(a\hat{y},\tau) + 2\phi_x(a\hat{z},\tau) + \phi_y(0,\tau) + \\ \phi_z(0,\tau) - 2\pi(\hat{z}-\hat{y})\cdot\vec{\psi}_{ex}, \qquad (8.39a)$$

$$O_J[\phi_y(0,\tau)] = +\phi_x(0,\tau) - 2\phi_x(a\hat{y},\tau) - 2\phi_y(0,\tau) + \\ 2\phi_y(a\hat{z},\tau) + \phi_z(0,\tau) + 2\pi(\hat{z}-\hat{x})\cdot\vec{\psi}_{ex}, \qquad (8.39b)$$

$$O_J[\phi_z(0,\tau)] = +\phi_x(0,\tau) - 2\phi_x(a\hat{z},\tau) + \phi_y(0,\tau) - 2\phi_y(a\hat{z},\tau) - \\ 2\phi_z(0,\tau) - 2\pi(\hat{y}-\hat{x})\cdot\vec{\psi}_{ex}, \qquad (8.39c)$$

$$O_J[\phi_x(a\hat{y},\tau)] = \phi_x(0,\tau) - 2\phi_x(a\hat{y},\tau) - \phi_y(0,\tau) + 2\pi\hat{z}\cdot\vec{\psi}_{ex}, \quad (8.39d)$$

$$O_J[\phi_x(a\hat{z},\tau)] = \phi_x(0,\tau) - 2\phi_x(a\hat{z},\tau) - \phi_z(0,\tau) - 2\pi\hat{y} \cdot \vec{\psi}_{ex}, \quad (8.39e)$$

$$O_J[\phi_y(a\hat{z},\tau)] = \phi_y(0,\tau) - 2\phi_y(a\hat{z},\tau) - \phi_z(0,\tau) + 2\pi\hat{x} \cdot \vec{\psi}_{ex}. \quad (8.39f)$$

The above set of equations will be solved numerically in the following section.

8.2.2. *A single cube*

A more symmetric structure is represented by a network consisting of six square elements forming a cube. In order to illustrate the labeling for junctions, fluxes and currents in this network, in Fig. 8.4 we depict it as the juxtaposition of two sub-networks: the first consisting of three square elements, as the one shown in Fig. 8.3, the second completing the cubic structure.

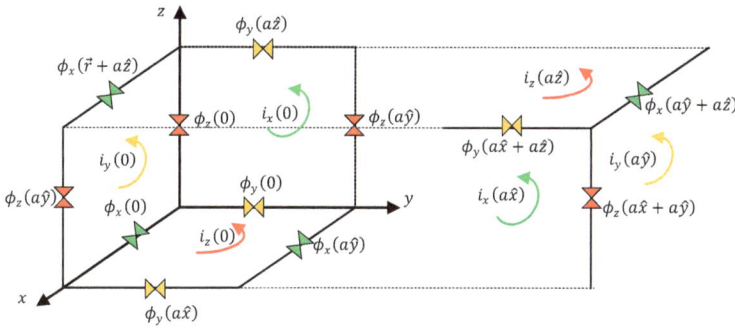

Fig. 8.4. Labeling in the three-dimensional Josephson junction network consisting of six square loops forming a cube. The whole system is shown as the juxtaposition of two sub-networks for clarity sake. All stationary superconducting phase differences and currents in the three-dimensional Josephson junction network are shown with the same convention of colors as in Fig. 8.3. All currents have positive sign if they circulate in the same direction of the shown arrow. The time dependence of all quantities has been omitted for simplicity.

The choice of splitting the network into two parts has been made merely to show more clearly the superconducting phase differences across each Josephson junction (JJ) in the network. In Fig. 8.4 the same choice of colors has been made as in Fig. 8.3. Moreover, the convention on the sign is the same adopted in Chapter 6, so that the minus-sign electrode precedes the plus-sign electrode in a junction as one moves along the positive direction of the axis on which the JJ lies.

As in the case of a network consisting of three orthogonal square loops, we start our analysis by considering Eq. (8.22a–c). These equations, in fact, are written in a general enough fashion and, as already stated, can be adopted for three-dimensional networks with cubic symmetry as, for example, the one described in Fig. 8.2. In the case of the single cube shown in Fig. 8.4, in particular, all junctions connect two sides of the structure, so that the procedure adopted for three orthogonal squares can be repeated only for the junctions closer to the origin. In this respect, we remark that the dynamic equations for the superconducting phase differences $\phi_x(0, \tau)$, $\phi_y(0, \tau)$, and $\phi_z(0, \tau)$ are still given by Eq. (8.27a), Eq. (8.27b), and Eq. (8.27c), respectively. The remaining nine equations are derived by means of the same type of reasoning adopted in getting to the latter three equations, therefore specializing Eq. (8.22a–c) to the case under exam.

Being the structure in Fig. 8.4 closed, it is instructive to prove that the sum of the magnetic fluxes threading the six loops in the network is zero; namely, that

$$-\sum_\xi \hat{\xi} \cdot \vec{\psi}(0) + \sum_\xi \hat{\xi} \cdot \vec{\psi}(a\hat{\xi}) = 0, \qquad (8.40)$$

where $\xi = x, y, z$, so that the first sum gives the net flux through the loops positioned at $\vec{r} = 0$, and the second sum represents the net flux through the remaining three faces of the cubic structure. The minus sign in front of the first term takes account of the particular convention adopted for the currents. By recalling Eq. (8.9), we may express the fluxes in terms of the superconducting phase differences. In this way, the first and the second sums in Eq. (8.40), which we can call S_1 and S_2, in the order, can be written as follows:

$$S_1 = -\sum_\xi \hat{\xi} \cdot \vec{\psi}(0) = \frac{1}{2\pi} \sum_\xi \hat{\xi} \cdot [\vec{\Delta}^{(+)} \times \vec{\phi}(0)], \qquad (8.41a)$$

$$S_2 = \sum_\xi \hat{\xi} \cdot \vec{\psi}(a\hat{\xi}) = -\frac{1}{2\pi} \sum_\xi \hat{\xi} \cdot [\vec{\Delta}^{(+)} \times \vec{\phi}(a\hat{\xi})]. \qquad (8.41b)$$

Let us thus separately calculate these sums. As for S_1, we may write:

$$S_1 = \frac{1}{2\pi}\sum_\xi \varepsilon_{\xi\mu\nu}\, \Delta_\mu \phi_\nu(0), \tag{8.42}$$

where $\varepsilon_{\xi\mu\nu}$ is the Levi-Civita tensor and the Greek letters represents the letters x, y, z. In this case we do not switch to numerical values for ξ, μ, ν, as done at the beginning of Section 8.2, since it is now clear the correspondence between the Latin letters i, j, k, and the Greek letters.

By expressing the operator Δ_μ in terms of the translation operator, as in Eq. (8.7a), we have:

$$S_1 = \frac{1}{2\pi}\sum_\xi \varepsilon_{\xi\mu\nu} [\phi_\nu(a\hat\mu) - \phi_\nu(0)]. \tag{8.43}$$

In the same way, the second sum can be expressed as follows:

$$S_2 = -\frac{1}{2\pi}\sum_\xi \varepsilon_{\xi\mu\nu} [\phi_\nu(a\hat\xi + a\hat\mu) - \phi_\nu(a\hat\xi)]. \tag{8.44}$$

First notice that the second addendum in the square brackets in S_1 does not depend on ξ and μ and that the first addendum in the square brackets in S_2 is symmetric for the exchange $\xi \leftrightarrow \mu$. Because of the antisymmetric properties of the tensor $\varepsilon_{\xi\mu\nu}$ and because of the invariance in the exchange of ξ and μ illustrated above, these terms give a null contribution to the sums S_1 and S_2, so that we may write the expression $S = S_1 + S_2$ as follows:

$$S = \frac{1}{2\pi}\sum_\xi \varepsilon_{\xi\mu\nu}\, \phi_\nu(a\hat\xi) + \frac{1}{2\pi}\sum_\xi \varepsilon_{\xi\mu\nu}\, \phi_\nu(a\hat\mu). \tag{8.45}$$

In the above expression we are summing not only on ξ, but also on all repeated indices, so that the two sums in Eq. (8.45) can be seen to be differ only in the sign, thus giving $S = 0$, as we wished to demonstrate.

We now illustrate how to generate the dynamic equations for $\phi_x(a\hat y, \tau)$, $\phi_x(a\hat z, \tau)$, and $\phi_x(a\hat y + a\hat z, \tau)$. The remaining equations, derived in a similar way, will only be displayed. The reader is advised to derive the latter and to prove that the same can be done by direct application of the constitutive equations for the network, namely, by means of Eq. (8.6a–c), without necessarily starting from Eq. (8.22a–c).

Naturally, we are here illustrating simple examples, but we would also like to attain some sort of generalization of our analysis, so that we insist in obtaining the dynamic equations of our system in making use of difference operators and, therefore, by starting from Eq. (8.22a–c).

Let us then consider Eq. (8.22a). For the single cube the following positions are allowed for $\phi_x(\vec{r}, \tau)$: $\vec{r} = a\hat{y}$, $\vec{r} = a\hat{z}$, $\vec{r} = a\hat{y} + a\hat{z}$. In this way, we have:

$$O_J[\phi_x(a\hat{y}, \tau)] = \left(\Delta_{-y}\Delta_y + \Delta_{-z}\Delta_z\right)\phi_x(a\hat{y}, \tau) -$$
$$\Delta_{-y}\Delta_x\phi_y(a\hat{y}, \tau) - \Delta_{-z}\Delta_x\phi_z(a\hat{y}, \tau) + B_x(a\hat{y}), \qquad (8.46a)$$

$$O_J[\phi_x(a\hat{z}, \tau)] = \left(\Delta_{-y}\Delta_y + \Delta_{-z}\Delta_z\right)\phi_x(a\hat{z}, \tau) -$$
$$\Delta_{-y}\Delta_x\phi_y(a\hat{z}, \tau) - \Delta_{-z}\Delta_x\phi_z(a\hat{z}, \tau) + B_x(a\hat{z}), \qquad (8.46b)$$

$$O_J[\phi_x(a\hat{y} + a\hat{z}, \tau)] = \left(\Delta_{-y}\Delta_y + \Delta_{-z}\Delta_z\right)\phi_x(a\hat{y} + a\hat{z}, \tau) -$$
$$\Delta_{-y}\Delta_x\phi_y(a\hat{y} + a\hat{z}, \tau) - \Delta_{-z}\Delta_x\phi_z(a\hat{y} + a\hat{z}, \tau) + B_x(a\hat{y} + a\hat{z}). \ (8.46c)$$

Before explicitly writing the difference operators in terms of the translation operators, let us substitute with $\mathbb{I}$ and $-\mathbb{I}$ the difference operators $\Delta_{-\xi}$ and Δ_ξ, respectively, which do not contribute to extra term in the above equations. In this way, by also making use of the commutation properties of the difference operators, we may recast Eq. (8.46a–c) in the following way:

$$O_J[\phi_x(a\hat{y}, \tau)] = \left(-\mathbb{I}\Delta_{-y} + \mathbb{I}\Delta_z\right)\phi_x(a\hat{y}, \tau) - \Delta_x\Delta_{-y}\phi_y(a\hat{y}, \tau) -$$
$$\mathbb{I}\Delta_x\phi_z(a\hat{y}, \tau) - 2\pi\left(\Delta_{-y}\psi_{ex}^{(z)}(a\hat{y}) - \mathbb{I}\psi_{ex}^{(y)}(a\hat{y})\right), \qquad (8.47a)$$

$$O_J[\phi_x(a\hat{z}, \tau)] = \left(\mathbb{I}\Delta_y - \mathbb{I}\Delta_{-z}\right)\phi_x(a\hat{z}, \tau) - \mathbb{I}\Delta_x\phi_y(a\hat{z}, \tau) -$$
$$\Delta_x\Delta_{-z}\phi_z(a\hat{z}, \tau) - 2\pi\left(\mathbb{I}\psi_{ex}^{(z)}(a\hat{z}) - \Delta_{-z}\psi_{ex}^{(y)}(a\hat{z})\right), \qquad (8.47b)$$

$$O_J[\phi_x(a\hat{y} + a\hat{z}, \tau)] = \left(-\mathbb{I}\Delta_{-y} - \mathbb{I}\Delta_{-z}\right)\phi_x(a\hat{y} + a\hat{z}, \tau) +$$
$$\Delta_x\phi_y(a\hat{z}, \tau) + \Delta_x\phi_z(a\hat{y}, \tau) - 2\pi\left(\Delta_{-y}\psi_{ex}^{(z)}(a\hat{y} + a\hat{z}) -\right.$$
$$\left.\Delta_{-z}\psi_{ex}^{(y)}(a\hat{y} + a\hat{z})\right), \qquad (8.47c)$$

where we have left the operator $\mathbb{I}$ in place of the elided difference operators, to give to the reader the possibility to trace back the logical process behind the derivation of the dynamic equations. Just to give an example of how to decide which difference operator should be replaced by $+\mathbb{I}$ or $-\mathbb{I}$, consider the first term on the right-hand side of Eq. (8.46a) $\Delta_{-y}\Delta_y\phi_x(a\hat{y},\tau)$. If we applied the fully defined difference operator Δ_y to $\phi_x(a\hat{y},\tau)$, we would obtain two terms, one term being $\phi_x(2a\hat{y},\tau)$, the second being the same $\phi_x(a\hat{y},\tau)$. Clearly, the first term does not correspond to any junction in the network, so that only the second term survives, with a minus sign in front. Therefore, in the first term on the right-hand side of Eq. (8.47a) we have reported the expression $-\mathbb{I}\Delta_{-y}\phi_x(a\hat{y},\tau)$. In going from Eq. (8.46c) to Eq. (8.47c), on the other hand, we have replaced, for the same reason, the term $\Delta_x\Delta_{-y}\phi_y(a\hat{y}+a\hat{z},\tau)$ with $-\Delta_x\phi_y(a\hat{z},\tau)$ and the term $\Delta_x\Delta_{-z}\phi_z(a\hat{y}+a\hat{z},\tau)$ with $-\Delta_x\phi_z(a\hat{y},\tau)$.

Let us now go one step further and substitute the difference operators with their expression in terms of the translations operators, as in Eq. (8.7a–b). Therefore, by now omitting the identity operator $\mathbb{I}$ where its appearance is not necessary, we have:

$$O_J[\phi_x(a\hat{y},\tau)] = \left(\hat{T}_{-y} + \hat{T}_z - 2\mathbb{I}\right)\phi_x(a\hat{y},\tau) + \left(\hat{T}_x - \mathbb{I}\right)\hat{T}_{-y}\phi_y(a\hat{y},\tau) - \left(\hat{T}_x - \mathbb{I}\right)\phi_z(a\hat{y},\tau) + 2\pi(\hat{z} + \hat{y})\cdot\vec{\psi}_{ex}, \tag{8.48a}$$

$$O_J[\phi_x(a\hat{z},\tau)] = \left(\hat{T}_y + \hat{T}_{-z} - 2\mathbb{I}\right)\phi_x(a\hat{z},\tau) - \left(\hat{T}_x - \mathbb{I}\right)\phi_y(a\hat{z},\tau) - \left(\hat{T}_x - \mathbb{I}\right)\hat{T}_{-z}\phi_z(a\hat{z},\tau) - 2\pi(\hat{z} + \hat{y})\cdot\vec{\psi}_{ex}, \tag{8.48b}$$

$$O_J[\phi_x(a\hat{y} + a\hat{z},\tau)] = \left(\hat{T}_{-y} + \hat{T}_{-z} - 2\mathbb{I}\right)\phi_x(a\hat{y} + a\hat{z},\tau) + \left(\hat{T}_x - \mathbb{I}\right)\phi_y(a\hat{z},\tau) + \left(\hat{T}_x - \mathbb{I}\right)\phi_z(a\hat{y},\tau) + 2\pi(\hat{z} - \hat{y})\cdot\vec{\psi}_{ex}. \tag{8.48c}$$

In calculating the externally applied flux terms in Eq. (8.48a–c), we proceeded as illustrated in the following example. Consider the term appearing in Eq. (8.47a), namely:

$$B_x(a\hat{y}) = -2\pi\left(\Delta_{-y}\psi_{ex}^{(z)}(a\hat{y}) - \psi_{ex}^{(y)}(a\hat{y})\right). \tag{8.49}$$

After application of the difference operator, we have:

$$B_x(a\hat{y}) = -2\pi \left(\psi_{ex}^{(z)}(a\hat{y}) - \psi_{ex}^{(z)}(0) - \psi_{ex}^{(y)}(a\hat{y}) \right). \quad (8.50)$$

The first term $\psi_{ex}^{(z)}(a\hat{y})$ does not contribute the externally applied flux, since the z-component is translated along the y-axis, thus being parallel to the cube face belonging the $x - z$ plane at $y = a$. However, the third term $\psi_{ex}^{(y)}(a\hat{y})$ does contribute to the externally applied flux, since it is orthogonal to the same cube face on the $x - z$ plane at $y = a$ and equal to $\psi_{ex}^{(y)}(0)$. In this way, we may set:

$$B_x(a\hat{y}) = 2\pi \left(\psi_{ex}^{(z)}(0) + \psi_{ex}^{(y)}(0) \right) = +2\pi(\hat{z} + \hat{y}) \cdot \vec{\psi}_{ex}. \quad (8.51)$$

Finally, by letting the translator operators act on all superconducting phase differences in Eq. (8.48a–c), we may write:

$$O_J[\phi_x(a\hat{y}, \tau)] = +\phi_x(0, \tau) - 2\phi_x(a\hat{y}, \tau) + \phi_x(a\hat{y} + a\hat{z}, \tau) -$$
$$\phi_y(0, \tau) + \phi_y(a\hat{x}, \tau) + \phi_z(a\hat{y}, \tau) - \phi_z(a\hat{x} + a\hat{y}, \tau) +$$
$$2\pi(\hat{y} + \hat{z}) \cdot \vec{\psi}_{ex}. \quad (8.52a)$$

$$O_J[\phi_x(a\hat{z}, \tau)] = +\phi_x(0, \tau) - 2\phi_x(a\hat{z}, \tau) + \phi_x(a\hat{y} + a\hat{z}, \tau) +$$
$$\phi_y(a\hat{z}, \tau) - \phi_y(a\hat{x} + a\hat{z}, \tau) - \phi_z(0, \tau) + \phi_z(a\hat{x}, \tau) -$$
$$2\pi(\hat{z} + \hat{y}) \cdot \vec{\psi}_{ex}, \quad (8.52b)$$

$$O_J[\phi_x(a\hat{y} + a\hat{z}, \tau)] = \phi_x(a\hat{y}, \tau) + \phi_x(a\hat{z}, \tau) - 2\phi_x(a\hat{y} +$$
$$a\hat{z}, \tau) - \phi_y(a\hat{z}, \tau) + \phi_y(a\hat{x} + a\hat{z}, \tau) - \phi_z(a\hat{y}, \tau) +$$
$$\phi_z(a\hat{x} + a\hat{y}, \tau) + 2\pi(\hat{z} - \hat{y}) \cdot \vec{\psi}_{ex}. \quad (8.52c)$$

Notice that Eq. (8.52b) can be obtained by Eq. (8.52a) by simply exchanging y and z and by changing sign to the externally applied flux term. By a similar procedure, exchanging x and y in Eq. (8.52a–c) and by opportunely changing the sign of the externally applied flux term, the three equations for $\phi_y(a\hat{x}, \tau)$, $\phi_y(a\hat{z}, \tau)$, and $\phi_y(a\hat{x} + a\hat{z}, \tau)$ can be written as follows:

$$O_J[\phi_y(a\hat{x}, \tau)] = -\phi_x(0, \tau) + \phi_x(a\hat{y}, \tau) + \phi_y(0, \tau) -$$
$$2\phi_y(a\hat{x}, \tau) + \phi_y(a\hat{x} + a\hat{z}, \tau) - \phi_z(a\hat{x} + a\hat{y}, \tau) + \phi_z(a\hat{x}, \tau) -$$
$$2\pi(\hat{x} + \hat{z}) \cdot \vec{\psi}_{ex}. \quad (8.53a)$$

$$O_J[\phi_y(a\hat{z}, \tau)] = \phi_x(a\hat{z}, \tau) - \phi_x(a\hat{y} + a\hat{z}, \tau) + \phi_y(0, \tau) -$$
$$2\phi_y(a\hat{z}, \tau) + \phi_y(a\hat{x} + a\hat{z}, \tau) + \phi_z(a\hat{y}, \tau) - \phi_z(0, \tau) +$$
$$2\pi(\hat{x} + \hat{z}) \cdot \vec{\psi}_{ex}, \tag{8.53b}$$

$$O_J[\phi_y(a\hat{x} + a\hat{z}, \tau)] = -\phi_x(a\hat{z}, \tau) + \phi_x(a\hat{y} + a\hat{z}, \tau) +$$
$$\phi_y(a\hat{x}, \tau) + \phi_y(a\hat{z}, \tau) - 2\phi_y(a\hat{x} + a\hat{z}, \tau) - \phi_z(a\hat{x}, \tau) +$$
$$\phi_z(a\hat{x} + a\hat{y}, \tau) + 2\pi(\hat{x} - \hat{z}) \cdot \vec{\psi}_{ex}. \tag{8.53c}$$

Finally, by exchanging x and z in Eq. (8.52a-c) and by opportunely changing the sign of the externally applied flux term, the three equations for $\phi_z(a\hat{x}, \tau)$, $\phi_z(a\hat{y}, \tau)$, and $\phi_z(a\hat{x} + a\hat{y}, \tau)$ can be written as follows:

$$O_J[\phi_z(a\hat{x}, \tau)] = -\phi_x(0, \tau) + \phi_x(a\hat{z}, \tau) + \phi_y(a\hat{x}, \tau) -$$
$$\phi_y(a\hat{x} + a\hat{z}, \tau) - 2\phi_z(a\hat{x}, \tau) + \phi_z(0, \tau) + \phi_z(a\hat{x} + a\hat{y}, \tau) +$$
$$2\pi(\hat{x} + \hat{y}) \cdot \vec{\psi}_{ex}. \tag{8.54a}$$

$$O_J[\phi_z(a\hat{y}, \tau)] = \phi_x(a\hat{y}, \tau) - \phi_x(a\hat{y} + a\hat{z}, \tau) - \phi_y(0, \tau) +$$
$$\phi_y(a\hat{z}, \tau) + \phi_z(0, \tau) - 2\phi_z(a\hat{y}, \tau) + \phi_z(a\hat{x} + a\hat{y}, \tau) -$$
$$2\pi(\hat{x} + \hat{y}) \cdot \vec{\psi}_{ex}. \tag{8.54b}$$

$$O_J[\phi_z(a\hat{x} + a\hat{y}, \tau)] = -\phi_x(a\hat{y}, \tau) + \phi_x(a\hat{y} + a\hat{z}, \tau) -$$
$$\phi_y(a\hat{x}, \tau) + \phi_y(a\hat{x} + a\hat{z}, \tau) + \phi_z(a\hat{y}, \tau) + \phi_z(a\hat{x}, \tau) -$$
$$2\phi_z(a\hat{y} + a\hat{x}, \tau) - 2\pi(\hat{x} - \hat{y}) \cdot \vec{\psi}_{ex}. \tag{8.54c}$$

In the next section the equations of the superconducting phase dynamics for three orthogonal square elements and for single cubes and will be solved numerically.

8.3. Numerical analysis of simple networks

In the previous section we have derived the dynamic equations for the superconducting phase differences across the junctions of two simple networks, one consisting of three orthogonal square elements, one consisting of a cubic structure.

For the first type of network, we found the following six dynamic equations (8.39a–f). For the cubic structure, instead, the dynamic equations are the following: (8.27a–c); (8.52a–c); (8.53a–c); (8.54a–c).

In the following two subsections a numerical solution of the dynamic equations will be given for these two simple networks.

8.3.1. *Three orthogonal square elements*

Let us consider the system of six coupled first-order differential equations for the network represented in Fig. 8.3, namely, Eq. (8.39a–f). We solve numerically for the magnetic state realized in this network, by finding the stationary values of the superconducting phase differences as a function of $\vec{\psi}_{ex}$. In order to carry out this result, we start from zero-field-cooling conditions (null superconducting phase differences at $t = 0$). The amplitude ψ_{ex} of the externally applied normalized magnetic flux $\vec{\psi}_{ex}$ is then slowly raised from zero, so that intermediate metastable states, starting from $\psi_{ex} = 0$, can be recorded. Differently from previous analyses carried out in the present work, we expect that, because of the vector nature of $\vec{\psi}_{ex}$, the system will show different responses depending on the direction in which $\vec{\psi}_{ex}$ is oriented. We thus express $\vec{\psi}_{ex}$ in polar coordinates as follows:

$$\vec{\psi}_{ex} = \psi_{ex}(\sin\theta\cos\varphi\,\hat{x} + \sin\theta\sin\varphi\,\hat{y} + \cos\theta\,\hat{z}). \qquad (8.55)$$

In this way, once the angles θ and φ are fixed, so that a specific direction of $\vec{\psi}_{ex}$ is defined, the amplitude ψ_{ex} can be varied and the response of the system recorded in terms of this amplitude for that specific direction. As a first type of check, let us apply the field along the three coordinate axes, namely, x, y, z. In this way, when the external magnetic field is applied along the x −axis, we have $\theta = \pi/2$ and $\varphi = 0$. When it is applied along the y −axis, we have $\theta = \pi/2$ and $\varphi = \pi/2$. Finally, along the z −axis, we have $\theta = 0$. It should then be possible to recognize two different types of responses, one along the axis in which the field is applied, one on the orthogonal plane. Therefore, by applying the field along the x −axis, taking $\vec{\psi}_{ex} = \psi_{ex}\hat{x}$, we expect that the magnetic response of the system will be mainly determined by the junctions lying in the $y - z$ plane. Therefore, the flux $\psi_x(0)$ and the current $i_x(0)$ are the most relevant quantities in determining the magnetic behavior of the system. However, due to the lack of complete cubic symmetry, the quantities $\psi_y(0)$ and $i_y(0)$, and $\psi_z(0)$ and $i_z(0)$, do play a role. In this case, for what previously stated, one expects to have: $\psi_y(0) = \psi_z(0)$; $i_y(0) = i_z(0)$. This is only a partial check for our system. A complete check could be given by getting an identical magnetic behavior, *mutatis mutandis*, when considering $\vec{\psi}_{ex} = \psi_{ex}\hat{y}$ or $\vec{\psi}_{ex} = \psi_{ex}\hat{z}$.

Therefore, we start by providing this complete check to validate our numerical approach. Let us then take $\beta = 1.0$ and let us associate the blue color to the quantities $\psi_x(0)$ and $i_x(0)$, the red color to $\psi_y(0)$ and $i_y(0)$, and the black color to $\psi_z(0)$ and $i_z(0)$.

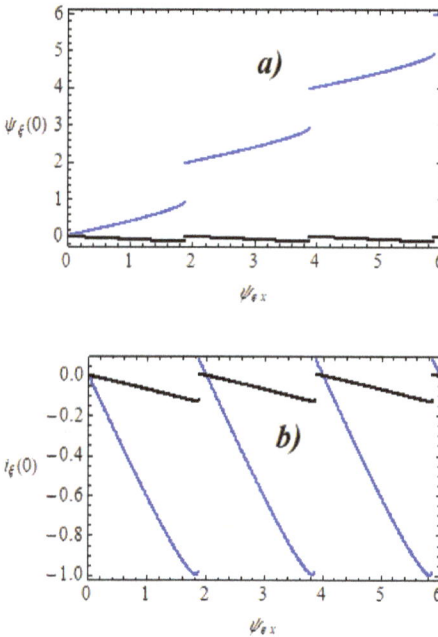

Fig. 8.5. Magnetic response of a network of JJs consisting of three orthogonal square elements for a magnetic applied in the x −direction. The stationary fluxes $\psi_\xi(0)$, and the stationary currents $i_\xi(0)$, where $\xi = x, y, z$, are shown in panel a) and panel b), respectively. As specified in the text, the blue curves represent the quantities $\psi_x(0)$ and $i_x(0)$. The black curves, on the other hand, represent the remaining quantities $\psi_y(0)$ and $i_y(0)$, and $\psi_z(0)$ and $i_z(0)$, since the red curves, representing the quantities $\psi_y(0)$ and $i_y(0)$, as specified in the text, here coincide with the black curves.

By applying the field first in the x−direction ($\vec{\psi}_{ex} = \psi_{ex}\hat{x}$) and by plotting the $\psi_\xi(0)$ vs. ψ_{ex} curves and $i_\xi(0)$ vs. ψ_{ex} curves ($\xi = x, y, z$) we should notice that blue curves determine the overall magnetic behavior of the system, in the way we shall later see, while red and black curves play a secondary role in this respect. This behavior is clearly visible in Fig. 8.5a–b, where the blue curves, representing the quantities $\psi_x(0)$ and $i_x(0)$, indeed play the major role in determining the magnetic response of the system. In fact, we may notice that the amplitudes of the black curves, which represent both the fluxes $\psi_y(0)$ and $\psi_z(0)$ in Fig. 8.5a and both the currents $i_y(0)$, and $i_z(0)$ in Fig. 8.5b, oscillate very close to the zero value.

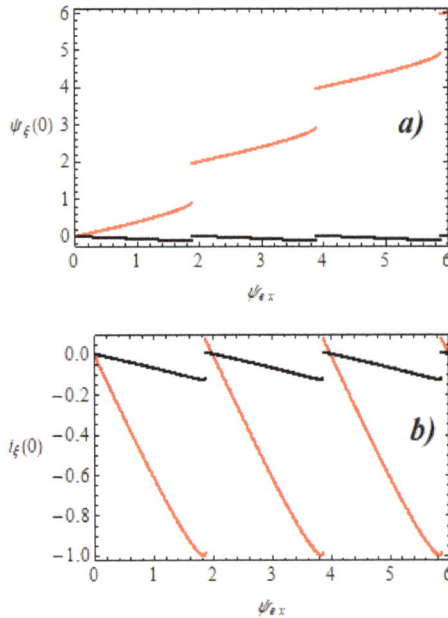

Fig. 8.6. Magnetic response of a network of JJs consisting of three orthogonal square elements for a field applied in the y−direction. The stationary fluxes $\psi_\xi(0)$, and the stationary currents $i_\xi(0)$, where $\xi = x, y, z$, are shown in panel a) and panel b), respectively. The red curves represent the quantities $\psi_y(0)$ and $i_y(0)$. The black curve, on the other hand, represent the remaining quantities $\psi_x(0)$ and $i_x(0)$, and $\psi_z(0)$ and $i_z(0)$, since the blue curves, representing the quantities $\psi_x(0)$ and $i_x(0)$, as specified in the text, here coincide with the black curves.

Moreover, as in two-dimensional systems, where attainment of the limiting value of the maximum Josephson current determines irreversible flux penetration, from Fig. 8.5b we see that this is possible only when $i_x(0) = -1$. The discontinuity in the $\psi_x(0)$ vs. ψ_{ex} blue curves in Fig. 8.5a confirms that this process occurs in the loop lying on the $y - z$ plane. The system thus behaves like a quasi-two-dimensional system. Notice however that, even for fields in the x −direction, magnetic flux is present in loops lying on the $x - z$ and $y - z$ planes.

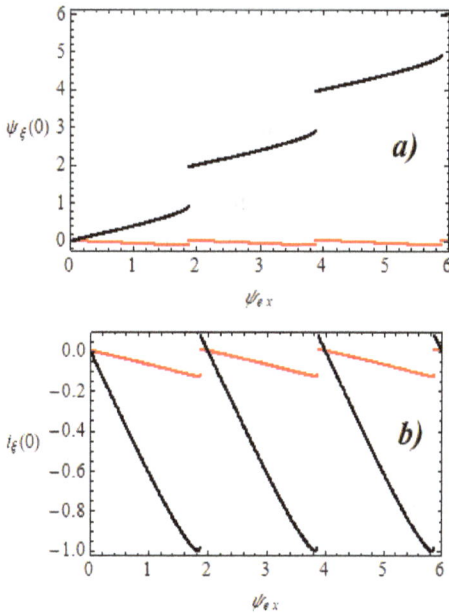

Fig. 8.7. Magnetic response of a network of JJs consisting of three orthogonal square elements for a field applied in the z −direction. The stationary fluxes $\psi_\xi(0)$, and the stationary currents $i_\xi(0)$, where $\xi = x, y, z$, are shown in panel a) and panel b), respectively. The black curves represent the quantities $\psi_z(0)$ and $i_z(0)$. The red curves, on the other hand, represent the remaining quantities $\psi_x(0)$ and $i_x(0)$, and $\psi_y(0)$ and $i_y(0)$, since the blue curves, representing the quantities $\psi_x(0)$ and $i_x(0)$, as specified in the text, here coincide with the red curves.

Let us now apply the magnetic field in the y −direction ($\vec{\psi}_{ex} = \psi_{ex}\hat{y}$). The $\psi_\xi(0)$ vs. ψ_{ex} curves and the $i_\xi(0)$ vs. ψ_{ex} curves are reported in

Fig. 8.6a and Fig. 8.6b, respectively. In this case we notice that red curves determine the overall magnetic behavior of the system, while blue and black curves play a secondary role in this respect. Notice also that the response of the system is identical to what obtained for a magnetic field applied in the x −direction.

Finally, by applying the magnetic field in the z −direction ($\vec{\psi}_{ex} = \psi_{ex}\hat{z}$), we obtain the $\psi_\xi(0)$ vs. ψ_{ex} curves and the $i_\xi(0)$ vs. ψ_{ex} curves shown in Fig. 8.7a and Fig. 8.7b, respectively. In this case we notice that black curves determine the overall magnetic behavior of the system. Once more, the response of the system is identical to what obtained for a magnetic field applied in the x −direction or in the y −direction.

Having validated the numerical routine for determining the magnetic states of the Josephson junction network shown in Fig. 8.3, we may now try to investigate how these states depend on the angles θ and φ or on the parameter β. Of course, the intrinsic complexity of the differential equations 8.39a–f, which govern the dynamics of the superconducting phase differences across the junctions in the network, will only allow us to numerically determine this dependence.

We shall then proceed as follows. We shall first fix the value of β to $\beta = 1.0$, as done before in plotting Fig.8.5a–b, Fig.8.6a–b, Fig.8.7a–b. By then considering three different values of the angle φ $(0, \pi/4, 3\pi/4)$, for each one of these values we let θ be $\pi/6, \pi/4, \pi/3$. Finally, by fixing $\varphi = 0$, we let the parameter β vary from $\beta = 0.5$ to $\beta = 3.0$ for $\theta = \pi/6, \pi/4, \pi/3$. At the end of this numerical analysis, we shall try to extract some general features of the magnetic states of the network under exam.

Let us then start by setting $\beta = 1.0$ and $\varphi = 0$. In Fig. 8.8a–c we report the $\psi_\xi(0)$ vs. ψ_{ex} curves for $\theta = \pi/6, \pi/4, \pi/3$, in the order. The $i_\xi(0)$ vs. ψ_{ex} curves, on the other hand, are shown, for $\theta = \pi/6, \pi/4, \pi/3$ in Fig. 8.9a–c, in the same order. As before, we have taken the blue, red, and black curves to represent the quantities $\psi_x(0)$, $\psi_y(0)$, and $\psi_z(0)$, respectively. Therefore, in Fig. 8.8a, obtained for $\theta = \pi/6$, we are able to detect the predominant values of $\psi_z(0)$ (black curve) over $\psi_x(0)$ (blue curve) and $\psi_y(0)$ (red curve). From Fig. 8.9a, however, obtained for the same value of θ, we may notice that, while the current $i_y(0)$ never reaches the minimum value -1, being it confined to the interval $[-0.2, 0]$, the current $i_x(0)$ does, so that irreversible flux penetration occurs both in the $x - y$ plane and in the $y - z$ plane.

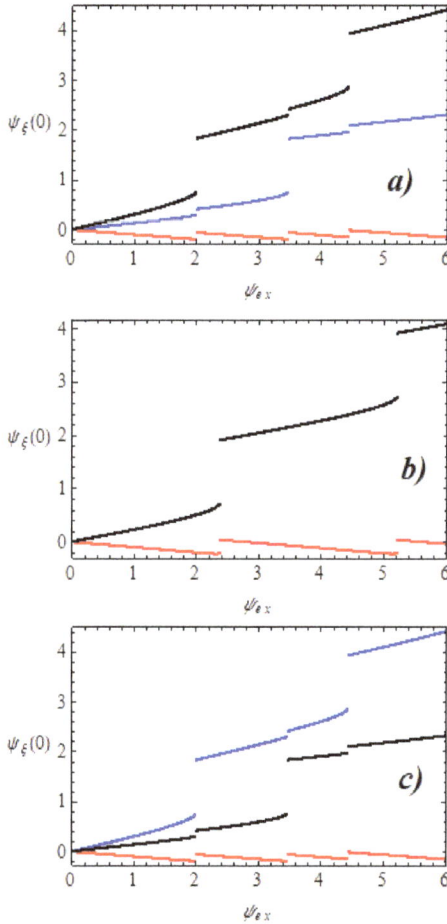

Fig. 8.8. Stationary normalized fluxes $\psi_\xi(0)$ as functions of ψ_{ex}. The blue, red, and black curves represent the quantities $\psi_x(0)$, $\psi_y(0)$, and $\psi_z(0)$, respectively. In all panels, we take $\beta = 1.0$ and $\varphi = 0$. The angle θ has been chosen as follows: $\theta = \pi/6$ in panel a), $\theta = \pi/4$ in panel b), $\theta = \pi/3$ in panel c). In panel b) the blue curve coincides with the black curve, so that it does not appear.

When considering Fig. 8.8b and 8.9b, obtained for $\theta = \pi/4$, we notice that the blue curves are not present.

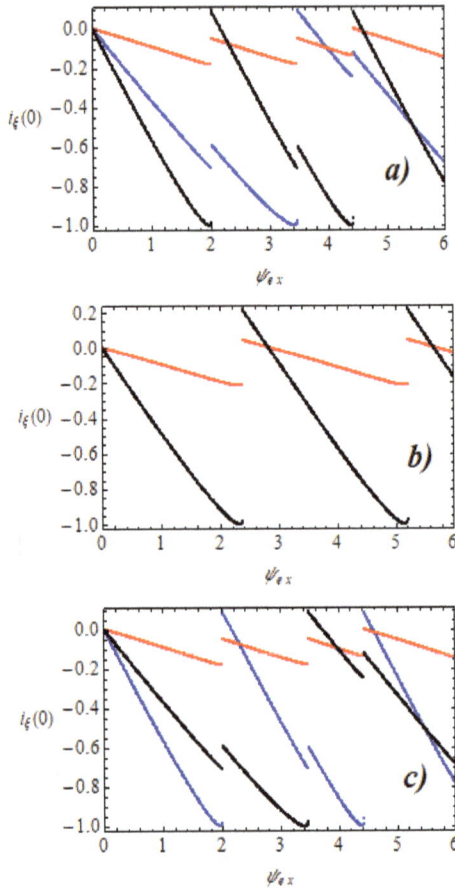

Fig. 8.9. Stationary normalized currents $i_\xi(0)$ as functions of ψ_{ex}. The blue, red, and black curves represent the quantities $i_x(0)$, $i_y(0)$, and $i_z(0)$, respectively. In all panels, we take $\beta = 1.0$ and $\varphi = 0$. The angle θ has been chosen as follows: $\theta = \pi/6$ in panel a), $\theta = \pi/4$ in panel b), $\theta = \pi/3$ in panel c). In panel b) the blue curve coincides with the black curve, so that it does not appear.

This occurs because the two loops, one lying on the $x - y$ plane and the other on the $y - z$ plane give an equal response to the applied field so that $\psi_x(0) = \psi_z(0)$ and $i_x(0) = i_z(0)$ and the blue and black curves are juxtaposed. We also notice that the blue curves and the black curves in Fig. 8.8c and 8.9c interchange their positions with respect to those in Fig. 8.8a and 8.9a, because the angles $\theta = \pi/6$ and $\theta' = \pi/3$ are complementary.

As a final remark, we may detect the values of ψ_{ex} at which first irreversible flux penetration occurs in the loops of the network, by either looking at Fig. 8.8a–c or at Fig. 8.9a–c. In fact, the first discontinuities with a jump $\Delta\psi_\xi(0) \approx 1$ in the $\psi_\xi(0)$ vs. ψ_{ex} curves in Fig. 8.8a–c correspond to first irreversible flux penetration in the system in the loop orthogonal to the ξ −direction. These flux transitions occur at $\psi_{ex} = \psi_1$, where ψ_1 is defined as in Chapter 6. Differently from the analysis carried out in Chapter 6 for two-dimensional systems, where the applied field was taken always orthogonal to the planar network of junctions, in three-dimensional networks we need to specify the loop, or the loops, in which this first irreversible flux transition occur. Corresponding to this principal transition, which might occur in one or more loops, depending on the particular symmetry introduced by the direction of the applied field, irreversible secondary transitions in adjacent loops with lower jumps can be detected.

More specifically, considering Fig. 8.8a and 8.9a, obtained for $\beta = 1, \varphi = 0$ and $\theta = \pi/6$, we may see that $\psi_1 \approx 2.0$. Notice that the principal flux transition, with a discontinuity $\Delta\psi_z(0) \approx 1$, occurs in the loop lying on the $x - y$ plane, while secondary transitions, with a much lower jump, occur in the remaining two loops. Notice also that at $\psi_{ex} \approx 3.4$ a discontinuity $\Delta\psi_x(0) \approx 1$ is detectable in Fig. 8.8a, so that, in this case, a principal transition occurs in the loop lying on the $y - z$ plane and only a secondary transition occurs in the loop lying on the $x - y$ plane. Therefore, the combined analysis of the $\psi_\xi(0)$ vs. ψ_{ex} and of the $i_\xi(0)$ vs. ψ_{ex} curves could provide a complete vision of the flux transition mechanisms in three-dimensional networks in much the same way as in two-dimensional systems.

Continuing our analysis, from Fig. 8.8b and 8.9b we may argue that, for $\beta = 1, \varphi = 0$, and $\theta = \pi/4$, we have $\psi_1 \approx 2.3$. This first irreversible flux transition occurs both in the loops lying on the $x - y$ plane and on the $y - z$ plane, because the $x -$ and $z -$ components of $\vec{\psi}_{ex}$ are equal. Moreover, a secondary transition occurs in the superconducting loop lying on the $x - z$ plane. Similar flux transitions occur at higher values of ψ_{ex}, as it can be seen from Fig. 8.8b and 8.9b.

Finally, from Fig. 8.8c and 8.9c, obtained for $\beta = 1$, $\varphi = 0$ and $\theta = \pi/3$, we have previously noticed that it is possible to obtain the magnetic states of the system by interchanging the components x and z of both $\psi_\xi(0)$ and $i_\xi(0)$. In this way, we argue that the principal flux transition, with a discontinuity $\Delta\psi_x(0) \approx 1$, occurs in the loop lying on the $y - z$ plane at $\psi_{ex} = \psi_1 \approx 2.0$.

In order to detect irreversibility generated by either principal or secondary flux transitions, we may integrate the differential equation first for increasing values of ψ_{ex}, starting from zero and going up to a maximum value $\psi_{ex}^{(max)}$, then for decreasing values of ψ_{ex}, from $\psi_{ex}^{(max)}$ to zero again. In Fig. 8.10a–b we show the results of this analysis for $\beta = 1$, $\varphi = 0$ and $\theta = \pi/6$. In order to retain the choice of color made before, we represented $\psi_x(0)$ with a blue curve, $\psi_y(0)$ with a red curve, and $\psi_z(0)$ with a black curve when raising the value of ψ_{ex} from zero to $\psi_{ex}^{(max)} = 6$. On the other hand, we have chosen to represent these same curves, when varying ψ_{ex} from $\psi_{ex}^{(max)}$ to zero, with cyan, orange, and gray color, in the order. In those intervals in which curves of the latter colors are not present (for example, where the blue and cyan curves coincide, so that the latter does not appear), the system behaves reversibly. By closely considering Fig. 8.10a–c, we might argue that, although the secondary transitions occur with a lower value of the discontinuity ($\Delta\psi_\xi(0) \ll 1$), they still indicate the presence of irreversible metastable magnetic states in the system. In fact, by considering Fig. 8.10a, we notice that in the ψ_{ex} region close to secondary flux transition occurring at $\psi_{ex} \approx 2.0$ and $\psi_{ex} \approx 4.6$ both cyan and blue curves coexist. Similarly, in Fig. 8.10b and 8.10c, the same can be noticed for all secondary transitions.

We now proceed in finding the magnetic stationary states for $\varphi = \pi/4$ and for $\theta = \pi/6, \pi/4, \pi/3$. Taking a first look at Fig. 8.11a–c and Fig. 8.12a–c, where the $\psi_\xi(0)$ vs. ψ_{ex} curves and the $i_\xi(0)$ vs. ψ_{ex} curves are represented, respectively, we soon notice the absence of the blue curves corresponding to $\psi_x(0)$ in Fig. 8.11a–c and to $i_x(0)$ in Fig. 8.12a–c. This absence is due to the following equalities, which are true for this particular case: $\psi_x(0) = \psi_y(0)$; $i_x(0) = i_y(0)$.

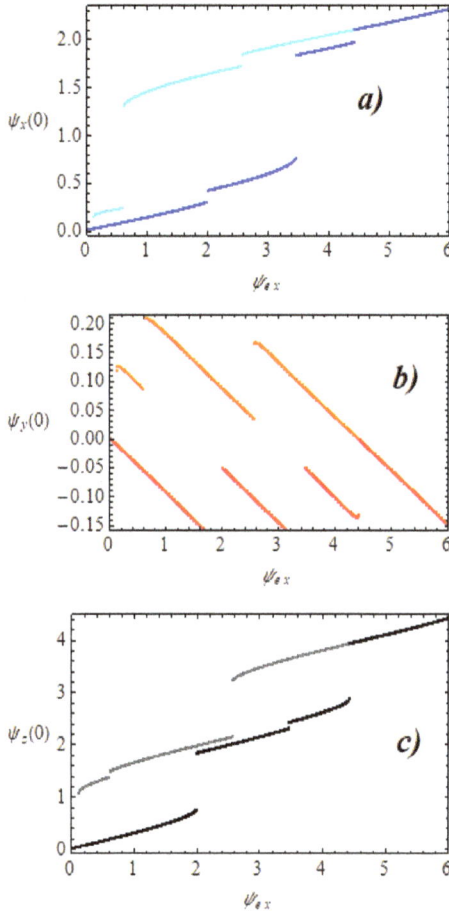

Fig. 8.10 Stationary normalized fluxes $\psi_\xi(0)$ as functions of ψ_{ex} both for increasing and decreasing field. In all panels, we take $\beta = 1.0$, $\varphi = 0$, $\theta = \pi/6$ In panel a) $\psi_x(0)$ is shown for increasing (blue curve) and decreasing (cyan curve) values of ψ_{ex}. In panel b) $\psi_y(0)$ is shown for increasing (red curve) and decreasing (orange curve) values of ψ_{ex}. In panel c) $\psi_z(0)$ is shown for increasing (black curve) and decreasing (gray curve) values of ψ_{ex}. In the intervals of ψ_{ex} for which the cyan, orange, and gray curves do not appear separately from the blue, red, and black curve, in the order, the system behaves reversibly.

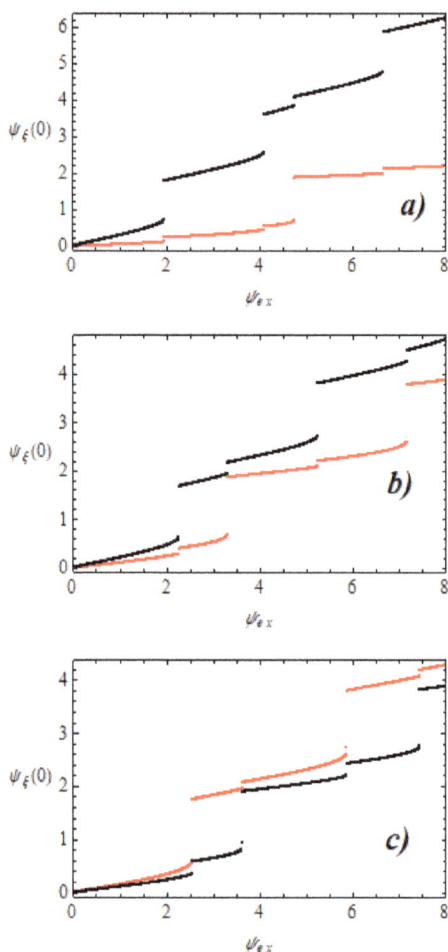

Fig. 8.11 Stationary normalized fluxes $\psi_\xi(0)$ as functions of ψ_{ex}. The red and black curves represent the quantities, $\psi_y(0)$, and $\psi_z(0)$, respectively. In all panels, we take $\beta = 1.0$ and $\varphi = \pi/4$. The angle θ has been chosen as follows: $\theta = \pi/6$ in panel a), $\theta = \pi/4$ in panel b), $\theta = \pi/3$ in panel c). In this case particular case $\psi_x(0) = \psi_y(0)$, so that the blue curve for $\psi_x(0)$ does not appear.

In fact, when the magnetic field is applied at $\varphi = \pi/4$, the superconducting loop lying on the $x - z$ plane is subject to the same forcing action which $\vec{\psi}_{ex}$ exerts on the loop lying on the $y - z$ plane.

Therefore, this particular orientation of $\vec{\psi}_{ex}$ determines an identical magnetic response of the two loops orthogonal to the $x-$ and $y-$direction. Let us now analyse the magnetic states for increasing values of θ.

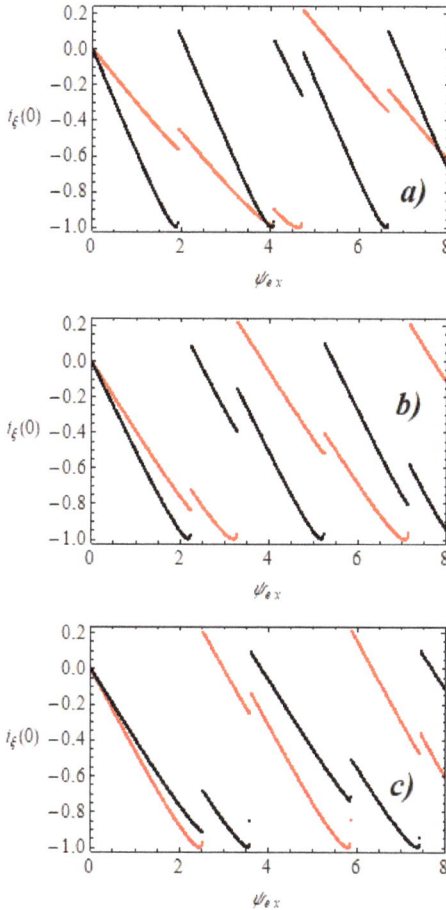

Fig. 8.12. Stationary normalized currents $i_\xi(0)$ as functions of ψ_{ex}. The red and black curves represent the quantities, $i_y(0)$, and $i_z(0)$, respectively. In all panels, we take $\beta = 1.0$ and $\varphi = \pi/4$. The angle θ has been chosen as follows: $\theta = \pi/6$ in panel a), $\theta = \pi/4$ in panel b), $\theta = \pi/3$ in panel c). In this case particular case $i_x(0) = i_y(0)$, so that the blue curve for $i_x(0)$ does not appear.

For $\theta = \pi/6$, the component of $\vec{\psi}_{ex}$ orthogonal to the $x - y$ plane is

$$\vec{\psi}_{ex} \cdot \hat{z} = \frac{\sqrt{3}}{2} \psi_{ex},\tag{8.56}$$

while the other two components are

$$\vec{\psi}_{ex} \cdot \hat{x} = \vec{\psi}_{ex} \cdot \hat{y} = \frac{\sqrt{2}}{4} \psi_{ex}.\tag{8.57}$$

Therefore, in this case the forcing term acting on the superconducting loop lying on the $x - y$ plane is greater than the corresponding term acting on the remaining two loops of the network. In this way, we expect, as confirmed by Fig. 8.11a, that the first irreversible transition occurs in the superconducting loop lying on the $x - y$ plane. Looking closely to Fig. 8.11a and 8.12a, we notice that this transition occurs at $\psi_{ex} = \psi_1 \approx 2.0$ and is accompanied by a secondary irreversible transition in the two remaining loops lying on the $x - z$ plane and on the $y - z$ plane, respectively.

For $\theta = \pi/4$ the component of $\vec{\psi}_{ex}$ orthogonal to the $x - y$ plane is

$$\vec{\psi}_{ex} \cdot \hat{z} = \frac{\sqrt{2}}{2} \psi_{ex},\tag{8.58}$$

while the other two components are

$$\vec{\psi}_{ex} \cdot \hat{x} = \vec{\psi}_{ex} \cdot \hat{y} = \frac{1}{2} \psi_{ex}.\tag{8.59}$$

Therefore, as for the preceding case, for $\theta = \pi/4$ we have $\vec{\psi}_{ex} \cdot \hat{z} > \vec{\psi}_{ex} \cdot \hat{x}$, so that the first irreversible transition is expected to occur in the superconducting loop lying on the $x - y$ plane. Notice, also, that this flux transition takes place at $\psi_{ex} = \psi_1 \approx 2.25$, as it can be estimated from Fig. 8.11b and Fig. 8.12b. The same principal transition is accompanied, as in the previous case, by a secondary irreversible transition in the two remaining loops.

For $\theta = \pi/3$, on the other hand, the component of $\vec{\psi}_{ex}$ orthogonal to the $x - y$ plane is

$$\vec{\psi}_{ex} \cdot \hat{z} = \frac{1}{2} \psi_{ex},\tag{8.60}$$

while the other two components are

$$\vec{\psi}_{ex} \cdot \hat{x} = \vec{\psi}_{ex} \cdot \hat{y} = \frac{\sqrt{6}}{4} \psi_{ex}. \tag{8.61}$$

Therefore, in this case we have $\vec{\psi}_{ex} \cdot \hat{x} > \vec{\psi}_{ex} \cdot \hat{z}$, so that the first irreversible transition is expected to occur in both the superconducting loop lying on the $x - z$ plane and on the $y - z$ plane. Notice, also, that this flux transition takes place at $\psi_{ex} = \psi_1 \approx 2.5$, as it can be estimated from Fig. 8.11c and Fig. 8.12c. The same principal transition is accompanied by a secondary irreversible transition, this time occurring in the superconducting loop lying on the $x - y$ plane.

Let us now consider the case in which, still setting $\beta = 1.0$, we take $\varphi = 3\pi/4$. In Fig. 8.13a–c we report the $\psi_\xi(0)$ vs. ψ_{ex} curves for $\theta = \pi/6, \pi/4, \pi/3$, in the order. The $i_\xi(0)$ vs. ψ_{ex} curves, on the other hand, are shown, for $\theta = \pi/6, \pi/4, \pi/3$ in Fig. 8.14a–c, respectively. The blue, red, and black curves in Fig. 8.13a–c represent the quantities $\psi_x(0)$, $\psi_y(0)$, and $\psi_z(0)$, in the order. In an analogous way, the blue, red, and black curves in Fig. 8.14a–c represent the quantities $i_x(0)$, $i_y(0)$, and $i_z(0)$, respectively.

For $\theta = \pi/6$ the component of $\vec{\psi}_{ex}$ are the following:

$$\vec{\psi}_{ex} \cdot \hat{x} = -\frac{\sqrt{2}}{4} \psi_{ex}, \tag{8.62a}$$

$$\vec{\psi}_{ex} \cdot \hat{y} = +\frac{\sqrt{2}}{4} \psi_{ex}, \tag{8.62b}$$

$$\vec{\psi}_{ex} \cdot \hat{z} = +\frac{\sqrt{3}}{2} \psi_{ex}. \tag{8.62c}$$

As shown in Fig. 8.13a, the negative sign of the x −component of the externally applied flux affects the sign of the fluxes in the loop lying on the $y - z$ plane. Therefore, if for $\varphi = \pi/4$ we detected positive values for $\psi_x(0)$, in the present case we have $\psi_x(0) < 0$. Corresponding to this sign reversal for $\psi_x(0)$, the current $i_x(0)$ is initially positive, at least up to the first irreversible flux transition, which is now to be understood very carefully. In fact, looking at Fig. 14a, we notice that no discontinuity is present in the current $i_y(0)$. Therefore, this current cannot determine first irreversible flux transition within the system.

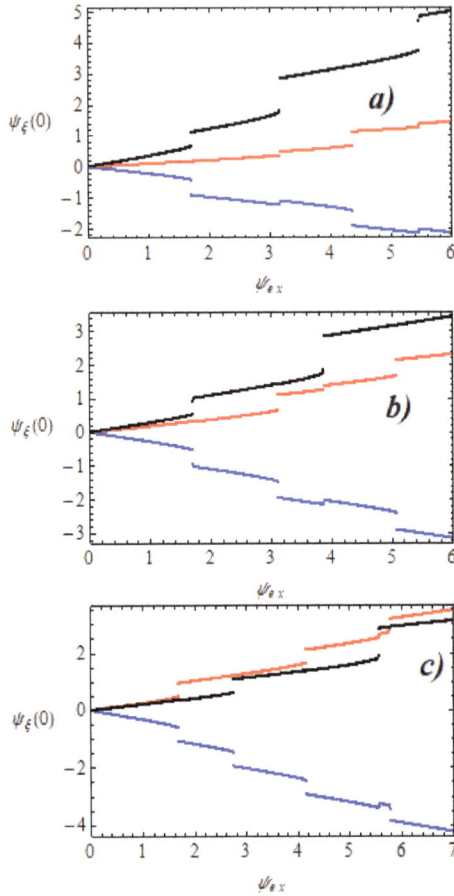

Fig. 8.13 Stationary normalized fluxes $\psi_\xi(0)$ as functions of ψ_{ex}. The blue, red, and black curves represent the quantities $\psi_x(0)$, $\psi_y(0)$, and $\psi_z(0)$, respectively. In all panels, we take $\beta = 1.0$ and $\varphi = 3\pi/4$. The angle θ has been chosen as follows: $\theta = \pi/6$ in panel a), $\theta = \pi/4$ in panel b), $\theta = \pi/3$ in panel c).

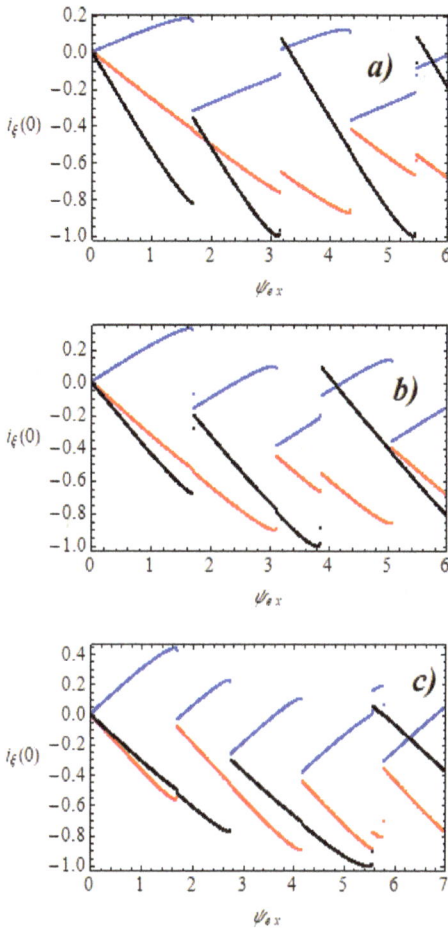

Fig. 8.14. Stationary normalized currents $i_\xi(0)$ as functions of ψ_{ex}. The blue, red, and black curves represent the quantities $i_x(0)$, $i_y(0)$, and $i_z(0)$, respectively. In all panels, we take $\beta = 1.0$ and $\varphi = 3\pi/4$. The angle θ has been chosen as follows: $\theta = \pi/6$ in panel a), $\theta = \pi/4$ in panel b), $\theta = \pi/3$ in panel c).

In order to justify this process, we thus need to consider the role played by the currents $i_x(0)$ and $i_z(0)$. The schematic representation given in Fig. 8.15 gives a rather clear picture of the current distribution in the system. In this representation the current $i_x(0)$ is seen to flow in the counterclockwise direction, as observed from a position on the positive x −axis. Therefore, the currents $i_x(0)$ and $i_z(0)$ flow in the same direction in the JJ on the $y-$axis characterized by the superconducting phase differences $\phi_y(0)$.

Similarly, the currents $i_x(0)$ and $i_y(0)$ flow in the same direction in the JJ on the $z-$axis characterized by the superconducting phase differences $\phi_z(0)$. From what above observed, in the first irreversible flux penetration process for $\theta = \pi/6$, both currents $i_x(0)$ and $i_z(0)$ concur in determining the phase slip of the JJ characterized by the superconducting phase differences $\phi_y(0)$. In this way, when the sum of the absolute values of $i_x(0)$ and $i_z(0)$ is equal to one, discontinuities in the $\psi_x(0)$ and $\psi_z(0)$ curve are noticed, as it can be seen from Fig. 8.13a for $\psi_{ex} \approx 1.7$. The second irreversible flux transition within the system occurs because the current $i_z(0)$ approaches the value $i_z(0) = -1$ at $\psi_{ex} \approx 3.1$. Moreover, in the third transition, at $\psi_{ex} \approx 4.4$, both currents $i_x(0)$ and $i_y(0)$ are directly involved in much the same way $i_x(0)$ and $i_z(0)$ concurred in determining the first irreversible flux transition. In this case, as we have already noticed, the junction involved in the flux transition is the one characterized by the superconducting phase difference $\phi_z(0)$.

In Fig. 8.13b, obtained for $\theta = \pi/4$, we detect a different pattern in the flux penetration mechanisms, which we shall here describe in some details. First of all, notice that the component of $\vec{\psi}_{ex}$ for $\theta = \pi/4$ are the following:

$$\vec{\psi}_{ex} \cdot \hat{x} = -\frac{1}{2}\psi_{ex}, \tag{8.64a}$$

$$\vec{\psi}_{ex} \cdot \hat{y} = +\frac{1}{2}\psi_{ex}, \tag{8.64b}$$

$$\vec{\psi}_{ex} \cdot \hat{z} = +\frac{\sqrt{2}}{2}\psi_{ex}. \tag{8.64c}$$

As far first irreversible flux transition is concerned, the current $i_y(0)$ does not contribute to the penetration process. In fact, only the currents $i_x(0)$ and $i_z(0)$ show relevant discontinuities at $\psi_{ex} \approx 1.6$ in Fig. 8.14b. Therefore, the junction involved in this flux transition is the one characterized by a superconducting phase difference $\phi_y(0)$, so that discontinuities in the fluxes $\psi_x(0)$ and $\psi_z(0)$ appear in Fig. 8.13b. The second irreversible flux transition occurring at $\psi_{ex} \approx 3.2$ can be justified by a phase slip of the JJ on the z −axis characterized by the superconducting phase differences $\phi_z(0)$. The currents $i_x(0)$ and $i_y(0)$ flowing in this junction, in fact, present a relevant discontinuity at this value of ψ_{ex}, as shown in Fig. 8.14b.

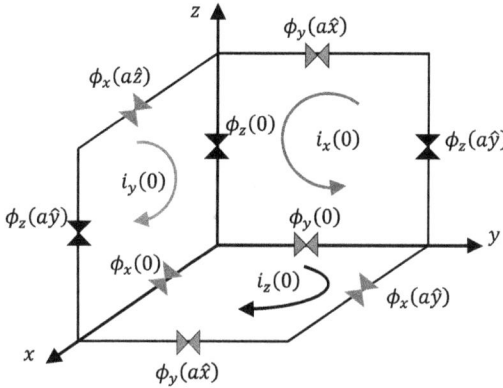

Fig. 8.15. Signs of the stationary normalized currents $i_\xi(0)$ for $\beta = 1.0$, $\varphi = 3\pi/4$ and for $0 < \theta < \pi/2$. Notice that in the junction characterized by a superconducting phase difference $\phi_y(0)$ both currents $i_x(0)$ and $i_z(0)$ flow in the same direction. The same happens for the currents $i_x(0)$ and $i_y(0)$ flowing in the junction across which a phase difference $\phi_z(0)$ is present.

The third irreversible flux transition occurring at $\psi_{ex} \approx 3.9$ is due to phase slip of the junctions characterized by the superconducting phase differences $\phi_x(a\hat{y})$ and $\phi_y(a\hat{x})$ and is caused by the fact that the current $i_z(0)$ approaches the value $i_z(0) = -1$ as the field increases on the left of the abscissa $\psi_{ex} \approx 3.9$.

Finally, in Fig. 8.13c, obtained for $\theta = \pi/3$, we detect still another pattern in the flux penetration mechanisms. The component of $\vec{\psi}_{ex}$ for $\theta = \pi/3$ are the following:

$$\vec{\psi}_{ex} \cdot \hat{x} = -\frac{\sqrt{6}}{4} \psi_{ex}, \tag{8.65a}$$

$$\vec{\psi}_{ex} \cdot \hat{y} = +\frac{\sqrt{6}}{4} \psi_{ex}, \tag{8.65b}$$

$$\vec{\psi}_{ex} \cdot \hat{z} = +\frac{1}{2}\psi_{ex}. \tag{8.65c}$$

As far as first irreversible flux transition is concerned, the current $i_z(0)$ does not contribute to the penetration process. In fact, only the currents $i_x(0)$ and $i_y(0)$ show relevant discontinuities at $\psi_{ex} \approx 1.7$ in Fig. 8.14c. Therefore, the junction involved in this flux transition is the one characterized by a superconducting phase difference $\phi_z(0)$, so that discontinuities in the fluxes $\psi_x(0)$ and $\psi_y(0)$ appear in Fig. 8.13c.

The second irreversible flux transition, occurring at $\psi_{ex} \approx 2.8$, is caused by a phase slip of the JJ on the y −axis characterized by the superconducting phase differences $\phi_y(0)$. The currents $i_x(0)$ and $i_z(0)$ flowing in this Josephson junction, in fact, present a relevant discontinuity at this value of ψ_{ex}, as shown in Fig. 8.14c.

The third irreversible flux transition occurring at $\psi_{ex} \approx 4.2$ is caused, as in the first irreversible flux penetration process, by phase slip of the junctions characterized by the superconducting phase difference $\phi_z(0)$, thus giving discontinuity in the currents $i_x(0)$ and $i_y(0)$ at $\psi_{ex} \approx 4.2$ in Fig. 8.14c and discontinuity in the fluxes $\psi_x(0)$ and $\psi_y(0)$ in Fig. 8.13c at the same value of ψ_{ex}.

We shall now take a look at the behavior of the network as the parameter β is varied. By fixing $\varphi = 0$ and taking $\theta = \pi/6$ and $\pi/4$, we consider the numerical solution of Eq. (8.39a–f) for $\beta = 0.50$, $\beta = 1.50$, $\beta = 3.00$.

In Fig. 8.16a–c the normalized magnetic fluxes $\psi_x(0)$, $\psi_y(0)$, and $\psi_z(0)$ are shown in terms of ψ_{ex} for $\theta = \pi/6$ and for the three values of β mentioned above. Correspondingly, in Fig. 8.17a–c the normalized currents $i_x(0)$, $i_y(0)$, and $i_z(0)$ are shown as functions of ψ_{ex}. A first characteristic feature of the network, evidenced by these curves, is that the first irreversible flux penetration process occurs at increasing values of ψ_{ex}, as β increases. In fact, from Fig. 8.16a and 8.17a, we may see that, for $\beta = 0.50$, $\psi_1 \approx 1.4$. On the other hand, from Fig. 8.16b and 8.17b, we may notice that, for $\beta = 1.50$, $\psi_1 \approx 2.6$. Finally, from Fig. 8.16c and 8.17c, we may argue that, for $\beta = 3.00$, $\psi_1 \approx 4.4$. The monotonically increasing behavior of ψ_1 with β had been already seen in a single superconducting loop interrupted by one or more Josephson junctions. In the very simple case of one junction interrupting a superconducting loop, the monotonically increasing function in Eq. (2.52) holds. In this case, due to the complexity of the system, we rely on the numerical evaluations of ψ_1.

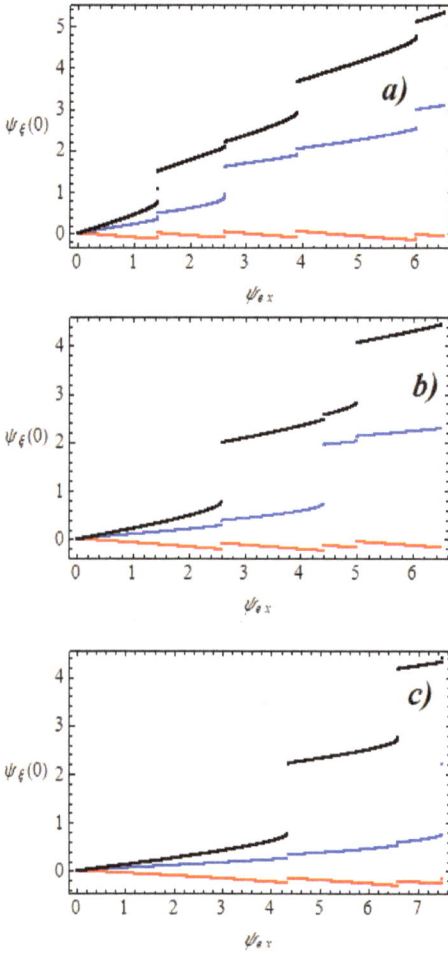

Fig. 8.16. Stationary normalized fluxes $\psi_\xi(0)$ as functions of ψ_{ex}. The blue, red, and black curves represent the quantities $\psi_x(0)$, $\psi_y(0)$, and $\psi_z(0)$, respectively. In all panels, we take $\theta = \pi/6$ and $\varphi = 0$. The parameter β has been chosen as follows: $\beta = 0.5$ in panel a), $\beta = 1.5$ in panel b), $\beta = 3.0$ in panel c).

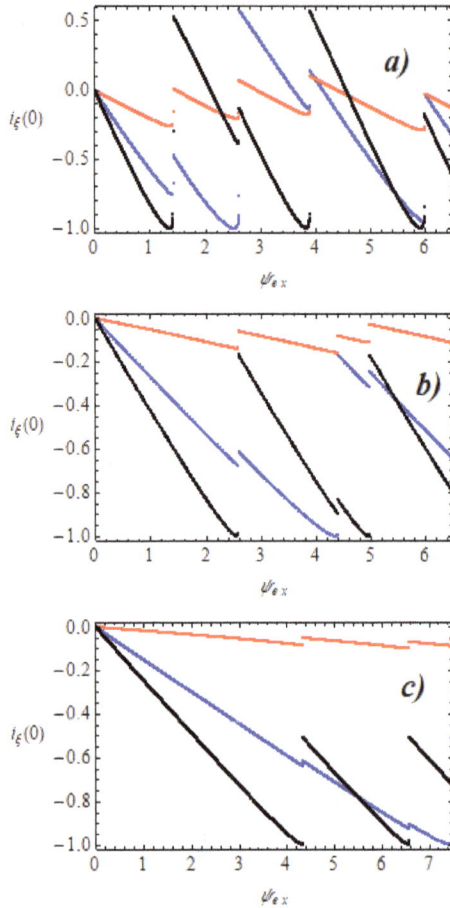

Fig. 8.17. Stationary normalized currents $i_\xi(0)$ as functions of ψ_{ex}. The blue, red, and black curves represent the quantities $i_x(0)$, $i_y(0)$, and $i_z(0)$, respectively. In all panels, we take $\theta = \pi/6$ and $\varphi = 0$. The parameter β has been chosen as follows: $\beta = 0.5$ in panel a), $\beta = 1.5$ in panel b), $\beta = 3.0$ in panel c).

In Fig. 8.18a–c the normalized magnetic fluxes $\psi_x(0)$, $\psi_y(0)$, and $\psi_z(0)$ are shown in terms of ψ_{ex} for $\theta = \pi/4$ and for the same three values of β as in Fig. 8.16a–c.

Correspondingly, in Fig. 8.19a–c the normalized currents $i_x(0)$, $i_y(0)$, and $i_z(0)$ are shown as functions of ψ_{ex}.

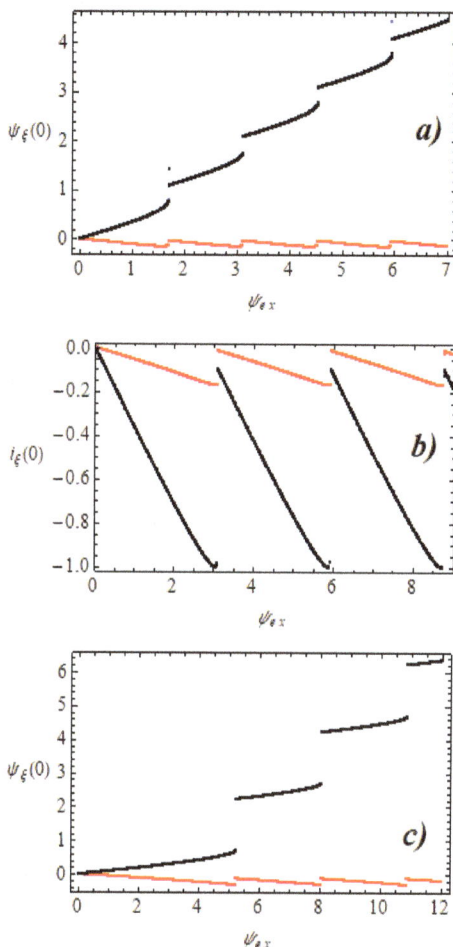

Fig. 8.18. Stationary normalized fluxes $\psi_\xi(0)$ as functions of ψ_{ex}. The red and black curves represent the quantities $\psi_y(0)$, and $\psi_z(0)$, respectively. In all panels, we take $\theta = \pi/4$ and $\varphi = 0$. The parameter β has been chosen as follows: $\beta = 0.5$ in panel a), $\beta = 1.5$ in panel b), $\beta = 3.0$ in panel c). The blue curve for $\psi_x(0)$ is not shown here, since it coincides with the black curve, given that $\psi_x(0) = \psi_z(0)$ in this case.

As for the first irreversible flux penetration process, it occurs at the following values of ψ_{ex}: from Fig. 8.18a and 8.19a, we have $\psi_1 \approx 1.7$ for $\beta = 0.50$; from Fig. 8.18b and 8.19b, we have $\psi_1 \approx 3.2$ for $\beta = 1.50$; finally, from Fig. 8.18c and 8.19c, we have $\psi_1 \approx 5.5$ for $\beta = 3.00$.

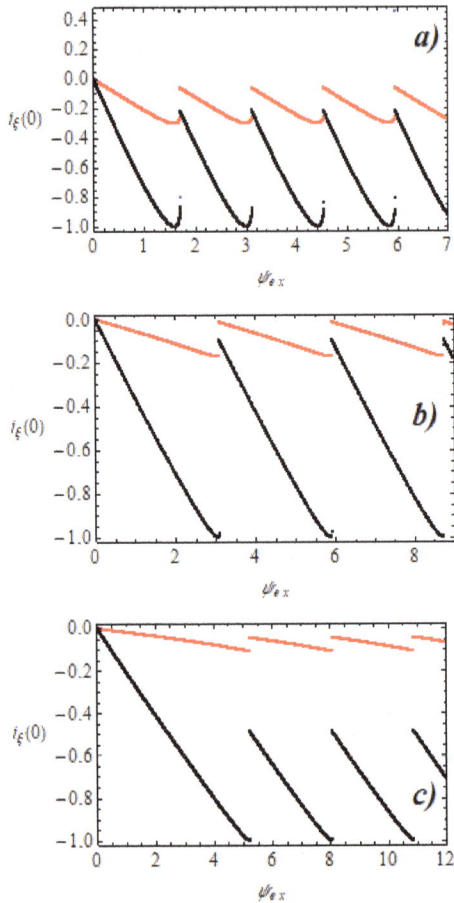

Fig. 8.19. Stationary normalized currents $i_\xi(0)$ as functions of ψ_{ex}. The red and black curves represent the quantities $i_y(0)$, and $i_z(0)$, respectively. In all panels, we take $\theta = \pi/4$ and $\varphi = 0$. The parameter β has been chosen as follows: $\beta = 0.5$ in panel a), $\beta = 1.5$ in panel b), $\beta = 3.0$ in panel c). The blue curves for $i_x(0)$ are not shown here, since they coincide with the black curves, given that $i_x(0) = i_z(0)$ in this case.

Also in this case, therefore, the monotonically increasing behavior of ψ_1 with β is preserved.

The detailed analysis carried out for $\beta = 1.0$ can be repeated in the cases we have considered above, varying the value of β and fixing $\varphi = 0$. After this rather lengthy discussion on the magnetic properties of the junction network consisting of three orthogonal loops, the reader may probably feel more confident in investigating more complex networks.

In the next subsection we shall in fact consider the response of a system with a higher degree of complexity, namely, the single cube.

8.3.2. *The single cube*

Let us consider the system schematically represented in Fig. 8.4: a network of twelve Josephson junctions located at the sides of a cube.

As compared to the network consisting of three orthogonal superconducting square loops, the cubic structure possesses a higher degree of complexity, given the higher number of nonlinear differential equations to be solved numerically, namely Eq. (8.27a–c), Eq. (8.52a–c), Eq. (8.53a–c), Eq. (8.54a–c). On the other hand, the system also shows useful symmetries, which allows us to make some simple considerations on its magnetic properties. In this first approach to the system, we shall only consider $\beta = 1$. As for the angles φ and θ, we first consider $\varphi = 0$ and $\theta = 0$. Successively, we investigate the cases $\varphi = 0$ and $\theta = \pi/6, \pi/4, \pi/3$, and $\varphi = \pi/6$ and $\theta = \pi/6, \pi/4, \pi/3$.

As in the case of a network consisting of three orthogonal square loops, we make a first run to validate the numerical procedure. The magnetic fluxes and the currents in the network are thus shown for $\beta = 1$, $\varphi = 0$, and $\theta = 0$ in Fig. 8.20a–b and Fig. 8.21a–b, respectively. In these curves we notice a regular behavior with increasing field of the quantities $\psi_z(0)$, $\psi_z(a\hat{z})$, $i_z(0)$, $i_z(a\hat{z})$ and the collapse to zero of the fluxes and currents in the lateral loops, as it was to be expected. Furthermore, the flux lines entering the network through the loop lying on the $x - y$ plane at the base of the cubic structure penetrate the structure vertically, so that they exit from the network from the upper cubic face orthogonal to the z −axis. This can be understood by noticing that, from Fig. 8.20a and 8.20b, the following equality holds: $\psi_z(0) = \psi_z(a\hat{z})$. As a consequence, as also shown in Fig. 8.21a and 8.21b, we have $i_z(0) = i_z(a\hat{z})$. As already seen when dealing with the total magnetic flux linked to the closed external surface of the network, we need to opportunely consider the choice of the sign convention that we made at the beginning of the present chapter. From the analysis of the

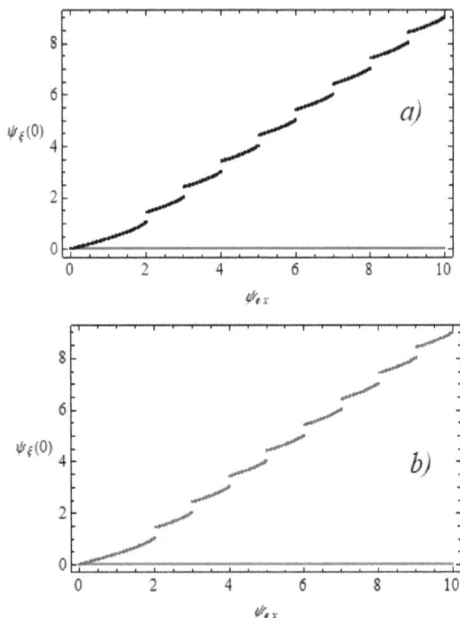

Fig. 8.20. Normalized fluxes as functions of ψ_{ex} in a cubic network for $\beta = 1.0$, $\theta = 0$, and $\varphi = 0$. In panel a) the stationary normalized fluxes $\psi_\xi(0)$ are show. The red and black curves represent the quantities $\psi_y(0)$, and $\psi_z(0)$, respectively. In panel b) the stationary normalized fluxes $\psi_\xi(a\hat{\xi})$ are shown. The orange and gray curves represent the quantities $\psi_y(a\hat{y})$, and $\psi_z(a\hat{z})$, respectively. The curves for $\psi_x(0)$ and $\psi_x(a\hat{x})$ are not shown here, since they coincide with the curves $\psi_y(0)$ and $\psi_y(a\hat{y})$, respectively. Notice the exact correspondence between the two sets of curves in panels a) and b).

properties of the system under a magnetic field orthogonal to the z−axis, we may generalize this behavior, *mutatis mutandis*, to the cases where $\vec{\psi}_{ex} = \psi_{ex}\,\hat{x}$ and $\vec{\psi}_{ex} = \psi_{ex}\,\hat{y}$. Finally, we notice that the first irreversible flux penetration process occurs at $\psi_{ex} \approx 2.2$.

This value can be predicted by applying Eq. (2.52) and by considering the prescription coming from Eq. (3.74), in which the normalized flux variable ψ, the normalized applied flux ψ_{ex}, and the parameter β are rescaled by a quantity $1/N$ for a single loop containing N junctions.

Having established the validity of the numerical approach adopted, we may proceed in determining the system response for the following cases: $\varphi = 0$ and $\theta = \pi/6, \pi/4, \pi/3$, and $\varphi = \pi/6$ and $\theta = \pi/6, \pi/4, \pi/3$.

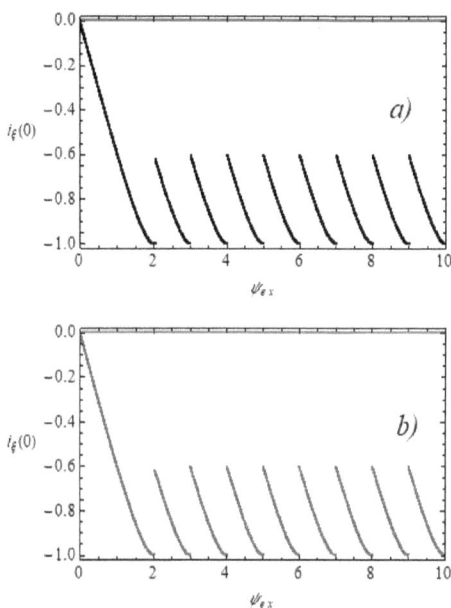

Fig. 8.21. Normalized currents as functions of ψ_{ex} in a cubic network for $\beta = 1.0$, $\theta = 0$, and $\varphi = 0$. In panel *a)* the stationary normalized currents $i_\xi(0)$ are show. The red and black curves represent the quantities $i_y(0)$, and $i_z(0)$, respectively. In panel *b)* the stationary normalized currents $i_\xi(a\hat{\xi})$ are shown. The orange and gray curves represent the quantities $i_y(a\hat{y})$, and $i_z(a\hat{z})$, respectively. The curves for $i_x(0)$ and $i_x(a\hat{x})$ are not shown here, since they coincide with the curves $i_y(0)$ and $i_y(a\hat{y})$, respectively. Notice the exact correspondence between the two sets of curves in panels *a)* and *b)*.

Additional properties of this network will be discussed in the next chapter.

In Fig. 8.22a–c and Fig. 8.23a–c the magnetic fluxes $\psi_\xi(0)$ and the currents $i_\xi(0)$ in the network are respectively shown for the following values of the parameters: $\beta = 1$, $\varphi = 0$, $\theta = \pi/6, \pi/4, \pi/3$. In some of these curves we see the reappearance of the blue curves for $\psi_x(0)$ and $i_x(0)$, not present in Fig. 20a–b and 21a–b. However, a zero value of the fluxes and currents in the superconducting loops lying on the $x - z$ plane can still be noticed. The curves for the magnetic fluxes $\psi_\xi(a\hat{\xi})$ and for the currents $i_\xi(a\hat{\xi})$ are not shown, since they are equal to the corresponding $\psi_\xi(0)$ and $i_\xi(0)$ curves. We now analyze these figures in detail.

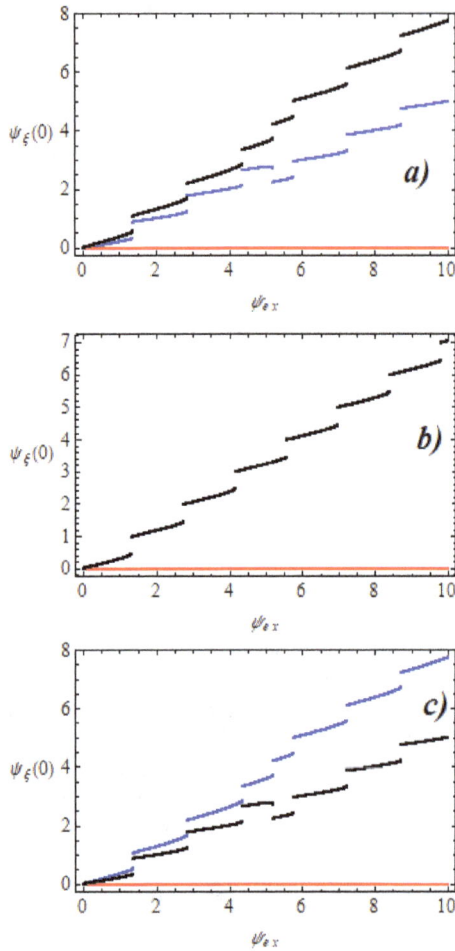

Fig. 8.22. Normalized magnetic flux as functions of ψ_{ex} in a cubic network for $\beta = 1.0$, and $\varphi = 0$. The blue, red, and black curves represent the quantities $\psi_x(0)$, $\psi_y(0)$, and $\psi_z(0)$, respectively. In panel a) the stationary values $\psi_\xi(0)$ are shown for $\theta = \pi/6$; in panel b) for $\theta = \pi/4$; in panel c) for $\theta = \pi/3$. In panel b) the blue curve coincides with the black curve and only the latter is shown. The quantities $\psi_\xi(a\hat{\xi})$ are equal to $\psi_\xi(0)$, so that it is not necessary to show them.

Starting by considering Fig. 8.22a, obtained for $\theta = \pi/6$, we may notice that the $\psi_z(0)$ vs. ψ_{ex} curve lies above the graphical representation of $\psi_x(0)$. As we have already mentioned, this can be understood by the fact that the component $\hat{z} \cdot \vec{\psi}_{ex} = \sqrt{3}/2$ is greater than $\hat{x} \cdot \vec{\psi}_{ex} = 1/2$. The opposite occurs in Fig. 8.22c, obtained for $\theta = \pi/3$, where the value of the components are interchanged, and where the $\psi_x(0)$ vs. ψ_{ex} curve lies above the graphical representation of $\psi_z(0)$.

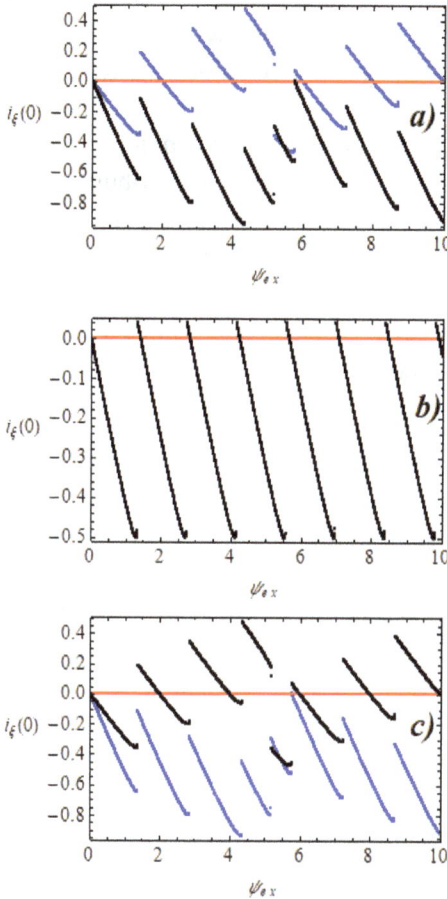

Fig. 8.23. Normalized currents $i_\xi(0)$ in the cubic network for $\beta = 1.0$, and $\varphi = 0$. The blue, red, and black curves represent the quantities $i_x(0)$, $i_y(0)$, and $i_z(0)$, respectively. In panel a) the stationary values $i_\xi(0)$ are shown for $\theta = \pi/6$; in panel b) for $\theta = \pi/4$; in panel c) for $\theta = \pi/3$. In panel b) the blue curve coincides with the black curve and only the latter is shown. The quantities $i_\xi(a\hat{\xi})$ are equal to $i_\xi(0)$, so that it is not necessary to show them.

These two curves are therefore dual and can be commented together. A particular feature in these curves breaks the regular behavior detected for $\theta = 0$. In fact, looking at Fig. 8.22a, for example, we see that the discontinuity of the $\psi_x(0)$ vs. ψ_{ex} curve at $\psi_{ex} \approx 5.2$ brings a decrement in the value of $\psi_x(0)$. It is thus interesting to understand this feature, not detectable, for instance, in Fig. 8.22b, where only incremental steps for each flux transition are present. In order to interpret this characteristic occurrence, we refer to Fig. 8.23a and to Fig. 8.24; in the former the currents $i_\xi(0)$ are shown for $\theta = \pi/6$, in the latter a schematic representation of the current distribution is given. By starting to consider the first irreversible transition, from Fig. 8.24 we notice that the currents $i_x(0)$ and $i_z(0)$ flow in the same sense in the JJs along the y −direction positioned at $\vec{r} = a\hat{x}$ and $\vec{r} = a\hat{z}$. Here we may consider $i_y(0)$ and currents $i_y(a\hat{y})$ to be zero. Notice also that the black arrows are longer than blue arrows.

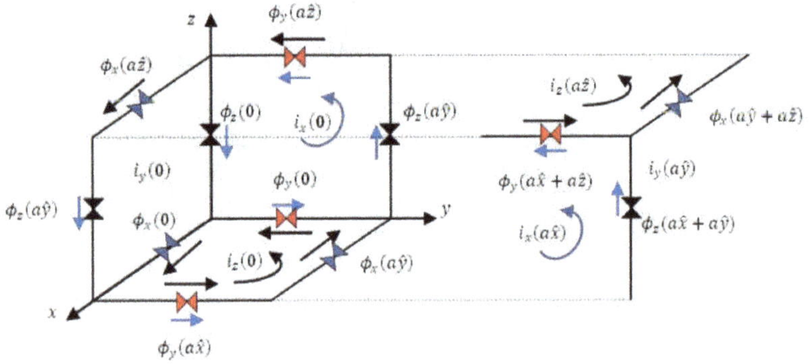

Fig. 8.24. Schematic representation of the normalized currents $i_\xi(\vec{r})$ in the cubic network for $\theta = \pi/6$. The blue and black arrows represent the quantities $i_x(\vec{r})$ and $i_z(\vec{r})$, respectively.

Therefore, the maximum Josephson current value is exceeded the first time in the JJs along the y −direction positioned at $a\hat{x}$ and $a\hat{z}$ at $\psi_{ex} \approx$ 1.37 (in this discussion we use three significant figures to specify the value at which transitions occur). As it appears from Fig. 8.22a and 8.23a, after the flux transition takes place at $\psi_{ex} \approx 1.37$, the absolute value of the currents $i_x(0)$ and $i_x(a\hat{x})$, and $i_z(0)$ and $i_z(a\hat{z})$ decreases, so that the absolute values of the sums $i_x(0) + i_z(a\hat{z})$ and $i_x(a\hat{x}) + i_z(0)$ do not exceed unity. The same JJs positioned at $\vec{r} = a\hat{x}$ and $\vec{r} = a\hat{z}$ are involved in the irreversible flux penetration process taking place at $\psi_{ex} \approx 2.85$ and at $\psi_{ex} \approx 4.35$. However, as seen in Fig. 6.23a after the latter transition there appears an anomalous one, occurring at $\psi_{ex} \approx 5.76$, by which the quantity $\psi_x(0)$ is seen to decrease in the network. In order to understand this apparent anomaly, let us first notice that the absolute values of the bias currents $i_x(0) + i_z(a\hat{z})$ and $i_x(a\hat{x}) + i_z(0)$ are less than one at $\psi_{ex} \approx 5.76$, as one can see by inspection from Fig. 8.23a. Therefore, the JJs involved in the flux penetration process occurring at $\psi_{ex} \approx 5.76$ cannot be the ones characterized by the superconducting phase differences $\phi_y(a\hat{x})$ and $\phi_y(a\hat{z})$, as it happened in the previously analyzed irreversible flux penetration processes.

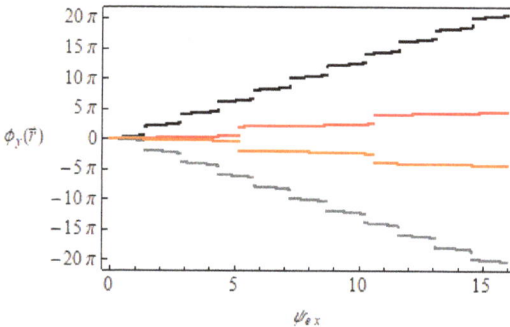

Fig. 8.25. Superconducting phase difference $\phi_y(\vec{r})$ across the Josephson junctions lying in the y −direction in a cubic network for $\beta = 1.0$, $\varphi = 0$, $\theta = \pi/6$. The red, grey, black, and orange curves represent the quantities $\phi_y(0)$, $\phi_y(a\hat{x})$, $\phi_y(a\hat{z})$, and $\phi_y(a\hat{x} + a\hat{z})$, respectively.

By then taking a second look at Fig. 8.24, where a representation of the circulating currents was exhibited to understand the first irreversible flux penetration process, we may utilize the same simplified view once more. In fact, we may adopt this schematic picture to analyze the present situation by reversing the sign of the current $i_x(0)$, i.e., by reversing the direction of the blue arrows. In this way, the junction subject to a higher value of the bias currents, involved in the flux penetration process, are those lying in the y−direction at position $\vec{r} = 0$ and $\vec{r} = a\hat{x} + a\hat{z}$. In order to verify the latter statement, in Fig. 8.25 we exhibit the behavior of the superconducting phase differences $\phi_y(0)$, $\phi_y(a\hat{x})$, $\phi_y(a\hat{z})$, and $\phi_y(a\hat{x} + a\hat{z})$ for $\beta = 1.0$, $\varphi = 0$, and $\theta = \pi/6$. From Fig. 8.25 we may argue both junction pairs, the first characterized by the superconducting phase differences $\phi_y(a\hat{x})$ and $\phi_y(a\hat{z})$, the second by $\phi_y(0)$ and $\phi_y(a\hat{x} + a\hat{z})$, suffer successive 2π phase slips as the externally applied flux increases. However, while the first pair is regularly forced into the resistive state by successive increases of ψ_{ex} of a given quantity $\Delta\psi_{ex}$, the interval over which the second pair suffers a 2π phase slip is about $3\Delta\psi_{ex}$. The juxtaposition of these two regular behaviors gives the apparently anomalous response in Fig. 8.23a and 8.23c.

As for Fig. 8.23b, we notice a rather simple behavior: successive 2π phase slips, at regular intervals of ψ_{ex} of the junctions lying in the y−direction, one located at $\vec{r} = a\hat{x}$ and another at $\vec{r} = a\hat{z}$.

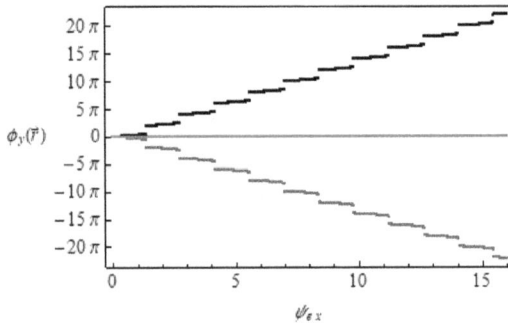

Fig. 8.26. Superconducting phase difference $\phi_y(\vec{r})$ across the Josephson junctions lying in the y−direction in a cubic network for $\beta = 1.0$, $\varphi = 0$, $\theta = \pi/4$. The grey, black, and orange curves represent the quantities $\phi_y(a\hat{x})$, $\phi_y(a\hat{z})$, and $\phi_y(a\hat{x} + a\hat{z})$, respectively. The curve for the phase difference $\phi_y(0)$ coincides with the orange curve, since $\phi_y(0) = \phi_y(a\hat{x} + a\hat{z})$ for $\varphi = 0$ and $\theta = \pi/4$.

This property can be confirmed by inspecting Fig. 8.26, where we report the ψ_{ex}−dependence of the superconducting phase differences $\phi_y(0)$, $\phi_y(a\hat{x})$, $\phi_y(a\hat{z})$, and $\phi_y(a\hat{x} + a\hat{z})$ for $\beta = 1.0$, $\varphi = 0$, and $\theta = \pi/4$. In fact, the superconducting phase differences $\phi_y(0)$ and $\phi_y(a\hat{x} + a\hat{z})$ are both seen to be zero and only the quantities $\phi_y(a\hat{x})$ and $\phi_y(a\hat{z})$ contribute to the dynamics of the whole system.

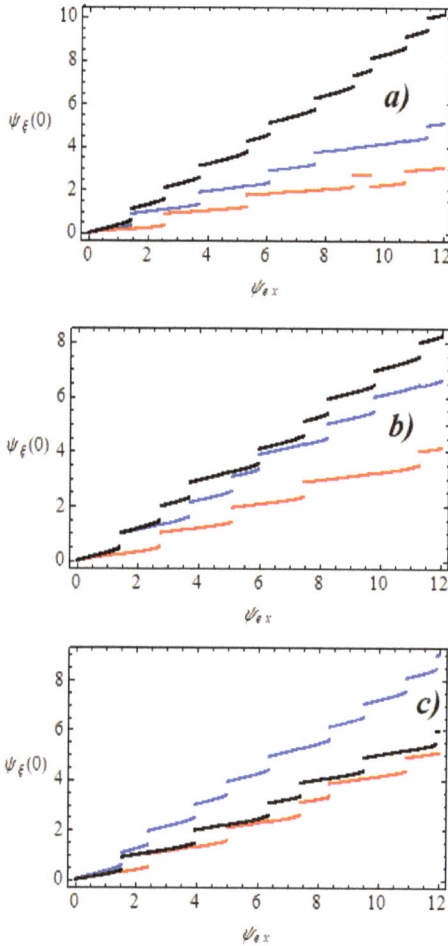

Fig. 8.27. Normalized magnetic flux as functions of ψ_{ex} in a cubic network for $\beta = 1.0$, and $\varphi = \pi/6$. The blue, red, and black curves represent the quantities $\psi_x(0)$, $\psi_y(0)$, and $\psi_z(0)$, respectively. In panel a) the stationary values $\psi_\xi(0)$ are shown for $\theta = \pi/6$; in panel b) for $\theta = \pi/4$; in panel c) for $\theta = \pi/3$. The quantities $\psi_\xi(a\hat{\xi})$ are equal to $\psi_\xi(0)$, so that it is not necessary to show them.

We can now proceed in determining the system response for $\varphi = \pi/6$ and $\theta = \pi/6, \pi/4, \pi/3$. In Fig. 8.27a–c and Fig. 8.28a–c the magnetic fluxes $\psi_\xi(0)$ and the currents $i_\xi(0)$ in the network are respectively shown for the following values of the parameters: $\beta = 1$, $\varphi = \pi/6$, $\theta = \pi/6, \pi/4, \pi/3$.

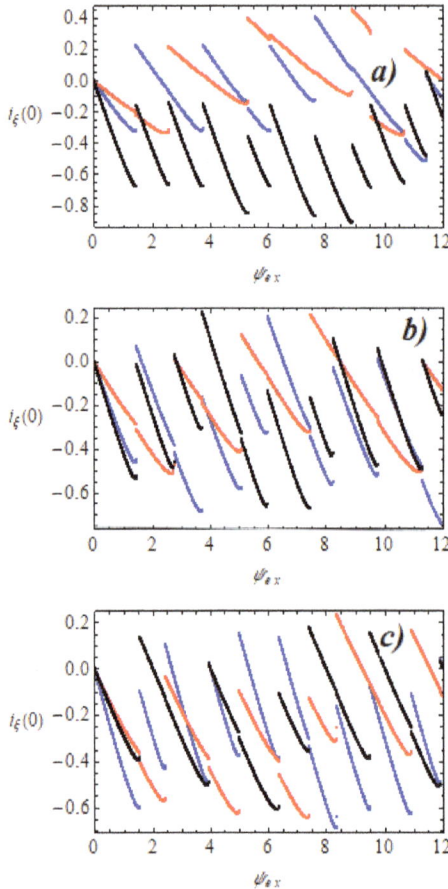

Fig. 8.28. Normalized currents $i_\xi(0)$ as functions of ψ_{ex} in the cubic network for $\beta = 1.0$, and $\varphi = \pi/6$. The blue, red, and black curves represent the quantities $i_x(0)$, $i_y(0)$, and $i_z(0)$, respectively. In panel a) the stationary values $i_\xi(0)$ are shown for $\theta = \pi/6$; in panel b) for $\theta = \pi/4$; in panel c) for $\theta = \pi/3$. The quantities $i_\xi(a\hat\xi)$ are equal to $i_\xi(0)$, so that it is not necessary to show them.

When analyzing the flux distribution in the network, we first of all notice that curves of three colors are present in all the graphical representations of Fig. 8.27a–c, meaning that fluxes of different values thread the superconducting perimeter of the three cubic faces located at $\vec{r} = 0$. Another important aspect, reported in the caption to Fig. 8.27a–c, is that the quantities $\psi_\xi(a\hat{\xi})$ are equal to $\psi_\xi(0)$. In this way, it is not necessary to show the normalized fluxes $\psi_\xi(a\hat{\xi})$ in separate figures.

Let us now try to describe the flux penetration mechanisms in the network for this orientation of the applied magnetic field, starting from $\theta = \pi/6$. Keeping in mind the pictorial representation of all currents flowing in the network in Fig. 8.29, we argue that flux penetration occurs because the two currents flowing in the same junction reach the maximum Josephson current value. In this way, they are seen to flow in the same direction in the branch where the junction lies. In Fig. 8.29 we have kept the sign of the currents positive, since only the relative sign counts.

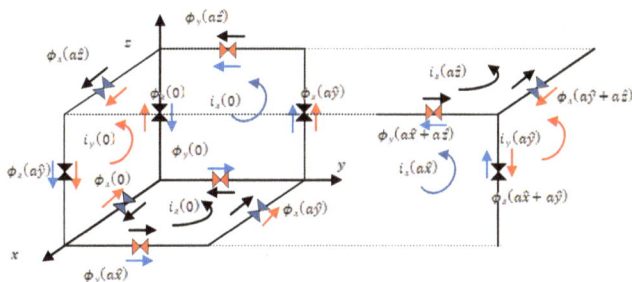

Fig. 8.29. Schematic representation of the normalized currents $i_\xi(\vec{r})$ in the cubic network for $\theta = \pi/6$. The blue, red, and black arrows represent the quantities $i_x(\vec{r})$, $i_y(\vec{r})$ and $i_z(\vec{r})$, respectively.

In this case the applied field components are the following:

$$\vec{\psi}_{ex} \cdot \hat{x} = \frac{\sqrt{3}}{4}\psi_{ex}, \tag{8.66a}$$

$$\vec{\psi}_{ex} \cdot \hat{y} = \frac{1}{4}\psi_{ex}, \tag{8.66b}$$

$$\vec{\psi}_{ex} \cdot \hat{z} = \frac{\sqrt{3}}{2}\psi_{ex}. \qquad (8.66c)$$

The z −component of the field is thus the highest, so that in Fig. 8.27a the $\psi_z(0)$ vs. ψ_{ex} curve lies above the blue and the red curves, representing $\psi_x(0)$ and $\psi_y(0)$, respectively. Therefore, once again, the junctions along the branches lying in the y −direction are responsible for flux penetration mechanisms. In particular, as for first irreversible flux penetration, occurring at $\psi_1 \approx 1.5$, the junctions suffering a 2π phase slip are those located at $\vec{r} = a\hat{x}$ and $\vec{r} = a\hat{z}$. In fact, a first current pair, $i_x(0)$ and $i_z(a\hat{z})$, and a second pair, $i_x(a\hat{x})$ and $i_z(0)$, are seen to flow in the same direction. Moreover, considering the absolute value of the individual currents in the pairs, it results greater than $i_y(0)$. Therefore, only the absolute value of the sum of the currents making up each of the above pairs approaches unity as $\psi_{ex} \rightarrow \psi_1$, as it can be also seen from Fig. 8.28a. Accordingly, from Fig. 8.27a, it can be seen that the fluxes $\psi_x(0)$ and $\psi_z(0)$ present a discontinuity at ψ_1.

Once first irreversible flux penetration has occurred, however, the current $i_x(0)$ and $i_x(a\hat{x})$ reverse their sign, so that, for higher values of the forcing term ψ_{ex}, only two pairs of currents are seen to flow in the same direction: the first pair is made by the currents $i_y(0)$ and $i_z(a\hat{z})$, the second by $i_y(a\hat{y})$ and $i_z(0)$. The junctions through which these current pairs flow are located at $\vec{r} = a\hat{y}$ and $\vec{r} = a\hat{z}$ and are directed along the x −direction. At $\psi_{ex} \approx 2.5$ these two junctions suffer a 2π phase slip and a second irreversible flux penetration process occurs, after which the currents $i_y(0)$ and $i_y(a\hat{y})$ reverse their sign. As a consequence, only the fluxes $\psi_y(0)$ and $\psi_z(0)$ present a discontinuity at $\psi_{ex} \approx 2.5$.

By continuing raising the value of the applied field, the individual currents in the original two pairs, namely, the pair $i_x(0)$ and $i_z(a\hat{z})$ and the pair $i_x(a\hat{x})$ and $i_z(0)$, flow again in the same direction. In this way, the junctions suffering a 2π phase slip at $\psi_{ex} \approx 3.8$ are again those lying along the y −direction and located at $\vec{r} = a\hat{x}$ and $\vec{r} = a\hat{z}$. Therefore, only the fluxes $\psi_x(0)$ and $\psi_z(0)$ present a discontinuity in correspondence with the field at which the third irreversible flux transition occurs.

Continuing now the analysis of flux penetration mechanisms, by considering the case $\theta = \pi/4$, we refer to Fig. 8.27b and 8.28b. Setting $\varphi = \pi/6$ and $\theta = \pi/4$ in Eq. (8.55), the applied field components are now seen to be the following:

$$\vec{\psi}_{ex} \cdot \hat{x} = \frac{\sqrt{6}}{4} \psi_{ex}, \tag{8.67a}$$

$$\vec{\psi}_{ex} \cdot \hat{y} = \frac{\sqrt{2}}{4} \psi_{ex}, \tag{8.67b}$$

$$\vec{\psi}_{ex} \cdot \hat{z} = \frac{\sqrt{2}}{2} \psi_{ex}. \tag{8.67c}$$

The z-component of $\vec{\psi}_{ex}$ is again the highest, being realized the following hierarchy in Eq. (8.67a–c): $\vec{\psi}_{ex} \cdot \hat{y} < \vec{\psi}_{ex} \cdot \hat{x} < \vec{\psi}_{ex} \cdot \hat{z}$. Therefore, as seen in Fig. 8.27b, the black curve lies on top of the blue and red curves. The flux penetration mechanisms can be reconstructed in much the same way as done for the case $\theta = \pi/6$, by means of the information provided by Fig. 8.27b, Fig. 8.28b and Fig. 8.29.

Finally, let us consider the case $\theta = \pi/3$, for which the applied field components are now seen to be the following:

$$\vec{\psi}_{ex} \cdot \hat{x} = \frac{3}{4} \psi_{ex}, \tag{8.68a}$$

$$\vec{\psi}_{ex} \cdot \hat{y} = \frac{\sqrt{3}}{4} \psi_{ex}, \tag{8.68b}$$

$$\vec{\psi}_{ex} \cdot \hat{z} = \frac{1}{2} \psi_{ex}. \tag{8.68c}$$

The hierarchy in the field components arising from Eq. (8.68a–c) is the following: $\vec{\psi}_{ex} \cdot \hat{y} < \vec{\psi}_{ex} \cdot \hat{z} < \vec{\psi}_{ex} \cdot \hat{x}$. Therefore, as seen in Fig. 8.27c, the blue curve lies on top of the black and red curves. As appearing from Fig. 8.28c, the current $i_x(0)$ plays much the same role that the current $i_z(0)$ played in the previous case ($\theta = \pi/4$). However, in the field interval $0 < \psi_{ex} < 12$, in which we have monitored the magnetic behavior of the cubic system, a distinctive transition occurs at $\psi_{ex} \approx 7.5$ where we do not register any discontinuity in the $\psi_x(0)$ vs. ψ_{ex} curve in Fig. 8.27c. Rather unusually, in the same graph a combined transition in the quantities $\psi_y(0)$ and $\psi_z(0)$ is seen to occur. This can be confirmed by looking at Fig. 8.28c, where a relatively large discontinuity for the red and black curves at $\psi_{ex} \approx 7.5$ can be detected. In order to reconstruct the penetration mechanism, we may look at the all possible combinations of currents which may give a 2π phase slip in junctions in which currents in the same direction flow. Among those, we need to choose the ones on which black and red arrows are present. Therefore, these junctions are located at $\vec{r} = a\hat{y}$ and at $\vec{r} = a\hat{z}$ and are directed along the x −direction.

With these examples we only intended to describe how to interpret the magnetic states and the flux penetration processes in the cubic system in the presence of the forcing term $\vec{\psi}_{ex}$ along a specific direction in space. In the following chapter, we shall investigate the behavior of the cubic system not only in the presence of a normalized applied flux $\vec{\psi}_{ex}$, but also in the presence of a bias current, injected at one vertex and derived at a second different vertex of the network. The task of interpreting the magnetic response of the same system will therefore become more difficult, given the presence of an additional forcing term.

Chapter 9

A Three-dimensional
Superconducting Interferometer

9.1. The system

In Chapter 8, the dynamic equations of Josephson junctions in three-dimensional networks were written and the magnetic states of some simple systems were defined. In analyzing the static magnetic properties of a cubic network of Josephson junctions, for instance, we only considered one forcing term, namely, the normalized externally applied flux vector $\vec{\psi}_{ex}$. In the present chapter we shall consider an additional forcing term: a current bias I_B injected at one vertex of the cubic structure and drawn from a second vertex. The network can thus be considered to be a starting point for conceiving a three-dimensional superconducting quantum interference device (3D SQUID). This system has been studied in the recent past (Oppenländer J. H., 1999) (De Luca R. D., 2000) and presents some promising features as a magnetic field sensor, much in the same way a d.c. SQUID (Barone, 1982) (Clarke, 2004) can be used to detect the local magnetic field component, orthogonal to the plane of the device. In the case of a 3D SQUID, all magnetic field components could be measured, if the system would be opportunely implemented. These preliminary studies and the analysis developed in previous chapters could be useful to describe the general properties of this innovative electronic device. Much more investigation on this topic, however, besides the seminal content of the present chapter, will be necessary.

In order to get a more precise idea of the way this magnetic field sensor can be devised and its properties found, we need to determine the dynamic equations for all phase differences in the network, as done in the previous chapter. Therefore, by considering Fig. 9.1, where the normalized bias current $i_B = I_B/I_{J0}$ enters the network at $\vec{r}_{in} = a\hat{x} + a\hat{z}$ and flows out from $\vec{r}_{out} = a\hat{y}$, we adopt branch currents to describe the magnetic states of the system. This is done for reasons which will be clear in what follows. For the moment being it suffices to say that this approach allows a more direct access to the voltage across each branch.

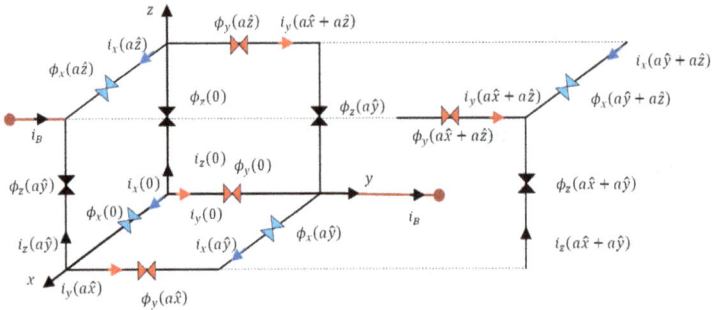

Fig. 9.1. Schematic representation of the stationary branch currents in a cubic network of Josephson junctions. In this system a normalized bias current i_B is injected at the node located at $\vec{r}_{in} = a\hat{x} + a\hat{z}$. The bias current flows out from the network at $\vec{r}_{out} = a\hat{y}$. Notice that each branch current carries on the same type of labeling and the same sign of the superconducting phase difference lying on the same branch.

This can be understood by noticing that the normalized branch currents $i_\xi(\vec{r}, \tau)$ carry on the same indices of the superconducting phase variables $\phi_\xi(\vec{r}, \tau)$ and are taken to flow from negative to positive junction electrodes. A negative branch current, therefore, flows in the opposite direction with respect to the previously described positive sense. In defining the time-dependence of these quantities, we use the time normalization

$$\tau = \frac{2\pi R I_{J0}}{\Phi_0} t, \qquad (9.1)$$

where t is the laboratory time, Φ_0 is the elementary flux quantum, I_{J0} and R are the maximum Josephson current and the resistive parameter for each junction in the network, respectively. We remark that the scaling factor for time normalization is different from the one used in Chapter 8.

Recalling the dynamic equations for an over-damped Josephson junction in Eq. (2.44), the superconducting phase difference $\phi_\xi(\vec{r}, \tau)$ must obey the following differential equation:

$$\frac{d\phi_\xi(\vec{r},\tau)}{d\tau} + \sin\phi_\xi(\vec{r},\tau) = i_\xi(\vec{r},\tau). \tag{9.2}$$

On the other hand, the superconducting phases $\phi_\xi(\vec{r}, \tau)$ satisfy the fluxoid quantization relation (8.6a), which, under zero-field cooling conditions, can be written as a vector product as follows:

$$\vec{\psi}(\vec{r},\tau) = -\frac{1}{2\pi}\vec{\Delta}^{(+)} \times \vec{\phi}(\vec{r},\tau). \tag{9.3}$$

The above compact expression has been derived in Chapter 8, where all quantities have been introduced. The same equation has been reported in Eq. (8.9). In order to briefly recall the definition of all terms, we start with the three-component vectors $\vec{\psi}(\vec{r}, \tau)$, by which we define the fluxes threading the six different loops in the network. These fluxes are then represented by the three components of $\vec{\psi}(0, \tau)$ and by the additional three quantities $\psi_\xi(a\hat{\xi}, \tau)$, with $\xi = x, y, z$. The three-component vectors $\vec{\phi}(\vec{r}, \tau)$ represent the superconducting phase variables across the twelve junctions in the network. Finally, the ξ −component of the forward difference operator $\vec{\Delta}^{(+)}$ is defined in such a way that, when applied on a scalar function $f(\vec{r}, \tau)$, gives:

$$\Delta_\xi^{(+)}f(\vec{r},\tau) = f(\vec{r} + a\hat{\xi},\tau) - f(\vec{r},\tau). \tag{9.4}$$

In the particular case of a single cube, we would only need to consider five independent quantities for $\vec{\psi}(\vec{r}, \tau)$. In fact, in Chapter 8 we have shown that the sum of the magnetic fluxes threading the six loops in the network is zero; this statement, which is a direct consequence of Maxwell's equations, is resumed in Eq. (8.40). However, in what follows we shall also consider the magnetic flux $\psi_z(a\hat{z}, \tau)$, in such a way that a six-component column vector Φ, shown in its transposed form, can be defined as follows:

$$\Phi = \left[\psi_x(0), \psi_x(a\hat{x}), \psi_y(0), \psi_y(a\hat{y}), \psi_z(0), \psi_z(a\hat{z}) \right]^{\mathrm{T}}, \quad (9.5)$$

where the τ-dependence of the components has been omitted for brevity. We might also define a twelve-component column vector φ describing all superconducting phase variables across the twelve junctions in the network, namely, $\phi_x(0), \phi_x(a\hat{y}), \phi_x(a\hat{z}), \phi_x(a\hat{y} + a\hat{z})$ in the x –direction, $\phi_y(0), \phi_y(a\hat{x}), \phi_y(a\hat{z}), \phi_y(a\hat{x} + a\hat{z})$ in the y –direction, and $\phi_z(0), \phi_z(a\hat{x}), \phi_z(a\hat{y}), \phi_z(a\hat{x} + a\hat{y})$ in the z –direction, taken in the order shown. Therefore, Eq. (9.3) can be written in the following form:

$$\Phi = -\frac{1}{2\pi} \mathbf{F}\varphi, \quad (9.6)$$

where $\mathbf{F}$ is the following 6×12 matrix, whose elements are defined by Eq. (9.3):

$$\mathbf{F} = \begin{pmatrix} 0 & 0 & 0 & 0 & 1 & 0 & -1 & 0 & -1 & 0 & 1 & 0 \\ 0 & 0 & 0 & 0 & 0 & 1 & 0 & -1 & 0 & -1 & 0 & 1 \\ -1 & 0 & 1 & 0 & 0 & 0 & 0 & 0 & 1 & -1 & 0 & 0 \\ 0 & -1 & 0 & 1 & 0 & 0 & 0 & 0 & 0 & 0 & 1 & -1 \\ 1 & -1 & 0 & 0 & -1 & 1 & 0 & 0 & 0 & 0 & 0 & 0 \\ 0 & 0 & 1 & -1 & 0 & 0 & -1 & 1 & 0 & 0 & 0 & 0 \end{pmatrix}. (9.7)$$

This matrix will prove to be of paramount importance for the following analysis.

It is now possible to express the fluxes in terms of the branch currents $i_\xi(\vec{r}, \tau)$ by means of the flux-current relations. In doing this, we shall take account of mutual inductance coefficients between opposite parallel sides in the cubic network, neglecting the inductance of orthogonal cubic faces. In this way, we shall give a simple example of how mutual inductance could be treated in the context of cubic networks of Josephson junctions.

Let us then first determine the flux $\psi_x(0)$, where we again omit the time dependence for the sake of brevity. In writing down this quantity, we need to consider the contribution, with the correct sign, of all branch currents flowing in the loop lying in the $y - z$ plane at $x = 0$. Therefore, we need to consider the following terms:

$$\beta \, \mathcal{I}_{yz}(0) = \beta\big[i_y(0) - i_y(a\hat{z}) + i_z(a\hat{y}) - i_z(0)\big], \qquad (9.8)$$

where $\mathcal{I}_{yz}(0)$ is the current term (in brackets) associated to the loop in the $y - z$ plane at $x = 0$ and $\beta = LI_{J0}/\Phi_0$, L being the inductance coefficient related to the current loop. Notice that the signs of all branch currents indicate a circulating character of $\mathcal{I}_{yz}(0)$. In this respect, we may think of this current term as a circulating current. The branch currents flowing in the x −direction do not give any contribution to the normalized magnetic flux $\psi_x(0)$. As specified above, here we choose to neglect the mutual inductance coefficient of orthogonal loops. A more complete analysis should take account of this additional term, due to the presence, for instance, of a common branch between any two orthogonal loops. This, for example, could be a task to accomplish, in order to make future progress. The currents flowing in the loop lying in the $y - z$ plane at $x = a$ do give a contribution to $\psi_x(0)$. To each current branch, in fact, we can associate magnetic field lines threading the loop in the $y - z$ plane at $x = 0$. Since, by symmetry, each current gives the same contribution, we may write the second term of $\psi_x(0)$ as follows:

$$m \, \mathcal{I}_{yz}(a\hat{x}) = m\big[i_y(a\hat{x}) - i_y(a\hat{x} + a\hat{z}) + i_z(a\hat{x} + a\hat{y}) - i_z(a\hat{x})\big], \quad (9.9)$$

where $\mathcal{I}_{yz}(a\hat{x})$ is a current term (in brackets) associated to the loop in the $y - z$ plane at $x = a$ and $m = MI_{J0}/\Phi_0$, M being the mutual inductance coefficient between the two loops lying in the $y - z$ plane, one at $x = 0$, one at $x = a$. The mutual inductance coefficient can be calculated as follows (Jackson, 1975):

$$M = \frac{\mu_0}{4\pi} \oint \oint \frac{d\vec{l}_1 \cdot d\vec{l}_2}{|\vec{r}_1 - \vec{r}_2|}, \qquad (9.10)$$

where $d\vec{l}_1$ and $d\vec{l}_2$ are elementary paths on the loop at $x = 0$ and at $x = a$, respectively. These elements are oriented in such a way to go counterclockwise around the two loops. Furthermore, in Eq. (9.10) the quantity $|\vec{r}_1 - \vec{r}_2|$ represents the distance between the elements $d\vec{l}_1$ and $d\vec{l}_2$. Taking advantage of the symmetry of the problem, one can consider only one of the four contributions of $d\vec{l}_1$ (a single branch in the first loop) and write

$$M = \frac{\mu_0}{\pi} \int_{y_1=0}^{a} \oint \frac{dy_1(\hat{y} \cdot d\vec{l}_2)}{|y_1\hat{y} - \vec{r}_2|},$$

(9.11)

so that the meaning of M is now more explicit. By the particular form of Eq. (9.11), we may see that only parallel branches distant $d = a$ or $d = \sqrt{2}a$ contribute, with different signs, to the integral. Therefore, we may write:

$$M = \frac{\mu_0}{\pi} \left[\int_0^a \int_0^a \frac{dy_1 dy_2}{\sqrt{(y_1-y_2)^2+a^2}} - \int_0^a \int_0^a \frac{dy_1 dy_2}{\sqrt{(y_1-y_2)^2+2a^2}} \right].$$

(9.12)

By denoting each integral in Eq. (9.12) by $f(d)$, where $d = a$ in the first and $d = \sqrt{2}a$ in the second, after some calculations we find:

$$f(d) = 2 \left[a \operatorname{arcsinh}\left(\frac{a}{d}\right) - \sqrt{a^2 + d^2} + d \right],$$

(9.13)

so that:

$$M = \frac{\mu_0}{\pi} \left[f(a) - f(\sqrt{2}a) \right].$$

(9.14)

We may finally express the normalized flux $\psi_x(0)$ as follows:

$$\psi_x(0) = \beta \, \mathfrak{I}_{yz}(0) + m \, \mathfrak{I}_{yz}(a\hat{x}) + \hat{x} \cdot \vec{\psi}_{ex},$$

(9.15)

where the last addendum is the contribution given by the externally applied flux.

Let us now set our attention on the quantity $\psi_x(a\hat{x})$. By choosing the appropriate sign for all branch currents flowing in the loop lying in the $y - z$ plane at $x = a$ and by considering mutual inductance effects due to the parallel face close to the origin, it is not difficult to argue that $\psi_x(a\hat{x})$ should have the following form:

$$\psi_x(a\hat{x}) = m \, \mathfrak{I}_{yz}(0) + \beta \, \mathfrak{I}_{yz}(a\hat{x}) + \hat{x} \cdot \vec{\psi}_{ex}.$$

(9.16)

In fact, when computing the term $\mathfrak{I}_{yz}(a\hat{x})$ as in Eq. (9.9), we realize that it must be multiplied by the normalized self-inductance coefficient β. As for mutual inductance, the opposite loop, lying in the $y - z$ plane and positioned at $x = 0$, plays the same role that the loop at $x = a$ played for the quantity $\psi_x(0)$. Finally, the forcing term is the same, so that we can justify Eq. (9.16). How do we go about now in finding $\psi_y(y)$ and $\psi_y(a\hat{y})$? We simply make the exchange of position variables $x \leftrightarrow y$ in Eq. (9.15) and Eq. (9.16), and choose the opportune signs when doing this in Eq. (9.8) and Eq. (9.9), obtaining:

$$\psi_y(0) = \beta \, \mathfrak{I}_{xz}(0) + m \, \mathfrak{I}_{xz}(a\hat{y}) + \hat{y} \cdot \vec{\psi}_{ex}, \qquad (9.17a)$$

$$\psi_y(a\hat{y}) = m \, \mathfrak{I}_{xz}(0) + \beta \, \mathfrak{I}_{xz}(a\hat{y}) + \hat{y} \cdot \vec{\psi}_{ex}, \qquad (9.17b)$$

with

$$\mathfrak{I}_{xz}(0) = -i_x(0) + i_x(a\hat{z}) - i_z(a\hat{x}) + i_z(0), \qquad (9.18a)$$

$$\mathfrak{I}_{xz}(a\hat{y}) = -i_x(a\hat{y}) + i_x(a\hat{y} + a\hat{z})$$
$$-i_z(a\hat{x} + a\hat{y}) + i_z(a\hat{y}). \qquad (9.18b)$$

We notice that a sign change occurs in all terms of Eq. (9.18a) and Eq. (9.18b) when, starting from Eq. (9.8) and Eq. (9.9), respectively, we make the substitution $x \leftrightarrow y$. And how do we find $\psi_z(0)$ and $\psi_z(a\hat{z})$? This time we make the exchange of position variables $y \leftrightarrow z$ in Eq. (9.17a–b) and Eq. (9.18a–b), with a change in sign of all terms in the latter, obtaining:

$$\psi_z(0) = \beta \, \mathfrak{I}_{xy}(0) + m \, \mathfrak{I}_{xy}(a\hat{z}) + \hat{z} \cdot \vec{\psi}_{ex}, \qquad (9.19a)$$

$$\psi_z(a\hat{z}) = m \, \mathfrak{I}_{xy}(0) + \beta \, \mathfrak{I}_{xy}(a\hat{z}) + \hat{z} \cdot \vec{\psi}_{ex}, \qquad (9.19b)$$

with

$$\mathfrak{I}_{xy}(0) = i_x(0) - i_x(a\hat{y}) + i_y(a\hat{x}) - i_y(0), \qquad (9.20a)$$

$$\mathfrak{I}_{xy}(a\hat{z}) = i_x(a\hat{z}) - i_x(a\hat{y} + a\hat{z}) + i_y(a\hat{x} + a\hat{z}) - i_y(a\hat{z}). \qquad (9.20b)$$

We have thus found the flux-current relations, including the mutual inductance effects, in the cubic system. These relations are represented by Eq. (9.15), Eq. (9.16), Eq. (9.17a–b) and Eq. (9.19a–b). We need to remark that only five independent fluxes and currents are present in the system. In fact, by Eq. (8.40) we may write:

$$\psi_x(0) + \psi_y(0) + \psi_z(0) = \psi_x(a\hat{x}) + \psi_y(a\hat{y}) + \psi_z(a\hat{z}), \quad (9.21)$$

so that, from Eq. (9.15), Eq. (9.16), Eq. (9.17a–b) and Eq. (9.19a–b), we have:

$$\mathcal{I}_{yz}(0) + \mathcal{I}_{xz}(0) + \mathcal{I}_{xy}(0) = \mathcal{I}_{yz}(a\hat{x}) + \mathcal{I}_{xz}(a\hat{y}) + \mathcal{I}_{xy}(a\hat{z}). \quad (9.22)$$

The above equation would thus allow solution for $\mathcal{I}_{xy}(a\hat{z})$, for instance, in terms of the five independent loop currents corresponding to five independent fluxes.

We are now ready to define the column vector $\mathcal{J}$, shown in its transposed form, as follows:

$$\mathcal{J} = \left[\mathcal{I}_{yz}(0), \mathcal{I}_{yz}(a\hat{x}), \mathcal{I}_{xz}(0), \mathcal{I}_{xz}(a\hat{y}), \mathcal{I}_{xy}(0), \mathcal{I}_{xy}(a\hat{z})\right]^{\mathrm{T}}. \quad (9.23)$$

Furthermore, we define:

$$\Gamma = \left[\hat{x} \cdot \hat{\psi}_{ex}, \hat{x} \cdot \hat{\psi}_{ex}, \hat{y} \cdot \hat{\psi}_{ex}, \hat{y} \cdot \hat{\psi}_{ex}, \hat{z} \cdot \hat{\psi}_{ex}, \hat{z} \cdot \hat{\psi}_{ex}\right]^{\mathrm{T}}. \quad (9.24)$$

By means of matrix notation, we may express the five independent flux-current equations, namely, Eq. (9.15), Eq. (9.16), Eq. (9.17a–b), and Eq. (9.19a–b), in the following compact form:

$$\Phi = \mathbf{G}\mathcal{J} + \psi_{ex}\Gamma, \quad (9.25)$$

where the 6×6 matrix $\mathbf{G}$ is given by:

$$\mathbf{G} = \beta \begin{pmatrix} 1 & b & 0 & 0 & 0 & 0 \\ b & 1 & 0 & 0 & 0 & 0 \\ 0 & 0 & 1 & b & 0 & 0 \\ 0 & 0 & b & 1 & 0 & 0 \\ 0 & 0 & 0 & 0 & 1 & b \\ 0 & 0 & 0 & 0 & b & 1 \end{pmatrix} = \beta \mathbf{D}, \quad (9.26)$$

where $b = m/\beta$. The above matrix is non-singular, as long as $m \neq \beta$, its determinant Δ_{G} being:

$$\Delta_{\mathrm{G}} = \beta^6 (1 - b^2)^3. \quad (9.27)$$

With Eq. (9.2), Eq. (9.6), and Eq. (9.25) we have fully described the constitutive equations of the network. In the following section, we shall explore the dynamic properties of the system, by opportunely combining these equations.

9.2. Dynamic equations

In the previous section, we have written down the constitutive equations for the cubic system in fig. 9.1. We shall now first summarize these equations and then we shall obtain the dynamic equations for all superconducting phase differences $\phi_\xi(\vec{r}, \tau)$ in the network. Let us then write Eq. (9.2) in matrix form and let us combine Eq. (9.6) and (9.25) to get:

$$\frac{d}{d\tau}\varphi + S(\varphi) = I, \tag{9.28a}$$

$$\mathcal{J} = -\frac{1}{\beta}\left(\frac{1}{2\pi}\mathbf{D}^{-1}\mathbf{F}\varphi + \psi_{ex}\mathbf{D}^{-1}\Gamma\right), \tag{9.28b}$$

where we assume $m \neq \beta$ ($b \neq 1$) and take $S(\varphi)$ to be a column vector whose components are the quantities $\sin\phi_\xi(\vec{r}, \tau)$ ordered according to the components of φ. Moreover, in Eq. (9.28a) the column vector I has components $i_\xi(\vec{r}, \tau)$, taken in the same order as the components of φ. We now realize, looking closely at Eq. (9.8), Eq. (9.9), Eq. (9.18a–b), and Eq. (9.19a–b), that the loop current vector $\mathcal{J}$ and the branch current vector I are linked in a way similar to the independent magnetic flux vector Φ and the superconducting phase vector φ. In fact, we can write:

$$\mathcal{J} = \mathbf{F}I. \tag{9.29}$$

This relation should not surprise, since the components of the branch current vector I have the same indices of the superconducting phase differences across the junctions. Furthermore, the loop current vector $\mathcal{J}$ is related to the magnetic flux vector Φ through Eq. (9.25) and keeps track of the circulating currents in the same way Φ accounts for the superconducting phase differences along the loop. The simple relation in Eq. (9.29) could also be derived by means of standard network analysis, through the notion of the tie-set matrix (Ghosh, 2010). This same equation will prove to be useful in the following analysis.

In the previous description of the characteristics of the cubic network we have not yet considered Kirchhoff's current law at the nodes of the network. Moreover, the normalized bias current i_B has not yet appeared in the summarized equations (9.28a–b), despite the fact that the column vector I appears on the right-hand side of the differential equation governing the dynamics of the superconducting phase differences. In order to completely account for these aspects, let us apply Kirchhoff's current law to the first node at the origin of the network, so that we may write:

$$i_x(0) + i_y(0) + i_z(0) = 0. \tag{9.30}$$

By considering a translation on the network of this equation along the x −axis, for the node at $a\hat{x}$ we write:

$$-i_x(0) + i_y(a\hat{x}) + i_z(a\hat{x}) = 0. \tag{9.31}$$

On the other hand, for a translation along the z −axis from the origin, for the node at $a\hat{z}$ we have:

$$i_x(a\hat{z}) + i_y(a\hat{z}) - i_z(0) = 0. \tag{9.32}$$

At the node positioned at $a\hat{x} + a\hat{z}$, where the normalized current is injected in the network, we have:

$$-i_B + i_x(a\hat{y} + a\hat{z}) + i_y(a\hat{z}) - i_z(a\hat{x}) = 0. \tag{9.33}$$

At the node positioned at $a\hat{y}$, where the normalized current is drawn from the network, we obtain:

$$i_B + i_x(a\hat{y}) - i_y(0) + i_z(a\hat{y}) = 0. \tag{9.34}$$

For the node positioned at $a\hat{y} + a\hat{z}$, we write:

$$i_x(a\hat{y} + a\hat{z}) - i_y(a\hat{z}) - i_z(a\hat{y}) = 0. \tag{9.35}$$

For the node positioned at $a\hat{x} + a\hat{y}$, we get:

$$-i_x(a\hat{y}) - i_y(a\hat{x}) + i_z(a\hat{x} + a\hat{y}) = 0. \tag{9.36}$$

Finally, for the node at $a\hat{x} + a\hat{y} + a\hat{z}$, we obtain:

$$-i_x(a\hat{y} + a\hat{z}) - i_y(a\hat{x} + a\hat{z}) - i_z(a\hat{x} + a\hat{y}) = 0. \qquad (9.37)$$

From Eq. (9.30) to Eq. (9.37) we have retained the plus sign for currents leaving from the node and the minus sign for currents entering the node, as conventionally done in network analysis. Eq. (9.30) - (9.37) define the so-called incidence matrix **A** for the network (Ghosh, 2010), whose elements are the following:

$$
\begin{pmatrix}
1 & 0 & 0 & 0 & 1 & 0 & 0 & 0 & 1 & 0 & 0 & 0 & 0 \\
-1 & 0 & 0 & 0 & 0 & 1 & 0 & 0 & 0 & 1 & 0 & 0 & 0 \\
0 & 1 & 0 & 0 & -1 & 0 & 0 & 0 & 0 & 0 & 1 & 0 & 1 \\
0 & -1 & 0 & 0 & 0 & -1 & 0 & 0 & 0 & 0 & 0 & 1 & 0 \\
0 & 0 & -1 & 0 & 0 & 0 & 0 & 1 & 0 & -1 & 0 & 0 & -1 \\
0 & 0 & 1 & 0 & 0 & 0 & 1 & 0 & -1 & 0 & 0 & 0 & 0 \\
0 & 0 & 0 & 1 & 0 & 0 & -1 & 0 & 0 & 0 & -1 & 0 & 0 \\
0 & 0 & 0 & -1 & 0 & 0 & 0 & -1 & 0 & 0 & 0 & -1 & 0
\end{pmatrix}. \qquad (9.38)
$$

By Kirchhoff current law (KCL) we are thus able to define the above matrix **A**, in which the number of rows is equal to the number of nodes and the number of columns is equal to the number of branch currents, including i_B. In matrix notation we can thus write:

$$\mathbf{A}\begin{pmatrix} I \\ i_B \end{pmatrix} = 0, \qquad (9.39)$$

Where the thirteen-component column vector on the right of the matrix **A** is formed by the twelve-component branch-current column vector I and by the bias current i_B. We can thus keep track of KCL at the nodes by the matrix **A**. However, in order to go from the loop-current vector J to the branch-current vector I, we need to define the fundamental-loop matrix for the network, in which an additional loop is added, carrying a link-current i_B which, in our case, enters at the node located at $a\hat{x} + a\hat{z}$, as specified in Eq. (9.33), and leaves at the node located at $a\hat{y}$, as indicated in Eq. (9.34). When we do this by standard network analysis methods, we find that the fundamental-loop matrix is equal to an augmented 7×13 matrix $\mathbf{F_A}$, expressed in terms of the core matrix **F** as follows:

$$\mathbf{F_A} = \begin{pmatrix} \mathbf{F} & 0 \\ \mathbf{H}^T & 1 \end{pmatrix}. \qquad (9.40)$$

where H is a column vector, expressed in its transposed form as

$$H = (0, -1, 0, 0, 0, 0, 0, +1, 0, 0, 0, -1)^T. \tag{9.41}$$

The transpose of F_A will thus link the thirteen-component column vector of the branch and bias currents appearing in Eq. (9.39) to the augmented seven-component loop current J_A, defined as follows:

$$J_A = \begin{pmatrix} J \\ J_7 \end{pmatrix}. \tag{9.42}$$

Therefore, from standard network analysis concepts (Ghosh, 2010), we may set

$$\begin{pmatrix} I \\ i_B \end{pmatrix} = F_A^T J_A = \begin{pmatrix} F^T & H \\ 0 & 1 \end{pmatrix} \begin{pmatrix} J \\ J_7 \end{pmatrix}. \tag{9.43}$$

In this way, we find the following two relations:

$$I = F^T J + J_7 H, \tag{9.44a}$$

$$i_B = J_7. \tag{9.44b}$$

Therefore, the sought relation between I and J is the following:

$$I = F^T J + i_B H. \tag{9.45}$$

Finally, by substituting Eq. (9.28b) into Eq. (9.45) we have:

$$I = -\frac{1}{2\pi\beta} F^T D^{-1} F\varphi - \frac{\psi_{ex}}{\beta} F^T D^{-1} \Gamma + i_B H. \tag{9.46}$$

With this procedure we are thus able to find the current vector I appearing on the right-hand side of Eq. (9.29a). The latter equation can be finally cast in the following form:

$$\frac{d}{d\tau}\varphi + S(\varphi) + \frac{1}{2\pi\beta} F^T D^{-1} F\varphi = i_B H - \frac{\psi_{ex}}{\beta} F^T D^{-1} \Gamma. \tag{9.47}$$

We remark that, while the matrix D depends on the parameter b, the matrix F does not. Another important aspect to point out is that, when we set $i_B = 0$ and $b = 0$ in Eq. (9.47), the matrix D reduces to the 6×6 identity matrix and the differential equation attains the same form of the set of equations written for a single cube in the presence of a magnetic field in Chapter 8. We shall devote the following subsection to prove the latter statement. Finally, we notice that the structure of Eq. (9.47) clearly shows that the forcing term splits in two parts: the first, i_BH, in which only the bias current term appears; the second, $-\psi_{ex}$F^{T}D$^{-1}/\beta$, in which only the externally applied magnetic field is present.

9.2.1. *Equivalence with the single cube equations*

In the present subsection we shall prove that Eq. (9.47), written for $i_B = 0$ and $b = 0$ is equivalent to the set of equations derived in Chapter 8 for the cubic network, namely, Eq. (8.27a–c), Eq. (8.52a–c), Eq. (8.53a–c), Eq. (8.54a–c). First of all, by setting $i_B = 0$ and $b = 0$ in Eq. (9.47) we obtain:

$$\frac{d}{d\tau}\varphi + S(\varphi) + \frac{1}{2\pi\beta}F^T F\varphi = -\frac{\psi_{ex}}{\beta}F^T\Gamma, \tag{9.48}$$

where, we recall, that the time normalization is given by Eq. (9.1). In order to compare the analysis done in this chapter to that of Chapter 8, we choose to deal with the laboratory time. In fact, as already stated, we have here made a different normalization choice with respect to Chapter 8. Therefore, going back to laboratory time and multiplying by $2\pi\beta$ both sides of Eq. (9.48), we rewrite this equation as follows:

$$\frac{L}{R}\frac{d}{dt}\varphi + 2\pi\beta\, S(\varphi) = -F^T F\varphi - 2\pi\psi_{ex}F^T\Gamma. \tag{9.49}$$

Recall that $L = 4L_0$ is the inductance parameter for the entire loop, so that β in Chapter 8 corresponds exactly to the parameter β in the present chapter. Let us then recall Eq. (8.27a–c), Eq. (8.52a–c), Eq. (8.53a–c), Eq. (8.54a–c) and notice that the non-linear operator O_J acts on all superconducting phase differences. By gathering all terms where the non-linear operator O_J appears, we may write:

$$O_J[\varphi] = \frac{4L_0}{R}\frac{d}{dt}\varphi + 2\pi\beta\, S(\varphi). \tag{9.50}$$

The above expression corresponds exactly to the left-hand side of Eq. (9.49). Next, by performing the matrix product $\mathbf{F}^T\Gamma$ in Eq. (9.49) and by using Eq. (9.25) and (9.48), we get:

$$-2\pi\psi_{ex}\,\mathbf{F}^T\Gamma = 2\pi\psi_{ex}\begin{pmatrix} (\hat{y}-\hat{z})\cdot\hat{\psi}_{ex} \\ (\hat{y}+\hat{z})\cdot\hat{\psi}_{ex} \\ -(\hat{y}+\hat{z})\cdot\hat{\psi}_{ex} \\ -(\hat{y}-\hat{z})\cdot\hat{\psi}_{ex} \\ -(\hat{x}-\hat{z})\cdot\hat{\psi}_{ex} \\ -(\hat{x}+\hat{z})\cdot\hat{\psi}_{ex} \\ (\hat{x}+\hat{z})\cdot\hat{\psi}_{ex} \\ (\hat{x}-\hat{z})\cdot\hat{\psi}_{ex} \\ (\hat{x}-\hat{y})\cdot\hat{\psi}_{ex} \\ (\hat{x}+\hat{y})\cdot\hat{\psi}_{ex} \\ -(\hat{x}+\hat{y})\cdot\hat{\psi}_{ex} \\ -(\hat{x}-\hat{y})\cdot\hat{\psi}_{ex} \end{pmatrix}. \tag{9.51}$$

By looking at the forcing terms in Eq. (8.27a–c), Eq. (8.52a–c), Eq. (8.53a–c), Eq. (8.54a–c), taken in the correct order, we notice that they correspond exactly to those in Eq. (9.51). We only need to prove that the remaining term in Eq. (9.49), namely, $-\mathbf{F}^T\mathbf{F}\varphi$, corresponds to the linear terms in the superconducting phase differences $\phi_\xi(\vec{r},t)$ in Eq. (8.27a–c), Eq. (8.52a–c), Eq. (8.53a–c), Eq. (8.54a–c), which can be expressed in matrix form as $\mathbf{C}\varphi$, where

$$\mathbf{C} = \begin{pmatrix}
-2 & 1 & 1 & 0 & 1 & -1 & 0 & 0 & 1 & -1 & 0 & 0 \\
1 & -2 & 0 & 1 & 1 & 1 & 0 & 0 & 0 & 0 & 1 & -1 \\
1 & 0 & -2 & 1 & 0 & 0 & 1 & -1 & -1 & 1 & 0 & 0 \\
0 & 1 & 1 & -2 & 0 & 0 & -1 & 1 & 0 & 0 & -1 & 1 \\
1 & -1 & 0 & 0 & -2 & 1 & 1 & 0 & 1 & 0 & -1 & 0 \\
-1 & 1 & 0 & 0 & 1 & -2 & 0 & 1 & 0 & 1 & 0 & -1 \\
0 & 0 & 1 & -1 & 1 & 0 & -2 & 1 & -1 & 0 & 1 & 0 \\
0 & 0 & -1 & 1 & 0 & 1 & 1 & -2 & 0 & -1 & 0 & 1 \\
1 & 1 & 0 & 0 & 1 & 0 & -1 & 0 & -2 & 1 & 1 & 0 \\
0 & 0 & -1 & 0 & 0 & 1 & 0 & -1 & 1 & -2 & 0 & 1 \\
0 & 0 & 0 & -1 & -1 & 0 & 1 & 0 & 1 & 0 & -2 & 1 \\
-1 & -1 & 1 & 1 & 0 & -1 & 0 & 1 & 0 & 1 & 1 & -2
\end{pmatrix}. \tag{9.52}$$

Therefore, in order to complete our proof, we need to show that $-\mathbf{F}^T\mathbf{F} = \mathbf{C}$ by a rather straightforward matrix multiplication. This can be done rather quickly using a mathematical software application, thus completing the proof.

9.3. General properties

In the previous section we have completely defined the dynamic equations, in a convenient matrix notation, for the three-dimensional superconducting quantum interference device. In order to dwell more deeply into the general properties of this system, which could be conceived as an ultra-sensitive flux-to-voltage transducer, we define the voltage across each branch of the cubic structure as follows:

$$V_\xi(\vec{r}, t) = \frac{\Phi_0}{2\pi}\frac{d}{dt}\phi_\xi(\vec{r}, t) + LI_{J0}\frac{d}{dt}i_\xi(\vec{r}, t). \tag{9.53}$$

By normalizing the voltage to the quantity RI_{J0}, defining $v_\xi(\vec{r}, \tau) = V_\xi(\vec{r}, t)/RI_{J0}$, and recalling the definition of the normalized time in Eq. (9.1), we may write:

$$v_\xi(\vec{r}, \tau) = \frac{d}{d\tau}\phi_\xi(\vec{r}, \tau) + 2\pi\beta\frac{d}{d\tau}i_\xi(\vec{r}, \tau). \tag{9.54}$$

The above equation can be written, more conveniently, in matrix notation as follows:

$$v(\tau) = \frac{d}{d\tau}\varphi(\tau) + 2\pi\beta\frac{d}{d\tau}I(\tau), \tag{9.55}$$

where $v(\tau)$ is the twelve-component voltage vector, the elements $v_\xi(\vec{r}, \tau)$ being ordered as done for the currents $i_\xi(\vec{r}, \tau)$ and for the superconducting phase differences $\phi_\xi(\vec{r}, \tau)$.

For constant forcing terms, the $v = 0$ state is realized, in this system, as in a two-junction superconducting quantum interferometer, by setting the time derivative of the vector $\varphi(\tau)$ equal to zero. In fact, by recalling Eq. (9.46), we may see that the time derivative of the current $I(\tau)$ is proportional to the time derivative of the vector $\varphi(\tau)$ in case ψ_{ex} and i_B do not depend on time. This aspect will be clarified in more details in what follows. Being the time derivative of the vector $\varphi(\tau)$ equal to zero, twelve constant phase differences, composing the vector φ_0, which depends both on i_B and on $\vec{\psi}_{ex}$, appear at the junctions electrodes. For a fixed magnetic field, the zero-voltage state (ZVS) survives up to a maximum value i_{BM} of the bias current i_B. The superconducting phase difference vector φ_0 is determined by the following non-linear vector equation:

$$S(\varphi_0) + \frac{1}{2\pi\beta}\mathbf{F}^T\mathbf{D}^{-1}\mathbf{F}\varphi_0 = i_B\mathrm{H} - \frac{\psi_{ex}}{\beta}\mathbf{F}^T\mathbf{D}^{-1}\Gamma. \qquad (9.56)$$

For a fixed value of $\vec{\psi}_{ex}$, when $i_B > i_{BM}$, the ZVS breaks down and a certain number of junctions in the network make a transition to a resistive state (RS). The resistive state can be partial (not all junctions are in the RS) or total (all junctions are in the RS), depending on the value of i_B and on the particular direction of $\vec{\psi}_{ex}$ chosen.

In the resistive state, the voltage is seen to show periodicity in time, the period being denoted by T. This property can be understood as follows. Relying on what seen in a two-junction interferometer, if the voltage vector $v(\tau)$ shows periodicity in time, the superconducting phase difference vector $\varphi(\tau)$ should be pseudo-periodic in τ, so that:

$$\varphi(\tau + T) = \varphi(\tau) + 2\pi\mathrm{K}, \qquad (9.57)$$

where K is a twelve-component vector whose elements are integers. By keeping the forcing terms constant, we may see that Eq. (9.47) allows, because of Eq. (9.57), the following equality:

$$\mathbf{F}^T\mathbf{D}^{-1}\mathbf{F}[\varphi(\tau) + 2\pi\mathrm{K}] = \mathbf{F}^T\mathbf{D}^{-1}\mathbf{F}\varphi(\tau), \qquad (9.58)$$

which can be simplified to give:

$$\mathbf{P}\mathrm{K} = \mathbf{0}, \qquad (9.59)$$

where $\mathbf{P} = \mathbf{F}^T\mathbf{D}^{-1}\mathbf{F}$ is a 12×12 matrix having determinant equal to zero. Being $\det\mathbf{P} = 0$, Eq. (9.59) has a non-null solution, because its null space is non-empty (Lang, 1987). The determinant of the matrix $\mathbf{P}$ can be proven to be zero more conveniently by a mathematical software application. Existence of a non-null solution for the vector K thus allows pseudo–periodicity of $\varphi(\tau)$. However, this condition is not sufficient to prove periodicity of $v(\tau)$, since we also need to show that the derivative of the current vector I is periodic with period T. The latter property, however, can be proven by considering the expression of the current vector I given by Eq. (9.28a). The proof goes as follows. First of all, assume that the forcing term f, defined from Eq. (9.47) and expressed as

$$f = i_B H - \frac{\psi_{ex}}{\beta} \mathbf{F}^T \mathbf{D}^{-1} \Gamma, \tag{9.60}$$

is constant. Therefore, we are in the condition of validity of Eq. (9.57). Consider now the expression for $I(\tau)$ given in Eq. (9.28a). Next substitute in the left-hand side of this equation the right-hand side of Eq. (9.57), thus obtaining an expression for $I(\tau + T)$. Then, by comparing the two expressions, we conclude that $I(\tau + T) = I(\tau)$. In this way, we note that the current vector $I(\tau)$ is periodic in time, thus completing our proof.

We should however bear in mind that instantaneous current and voltages are not directly observable quantities in electronic superconducting devices, so that their average value, calculated with respect to time, should be found. In this way, we may show that

$$\langle v \rangle = \frac{1}{T} \int_\tau^{\tau+T} v(\tau) \, d\tau = \frac{\varphi(\tau+T) - \varphi(\tau)}{T} = \frac{2\pi}{T} K, \tag{9.61}$$

where, as usual, we indicate with $\langle v \rangle$ the time-averaged value of $v(\tau)$. How can we interpret the result of Eq. (9.61)? By looking at the first and the last term in the chain of equalities, we notice that the non-null components of the vector K correspond to the components of the superconducting phase difference vector φ relative to the junctions in the running state. All the remaining junctions, if any, will be in a flat ZVS (where effectively the voltage is zero) or in effective ZVS (where only the average value of the voltage is zero).

Having established periodicity of $v(\tau)$ in the time domain, it is now possible to study the periodic behavior of the quantity $\langle v \rangle$ with respect to the amplitude ψ_{ex}, for a given direction of the normalized external magnetic flux $\vec{\psi}_{ex}$. Let us first notice that, by Eq. (9.47), we may write

$$\langle \frac{d}{d\tau} \varphi \rangle = f - \langle S(\varphi) \rangle - \frac{1}{2\pi\beta} P\langle\varphi\rangle, \qquad (9.62)$$

where f is the forcing-term vector given by Eq. (9.60). Let us now denote by φ^* the solution of Eq. (9.47) after an increment p of the forcing term, so that we write:

$$\langle \frac{d}{d\tau} \varphi^* \rangle = f + p - \langle S(\varphi^*) \rangle - \frac{1}{2\pi\beta} P\langle\varphi^*\rangle. \qquad (9.63)$$

If we require $v(\tau)$ to be invariant under the translation $f \to f + p$, by considering Eq. (9.57), we may write:

$$\varphi^* = \varphi + 2\pi \widetilde{K}, \qquad (9.64)$$

where the component of the vector $\widetilde{K}$ are integers. Let us now impose $\langle v^* \rangle = \langle v \rangle$. We notice that, because of Eq. (9.64), it suffices to set

$$\langle \frac{d}{d\tau} \varphi^* \rangle = \langle \frac{d}{d\tau} \varphi \rangle \Rightarrow p - \frac{1}{2\pi\beta} P(\langle\varphi^*\rangle - \langle\varphi\rangle) = 0. \qquad (9.65)$$

Therefore, the quantity p must satisfy the following relation:

$$p = \frac{1}{\beta} P \widetilde{K}. \qquad (9.66)$$

If the increase p in the forcing term is exclusively given by an increase $\Delta\psi_{ex}$ in the amplitude of the externally applied flux, leaving the direction of $\vec{\psi}_{ex}$ fixed, we have

$$p = -\frac{\Delta\psi_{ex}}{\beta} F^T D^{-1} \Gamma, \qquad (9.67)$$

where Γ has been defined in Eq. (9.24). Recalling now that $P = F^T D^{-1} F$, by combining Eq. (9.66) and Eq. (9.67), we have:

$$F^T D^{-1} [F\widetilde{K} + \Delta\psi_{ex}\Gamma] = 0. \qquad (9.68)$$

The 12×6 matrix $E = F^T D^{-1}$ has rank equal to five, i.e., rank(E) = 5, as it can be proven by means of direct calculation or by computer software applications. Therefore, the system of 12 equations in (9.68), although being always compatible, is redundant. In order to extract some useful information from this system, let us proceed as follows. First of all, let us consider the solutions of Eq. (9.68) not belonging to the null-space of the matrix **E**, so that, we may write:

$$\Delta\psi_{ex}\Gamma = -F\tilde{K}. \tag{9.69}$$

The right hand side of Eq. (9.69) can be written as a column vector whose components are integer numbers. By then expressing Eq. (9.69) in terms of its components, we can write:

$$\Delta\psi_{ex}\begin{pmatrix} \hat{x}\cdot\hat{\psi}_{ex} \\ \hat{x}\cdot\hat{\psi}_{ex} \\ \hat{y}\cdot\hat{\psi}_{ex} \\ \hat{y}\cdot\hat{\psi}_{ex} \\ \hat{z}\cdot\hat{\psi}_{ex} \\ \hat{z}\cdot\hat{\psi}_{ex} \end{pmatrix} = \begin{pmatrix} l \\ l' \\ m \\ m' \\ n \\ n' \end{pmatrix}. \tag{9.70}$$

Therefore, we need to have $l = l'$, $m = m'$, $n = n'$, so that Eq. (9.70) can be summarized in the following way:

$$\Delta\psi_{ex}\,\alpha = l; \quad \Delta\psi_{ex}\,\beta = m; \quad \Delta\psi_{ex}\,\gamma = n. \tag{9.71}$$

where $\alpha = \hat{x}\cdot\hat{\psi}_{ex}$, $\beta = \hat{y}\cdot\hat{\psi}_{ex}$, $\gamma = \hat{z}\cdot\hat{\psi}_{ex}$, with $\alpha^2 + \beta^2 + \gamma^2 = 1$. The periodicity of the $\langle v \rangle$ vs. ψ_{ex} curves for a fixed value of the bias current i_B and for a fixed direction in space of the applied magnetic field can be found by means of Eq. (9.71). The expression for $\Delta\psi_{ex}$ can be found by squaring and summing homologous sides in the relations in Eq. (9.71), so that:

$$\Delta\psi_{ex} = \sqrt{l^2 + m^2 + n^2}. \tag{9.72}$$

In what follows we shall consider some special cases, keeping in mind that our analysis might still not completely describe all properties of the system, since the solutions of Eq. (9.68) belonging to thc null-space of the matrix **E** have not yet considered. In the following analysis we shall assume that, for fixed values of i_B, the vector $\tilde{K}$ in Eq. (9.69) does not depend on ψ_{ex}.

a) Magnetic field applied along one of the axes.

To fix the ideas, let us say that the magnetic field is applied along the x −axis. In this case, we set: $\alpha = 1$, $\beta = \gamma = 0$. From Eq. (9.71) we have:

$$n = m = 0, \tag{9.73}$$

so that:

$$\Delta\psi_{ex} = |l|. \tag{9.74}$$

For $l = 1$ we recover the periodicity of a two-junction quantum interferometer, as studied in Chapter 3. The simple relation in Eq. (9.74) represents a strong similarity between two-junction quantum interferometers and current-biased cubic networks in the presence of a magnetic field along one of the three coordinate axes.

b) Magnetic field parallel to one of the coordinate planes $x - y$, $x - z$, $y - z$.

Let us take the magnetic field parallel to the $x - y$ plane. In this case, we have: $\alpha, \beta \neq 0$, and $\gamma = 0$. We may distinguish the following subcases:

  i) $\alpha = \beta$, and $\gamma = 0$;
  ii) $\alpha \neq \beta$, and $\gamma = 0$.

By considering first the case *i)*, the relations in Eq. (9.71) give:

$$l = m, \tag{9.75a}$$

$$n = 0. \tag{9.75b}$$

therefore, by Eq. (9.72), we have:

$$\Delta\psi_{ex} = \sqrt{2}\,|l|. \tag{9.76}$$

We notice that there are four different directions in space of $\hat{\psi}_{ex}$ all at $\pi/4$ with respect both to the x −axis and to the y −axis. By considering all possibilities, i.e., by taking $\alpha = 0$ or $\beta = 0$, we have twelve possible direction in space giving $\Delta\psi_{ex} = \sqrt{2}|l|$.

By considering now the case *ii)*, the relations in Eq. (9.71) give:

$$l \neq m, \tag{9.77}$$

$$n = 0, \tag{9.77b}$$

Therefore, by Eq. (9.72), we have:

$$\Delta\psi_{ex} = \sqrt{l^2 + m^2}. \tag{9.78}$$

In this case, the number of ways in which we may obtain the lowest value of $\Delta\psi_{ex}$ is greater that in the case *i)*. For example, let us consider the case $\Delta\psi_{ex} = \sqrt{5}\,|l|$. The value $\Delta\psi_{ex} = \sqrt{5}\,|l|$ can be obtained in eight ways for $\gamma = 0$, by considering either $|l| = 2|m|$, or $|m| = 2|l|$. Furthermore, by exchanging α with γ, or β with γ, in 16 additional ways, we obtain a total of 24 possible directions in space.

c) Magnetic field not parallel to any coordinate plane.

Let us take $\alpha, \beta, \gamma \neq 0$, distinguishing the following subcases:
  i) $\alpha = \beta = \gamma$;
  ii) $\alpha = \beta \neq \gamma$;
  iii) $\alpha \neq \beta \neq \gamma$.
By considering first the case *i)*, the relations in Eq. (9.71) give:

$$l = m = n. \tag{9.79}$$

Therefore, by Eq. (9.72), we have:

$$\Delta\psi_{ex} = \sqrt{3}\,|l|, \tag{9.80}$$

for eight directions in space, one for each octant of the $0xyz$ coordinate system.

In the second case we have:

$$l = m \neq n. \tag{9.81}$$

Therefore, by Eq. (9.72), we may write:

$$\Delta\psi_{ex} = \sqrt{2l^2 + n^2}. \tag{9.82}$$

For $|l| = 1$ and $|n| = 2$, we realize the lowest value of $\Delta\psi_{ex}$ in this case, namely, $\Delta\psi_{ex} = \sqrt{6}$. Considering all other possible ways of getting $\Delta\psi_{ex} = \sqrt{6}$, we count a total of 24 directions in space.

In the third case we set

$$l \neq m \neq n, \qquad (9.83)$$

Therefore, for this case we have Eq. (9.72) with integers l, m, and n all different from each-other. Moreover, in this case the lowest periodicity, $\Delta\psi_{ex} = \sqrt{14}$, can be obtained in eight different way by setting $|l| = 1$, $|m| = 2$, $|n| = 3$. Counting all possible permutations of the numbers $1, 2, 3$, we end up with a number of 48 directions in space giving periodicity $\Delta\psi_{ex} = \sqrt{14}$.

A similar analysis can be performed by considering the one-dimensional null-space of the operator **E** in Eq. (9.68), whose base is given by the following vector:

$$\mathbf{K_0} = (-1, +1, -1, +1, -1, +1,)^{\mathrm{T}}. \qquad (9.84)$$

In this case, in order to satisfy Eq. (9.68) we might set

$$\Delta\psi_{ex}\Gamma = \mathbf{K_0} - \mathbf{F}\widetilde{\mathbf{K}}. \qquad (9.85)$$

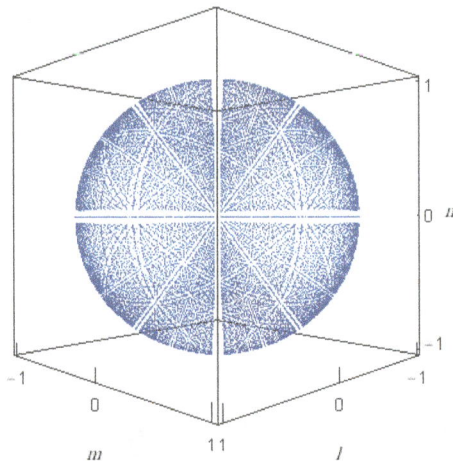

Fig. 9.2. Points on the unitary sphere describing directions in space for which periodic behavior of the $\langle v \rangle$ vs. ψ_{ex} curves of a three-dimensional superconducting interferometer can be detected. The integers l, m, n range from -15 to 15. Local lack of points on the sphere confirms the selection rule indicated by Gauss condition.

Being all components of the column vector on the right-hand side of Eq. (9.85) given by integer numbers, exactly the same cases, as before, are detectable. A more refined and lengthy analysis, in which a multiplicative factor of K_0 is considered, leads to the same conclusion.

Up to this point we have not mentioned that there could exist directions in space for which periodicity is not present. In fact, by Gauss condition (Hardy, 1979) by summing the squares of three integers, we cannot obtain the following numbers: $4^a(8b + 7)$, where a and b are non-negative integers. Therefore, as a consequence of Gauss condition, the relation in Eq. (9.72) imposes the following selection rule for the period $\Delta\psi_{ex}$:

$$\Delta\psi_{ex} \neq \sqrt{4^a(8b + 7)}, \qquad (9.86)$$

with a and b non-negative integers. In Fig. 9.2 we represent a collection of directions in space, on the unitary sphere, for which a periodicity in ψ_{ex} does exist for integers l, m, n ranging from -15 to 15.

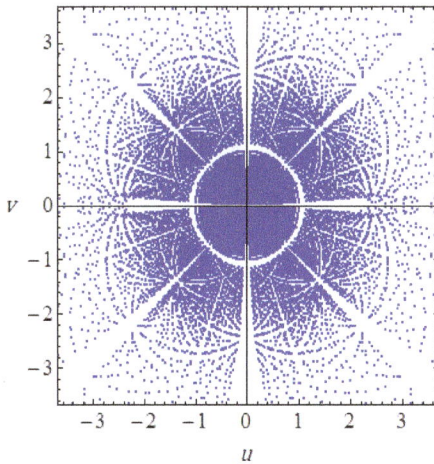

Fig. 9.3. Stereographic projection on the $u - v$ plane of points on the unitary sphere describing direction in space for which periodic behavior of the $\langle v \rangle$ vs. ψ_{ex} curves of a three-dimensional superconducting interferometer is detectable. The integers l, m, n range from -15 to 15. Points inside and outside the unitary circle $u^2 + v^2 = 1$ represent the

projection on the $u - v$ plane of points of the Southern and Northern hemisphere, respectively.

In Fig. 9.3 a stereographic projection of the points on the unitary sphere represented in Fig. 9.2 is given. We recall that a stereographic projection on the $u - v$ plane of a point of coordinates (x, y, z) in space, can be realized by means of the following transformations:

$$u = \frac{x}{1-z},$$ (9.87a)

$$v = \frac{y}{1-z}.$$ (9.87b)

In this way, the South Pole $(0, 0, -1)$ is mapped into the origin of the $u - v$ plane, while points in the vicinity of the North Pole $(0, 0, 1)$ are projected at infinity. We remark that the selection rule given by Gauss condition in Eq. (9.86) is confirmed by the presence of "equatorial gaps" in Fig. 9.2, which indicate absence of points on the sphere, for the range of the integers l, m, n going from -15 to 15. When this range increases, the equatorial gaps are expected to reduce in size.

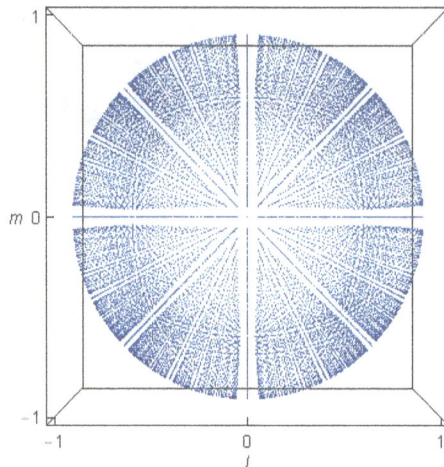

Fig. 9.4. Top view of the unitary sphere in fig. 9.2 describing the directions in space for which periodic behavior of the $\langle v \rangle$ vs. ψ_{ex} curves of a three-dimensional superconducting interferometer is detectable. The integers l, m, n range from -15 to 15.

In Fig. 9.4 we give a top view of Fig. 9.2, in order to make a correspondence between the principal equatorial gaps on the unitary sphere and those seen in the stereographic projection in Fig. 9.3. The gaps appearing at $m = \pm l$ in Fig. 9.4 are visible in the vicinities of the lines $u = \pm v$ in Fig. 9.3. The lack of points at $u = 0$ or $v = 0$ in Fig. 9.3, on the other hand, represent the equatorial gaps occurring at $l = 0$ or $m = 0$, respectively in Fig. 9.4. Secondary equatorial gaps can be detected along lines at $\pm \pi/6$ and $\pm \pi/3$ with respect to the m −axis. Corresponding thin regions, where absence of points can be noticed, appear on the $u - v$ plane in Fig. 9.3 at the same inclination angles with respect to the u −axis.

9.4. Numerical analysis

In order to numerically confirm some of the general properties investigated in the previous section, we shall here analyze the solution of Eq. (9.47) in various cases. Starting by considering the time dependence of the phase differences and voltages, we numerically determine the existence of a maximum Josephson current i_{BM} for the system. The transition from the Zero Voltage State (ZVS) to the Resistive State (RS) of the network is thus investigated. The periodicity in time of the voltage vector $v(\tau)$ is evidenced and the time-average $\langle v_\xi(\vec{r}) \rangle$ of the components $v_\xi(\vec{r}, \tau)$ is plotted against the amplitude ψ_{ex} of the externally applied flux $\vec{\psi}_{ex}$, taken in the direction of the applied magnetic field. The periodicity of the $\langle v_\xi \rangle$ vs. ψ_{ex} curves is investigated on the basis of the analytic properties derived in the previous section. The average voltage $\langle v_B \rangle$ across the two nodes, at which the bias current i_B is injected and drawn, is found and the i_B vs. $\langle v_B \rangle$ curves are determined for different orientations $\hat{\psi}_{ex}$ of the applied magnetic field and for various values of ψ_{ex}.

9.4.1. *Zero-voltage states vs. resistive states*

In the present subsection, we consider the vector dynamic equation (9.47), which we rewrite in a more compact form as follows:

$$\frac{d}{d\tau}\varphi + S(\varphi) + \frac{1}{2\pi\beta}\,\mathbf{EF}\varphi = i_B\mathbf{H} - \frac{\psi_{ex}}{\beta}\,\mathbf{E\Gamma}. \qquad (9.88)$$

where, we recall, $\mathbf{E} = \mathbf{F}^{\mathrm{T}}\mathbf{D}^{-1}$. In the above equation the bias current value i_B and the amplitude ψ_{ex} of the quantity $\vec{\psi}_{ex}$ appear explicitly. The vector H, on the other hand, as we have seen in Section 9.2, is determined by the position of the leads when biasing the circuit by means of a current i_B. In our case, we have chosen the following two nodes: node a positioned at $\vec{r} = a\hat{x} + a\hat{z}$, where i_B is injected, node b located at $\vec{r} = a\hat{y}$, where i_B is drawn. For this specific case we found H = $(0, -1, 0, 0, 0, 0, 0, +1, 0, 0, 0, -1)^{\mathrm{T}}$. Different choices of nodes will give different components for H, but will leave the dynamic equation unaltered in form.

By standard numerical routines, we shall proceed in solving the differential equation (9.47), fixing the value of the parameters β and m to 1.0 and 0.1, respectively. We shall first consider the case in which only the bias current is present, so that we take $i_B > 0$ and $\psi_{ex} = 0$. Successively, we shall apply the magnetic field along the z −direction, in order to see whether a magnetic behavior similar to a two-junction quantum interferometer can be detected. Moreover, we shall investigate the case in which the magnetic field is inclined of an angle θ with respect to the z −axis. Finally, we shall study the case in which an additional azimuth angular displacement to the field direction is given. The period T of the time-dependent voltage curves $v_\xi(\vec{r}, \tau)$ will be found in each case.

As a first example, then, let us consider the magnetic field along the z −axis and let us solve for the superconducting phase differences by starting from zero-field cooling conditions (all phases initially equal to zero) and by slowly raising the current i_B, up to a specific maximum value. In this starting analysis, we take $\psi_{ex} = 0$, in order to keep track of the current distribution inside the network in a rather simple way. When calculating the voltage vector $v(\tau)$ from Eq. (9.55), we assume time-independent forcing terms, so that, by recalling Eq. (9.46), we may write:

$$v(\tau) = (\mathbb{I} - \mathbf{EF})\frac{\mathrm{d}}{\mathrm{d}\tau}\varphi(\tau). \qquad (9.89)$$

In this way, we may set:

$$\langle v \rangle = (\mathbb{I} - \mathbf{EF})[\varphi(\tau_0 + T) - \varphi(\tau_0)], \qquad (9.90)$$

where τ_0 is an arbitrarily chosen time. We remark that Eq. (9.89) and (9.90) can be used only when the external field and the bias current do not depend explicitly on time.

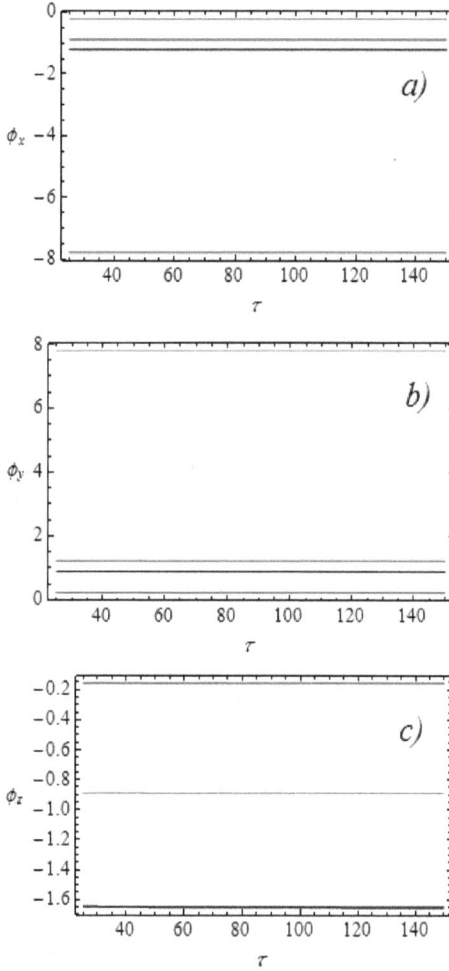

Fig. 9.5. Superconducting phase differences $\phi_\xi(\vec{r}, \tau)$ across the junctions in a current-biased cubic network with parameters $\beta = 1.0$ and $m = 0.1$, for $\psi_{ex} = 0$ and $i_B = 2.93$. Blue, red, black, and orange curves represent the quantities $\phi_\xi(\vec{r}, \tau)$ ($\xi = x, y, z$) in increasing hierarchical order. For the superconducting phases $\phi_x(\vec{r})$ in panel a) we have: $\phi_x(0)$, $\phi_x(a\hat{y})$, $\phi_x(a\hat{z})$, $\phi_x(a\hat{y} + a\hat{z})$. For the superconducting phases $\phi_y(\vec{r})$ in panel b) we have: $\phi_y(0)$, $\phi_y(a\hat{x})$, $\phi_y(a\hat{z})$, $\phi_y(a\hat{x} + a\hat{z})$. Finally, for branch currents $\phi_z(\vec{r})$ in panel c) the order is the following: $\phi_z(0)$, $\phi_z(a\hat{x})$, $\phi_z(a\hat{y})$, $\phi_y(a\hat{x} + a\hat{y})$. In the latter panel the red curve coincides with the black curve and they both appear in the bottom.

Not considering here time-dependent forcing terms, from now on we shall adopt this scheme. We should however remember that, each time we apply Eq. (9.90), full notion of the period T should be acquired.

In Fig. 9.5a–c the superconducting phases $\phi_\xi(\vec{r}, \tau)$, solutions of Eq. (9.88) for $\beta = 1.0$, $m = 0.1$, $\psi_{ex} = 0$, and $i_B = 2.93$, are shown. We notice that all superconducting phases are constant. This means that the system is in a ZVS. In fact, if we were to draw the curves for the voltages $v_\xi(\vec{r}, \tau)$ across each branch, we would see all of them to be zero, according to Eq. (9.89). At $i_B = 2.93$ we are, however, close to the transition to the resistive state. By taking a look at the constant values in Fig. 9.5a, where the superconducting phases $\phi_x(\vec{r}, \tau)$ are shown, we notice that they are all negative, meaning that the currents $i_x(\vec{r}, \tau)$, when the bias current i_B is injected at node a, all flow in the negative direction. In Fig. 9.5b, on the other hand, the superconducting phases $\phi_y(\vec{r}, \tau)$ are all positive. Finally, in Fig. 9.5c, the superconducting phases $\phi_z(\vec{r}, \tau)$ are again seen to be negative. This feature is common to the currents $i_\xi(\vec{r}, \tau)$, which are shown in Fig. 9.6a–c for the same values of the parameters and for $i_B = 2.93$. This current distribution can be understood as follows. The current i_B, injected in node a, divides in three branch current: $i_x(a\hat{z}, \tau)$ (black line in Fig. 9.6a), $i_y(a\hat{x} + a\hat{z}, \tau)$ (orange line in Fig. 9.6b), $i_z(a\hat{x}, \tau)$ (red line in Fig. 9.6c). The first current is negative and flows front-back away from the node; the second is positive and flows left-to-right away from the node; the third is negative and flows top-bottom away from the node. The sum of the absolute values of these currents should therefore be equal to $i_B = 2.93$ by KCL, as one can effectively deduce from Fig. 9.6a–c. In addition, the absolute values of the currents flowing on the cubic face parallel to the $y - z$ plane and passing through node a ($i_y(a\hat{x} + a\hat{z}, \tau)$ and $i_z(a\hat{x}, \tau)$) are almost equal to 1. The junctions lying on these branches would thus be subject to phase-slip processes if the value of the bias current were increased. The same would be true for the junctions close to node b through which the currents $i_x(a\hat{y}, \tau)$ (red line in Fig. 9.6a) and $i_z(a\hat{y}, \tau)$ (black line in Fig. 9.6c) flow.

In order to investigate the transition from the ZVS to the RS, we slowly further raise the bias current, up to $i_B = 3.20 > i_{BM}$. Therefore, the superconducting phases $\phi_x(\vec{r}, \tau)$, $\phi_y(\vec{r}, \tau)$, and $\phi_z(\vec{r}, \tau)$ are shown, respectively, in Fig. 9.7a–b, Fig. 9.8a–b and Fig. 9.9a–b.

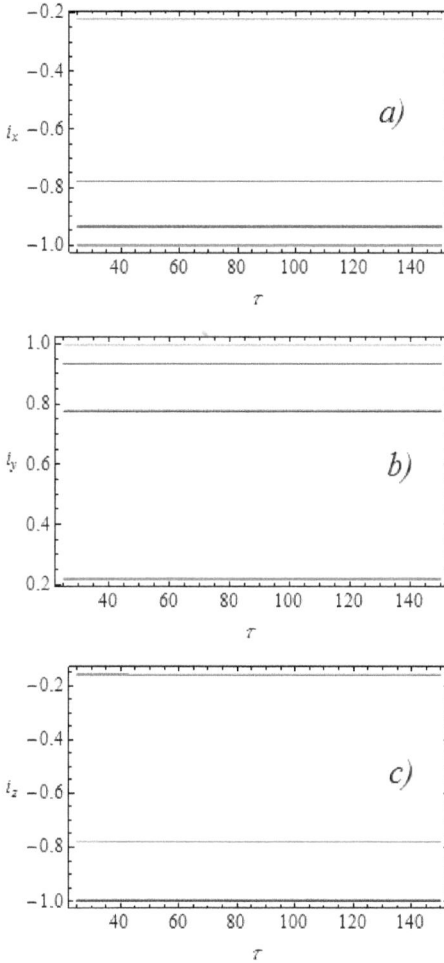

Fig. 9.6. Branch currents $i_\xi(\vec{r}, \tau)$ flowing through the junctions in a current-biased cubic network with parameters $\beta = 1.0$ and $m = 0.1$, for $\psi_{ex} = 0$ and $i_B = 2.93$. Blue, red, black, and orange curves represent the quantities $i_\xi(\vec{r}, \tau)$ ($\xi = x, y, z$) in increasing hierarchical order. For branch currents $i_x(\vec{r})$ in panel a) we have: $i_x(0)$, $i_x(a\hat{y})$, $i_x(a\hat{z})$, $i_x(a\hat{y} + a\hat{z})$. For branch currents $i_y(\vec{r})$ in panel b) we have: $i_y(0)$, $i_y(a\hat{x})$, $i_y(a\hat{z})$, $i_y(a\hat{x} + a\hat{z})$. Finally, for branch currents $i_z(\vec{r})$ in panel c) the order is the following: $i_z(0)$, $i_z(a\hat{x})$, $i_z(a\hat{y})$, $i_y(a\hat{x} + a\hat{y})$. In the latter panel the red curve coincides with the black curve and they both appear in the bottom.

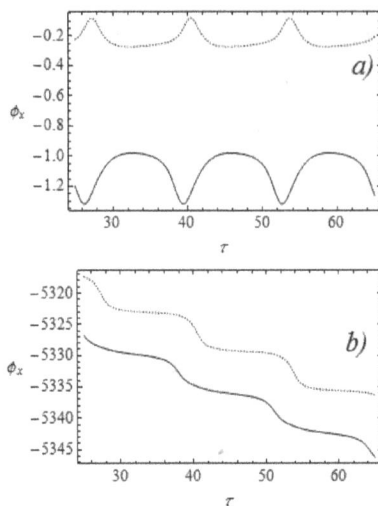

Fig. 9.7. Superconducting phase differences $\phi_x(\vec{r},\tau)$ in a current-biased cubic network with $\beta = 1.0$ and $m = 0.1$, for $\psi_{ex} = 0$ and $i_B = 3.20$. In panel a) $\phi_x(0)$ (full line), $\phi_x(a\hat{y} + a\hat{z})$ (dashed line) are shown. In panel b) $\phi_x(a\hat{y})$ (full line), $\phi_x(a\hat{z})$ (dashed line) are shown.

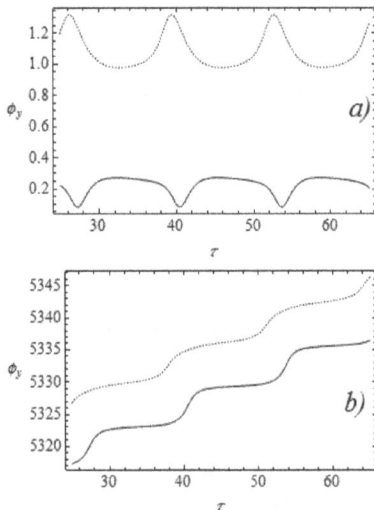

Fig. 9.8. Superconducting phase differences $\phi_y(\vec{r},\tau)$ in a current-biased cubic network with $\beta = 1.0$ and $m = 0.1$, for $\psi_{ex} = 0$ and $i_B = 3.20$. In panel a) $\phi_y(a\hat{x})$ (full line), $\phi_y(a\hat{z})$ (dashed line) are shown. In panel b) $\phi_y(0)$ (full line), $\phi_y(a\hat{x} + a\hat{z})$ (dashed line) are shown.

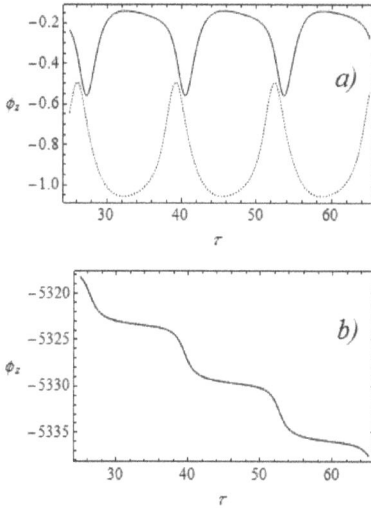

Fig. 9.9. Superconducting phase differences $\phi_z(\vec{r}, \tau)$ in a current-biased cubic network with $\beta = 1.0$ and $m = 0.1$, for $\psi_{ex} = 0$ and $i_B = 3.20$. In panel a) $\phi_z(0)$ (full line), $\phi_z(a\hat{y} + a\hat{z})$ (dashed line) are shown. In panel b) the two equal quantities $\phi_z(a\hat{x})$ and $\phi_z(a\hat{y})$ are shown.

This time the corresponding voltages $v_x(\vec{r}, \tau)$, $v_y(\vec{r}, \tau)$, and $v_z(\vec{r}, \tau)$ are shown in Fig. 9.10a–b, Fig. 9.11a–b and Fig. 9.12a–b, respectively.

Let us start by analyzing the superconducting phases in Fig. 9.7a–b, 9.8a–b, 9.9a–b. We notice that the typical behavior of successive phase-slip processes appears only in panels b) of all these figures. This means that only two junctions lying in the x −directions, two lying in the y −direction, and two in the z −direction are in the RS. The phase differences across these junctions are: $\phi_x(a\hat{y})$ (full line in Fig.9.7b), $\phi_x(a\hat{z})$ (dashed line in Fig.9.7b), $\phi_y(0)$ (full line in Fig.9.8b), $\phi_y(a\hat{x} + a\hat{z})$ (dashed line in Fig.9.8b), $\phi_z(a\hat{x})$ and $\phi_z(a\hat{x} + a\hat{y})$ (full line in Fig.9.9b). The above phase differences pertain to junctions which, for $i_B = 2.93$, in the previous computer run, showed absolute value of the branch currents flowing through them close to 1. The remaining junctions, for which the absolute value of the branch currents flowing through them was lower than 0.8 for $i_B = 2.93$, stay in the ZVS.

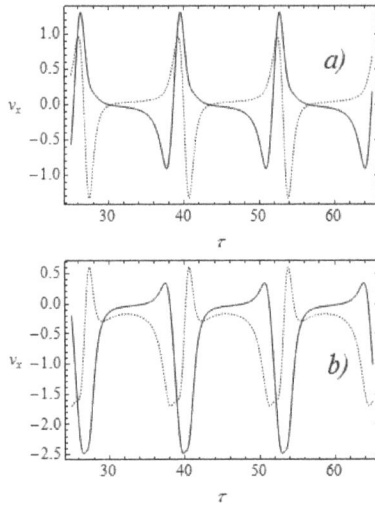

Fig. 9.10. Voltages $v_x(\vec{r}, \tau)$ across the branches of a current-biased cubic network with $\beta = 1.0$ and $m = 0.1$, for $\psi_{ex} = 0$ and $i_B = 3.20$. In panel a) $v_x(0)$ (full line), $v_x(a\hat{y} + a\hat{z})$ (dashed line) are shown. In panel b) $v_x(a\hat{y})$ (full line), $v_x(a\hat{z})$ (dashed line) are shown. The period can be estimated to be $T \approx 13.0$

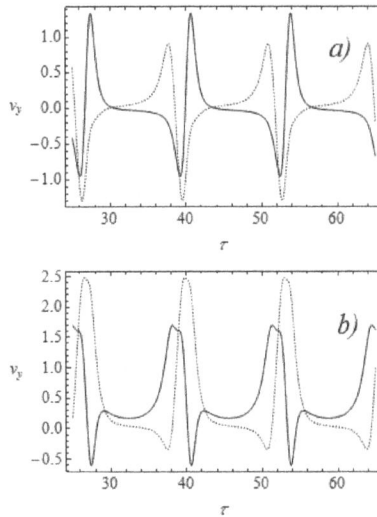

Fig. 9.11. Voltages $v_y(\vec{r}, \tau)$ across the branches of a current-biased cubic network with $\beta = 1.0$ and $m = 0.1$, for $\psi_{ex} = 0$ and $i_B = 3.20$. In panel a) $v_y(a\hat{x})$ (full line), $v_y(a\hat{z})$ (dashed line) are shown. In panel b) $v_y(0)$ (full line), $v_y(a\hat{x} + a\hat{z})$ (dashed line) are shown. The period can be estimated to be $T \approx 13.0$

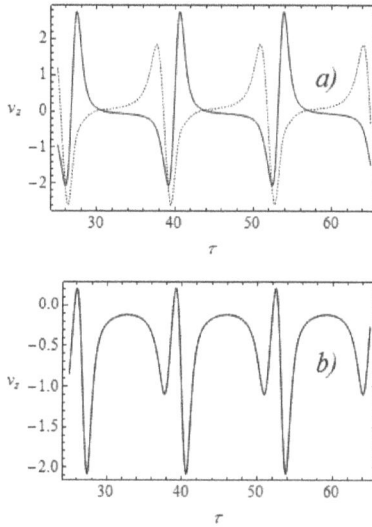

Fig. 9.12. Voltages $v_z(\vec{r}, \tau)$ in a current-biased cubic network with $\beta = 1.0$ and $m = 0.1$, for $\psi_{ex} = 0$ and $i_B = 3.20$. In panel a) $v_z(0)$ (full line), $v_z(a\hat{y} + a\hat{z})$ (dashed line) are shown. In panel b) the two equal quantities $v_z(a\hat{x})$ and $v_z(a\hat{y})$ are shown. The period can be estimated to be $T \approx 13.0$.

This can be argued by the properties of the voltages $v_x(\vec{r}, \tau)$, $v_y(\vec{r}, \tau)$, and $v_z(\vec{r}, \tau)$ shown in Fig. 9.10a–b, Fig. 9.11a–b and Fig. 9.12a–b, respectively. In fact, from these curves we may notice that the voltages $v_\xi(\vec{r}, \tau)$ in all panels a) oscillate periodically about a null value. For these voltages we may deduce that their time-averaged values $\langle v_\xi(\vec{r}, \tau) \rangle$ are zero. It is thus confirmed that the junctions characterized by phase differences $\phi_x(0)$, $\phi_x(a\hat{y} + a\hat{z})$, $\phi_y(a\hat{x})$, $\phi_y(a\hat{z})$, $\phi_z(0)$, $\phi_z(a\hat{y} + a\hat{z})$ remain in an effective ZVS when the bias current is brought to the value of $i_B = 3.20$. On the other hand, all remaining junctions are in the resistive state for $i_B = 3.20$. This can be argued by noticing, from Fig. 9.10b, Fig. 9.11b, and Fig. 9.12b, that the time-averaged values $\langle v_\xi(\vec{r}, \tau) \rangle$ are different from zero for the remaining junctions in the network. Notice also that, independently of the state (zero-voltage or resistive) of the junctions, the period T of the voltage is equal for all curves and can be estimated to be $T = 13.0$.

The question is now the following: can we make all junctions go into the RS by further raising the value of i_B?

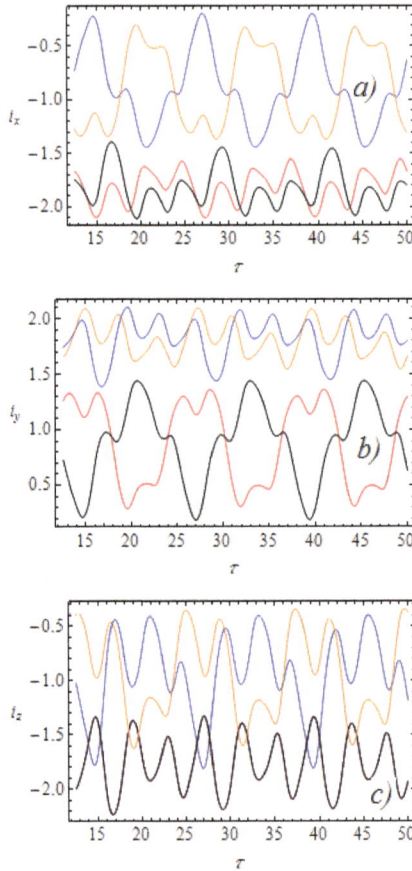

Fig. 9.13. Branch currents $i_\xi(\vec{r}, \tau)$ flowing through the junctions in a current-biased cubic network with parameters $\beta = 1.0$ and $m = 0.1$, for $\psi_{ex} = 0$ and $i_B = 5.40$. Blue, red, black, and orange curves represent the quantities $i_\xi(\vec{r}, \tau)$ ($\xi = x, y, z$) in increasing hierarchical order. For branch currents $i_x(\vec{r})$ in panel a) we have: $i_x(0)$, $i_x(a\hat{y})$, $i_x(a\hat{z})$, $i_x(a\hat{y} + a\hat{z})$. For branch currents $i_y(\vec{r})$ in panel b) we have: $i_y(0)$, $i_y(a\hat{x})$, $i_y(a\hat{z})$, $i_y(a\hat{x} + a\hat{z})$. Finally, for branch currents $i_z(\vec{r})$ in panel c) the order is the following: $i_z(0)$, $i_z(a\hat{x})$, $i_z(a\hat{y})$, $i_y(a\hat{x} + a\hat{y})$. In the latter panel the red curve coincides with the black curve and they both appear in the lowest part of the graph. The period can be estimated to be $T \approx 12.5$

We can answer by a computer run using the same parameters as in the previous two cases, this time setting $i_B = 5.40$. In this way, by using a value of the bias current well above i_{BM} for $\beta = 1.0$, $m = 0.1$ and $\psi_{ex} = 0$, we obtain the results shown in Fig. 9.13a–c for the currents $i_\xi(\vec{r}, \tau)$ in all branches. In particular, the branch currents $i_x(\vec{r}, \tau)$, $i_y(\vec{r}, \tau)$, and $i_z(\vec{r}, \tau)$ are shown in Fig. 9.13a, Fig. 9.13b, and Fig. 9.13c, respectively. In all curves we note oscillations of the currents which, for given values of the normalized time τ, give $i_\xi(\vec{r}, \tau) > 1$. This is indicative of the fact that all junctions are in the running state. In this way, we have seen that, by raising the value of i_B, we can go form a partial to a total RS in the network.

Having thoroughly investigated the properties of the cubic network at zero field ($\psi_{ex} = 0$), we now turn our attention to the dynamic properties of the superconducting phases, currents and voltages of the system in the presence of a magnetic field applied along the z −direction. We would like to anticipate the qualitative behavior of the system when $\psi_{ex} \neq 0$, based on what already seen in this chapter and on what already studied in the present book. How can we do this? Let us consider the response already studied in this subsection for $\psi_{ex} = 0$ and let us imagine that we can superimpose, to some extent, this same response to the effects of a magnetic field on the superconducting system. Naturally, being the dynamic equations (9.90) non-linear, this logical procedure would not always give the correct qualitative answer. Our scope, however, is only to guess the dynamic behavior of the system, before making one or more computer runs. Let us then consider the very first case studied, namely $i_B = 2.93$. In the absence of a magnetic field all junctions are in the ZVS. Let us then slowly apply a magnetic field in the z −direction, up to the point when a value of $\psi_{ex} = 0.2$ is reached. We may imagine that the network will try to shield the magnetic field, so that additional currents will circulate clockwise, as observed from above, on the faces of the cubic structure parallel to the $x - y$ plane. These circulating currents will add to the branch currents $i_\xi(\vec{r}, \tau)$. They will thus increase the amplitude of $i_\xi(\vec{r}, \tau)$ if flowing in the same direction on the branch, will decrease the amplitude of $i_\xi(\vec{r}, \tau)$ otherwise. For $i_B = 2.53$ and $\psi_{ex} = 0$ some junctions were already on the verge of making a transition from ZVS to RS. Therefore, a small additional current would make them go into a resistive case. We thus expect that some junctions will suffer phase-slip processes by application of a magnetic field in the z −direction even for $i_B = 2.93$.

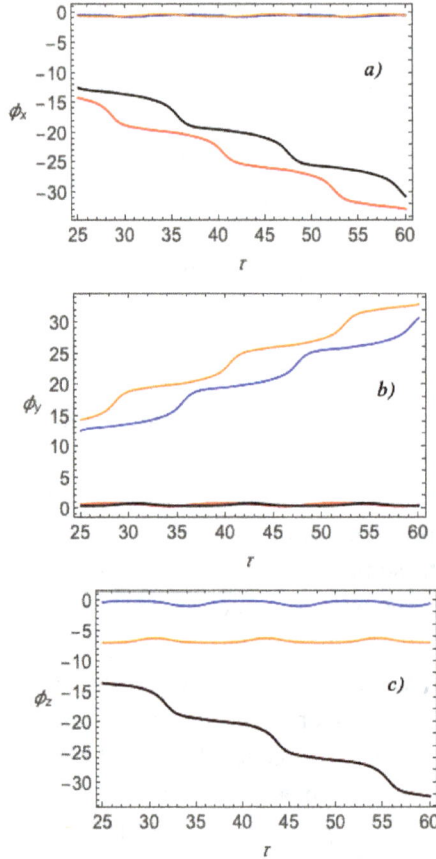

Fig. 9.14. Superconducting phase differences $\phi_\xi(\vec{r}, \tau)$ across the junctions in a current-biased cubic network with parameters $\beta = 1.0$ and $m = 0.1$, for $\vec{\psi}_{ex} = 0.2\,\hat{z}$ and $i_B = 2.93$. Blue, red, black, and orange curves represent the quantities $\phi_\xi(\vec{r}, \tau)$ ($\xi = x, y, z$) in increasing hierarchical order. For the superconducting phases $\phi_x(\vec{r})$ in panel a) we have: $\phi_x(0)$, $\phi_x(a\hat{y})$, $\phi_x(a\hat{z})$, $\phi_x(a\hat{y} + a\hat{z})$. For the superconducting phases $\phi_y(\vec{r})$ in panel b) we have: $\phi_y(0)$, $\phi_y(a\hat{x})$, $\phi_y(a\hat{z})$, $\phi_y(a\hat{x} + a\hat{z})$. Finally, for branch currents $\phi_z(\vec{r})$ in panel c) the order is the following: $\phi_z(0)$, $\phi_z(a\hat{x})$, $\phi_z(a\hat{y})$, $\phi_y(a\hat{x} + a\hat{y})$. In the latter panel the red curve coincides with the black curve and they both appear in the bottom of the graph.

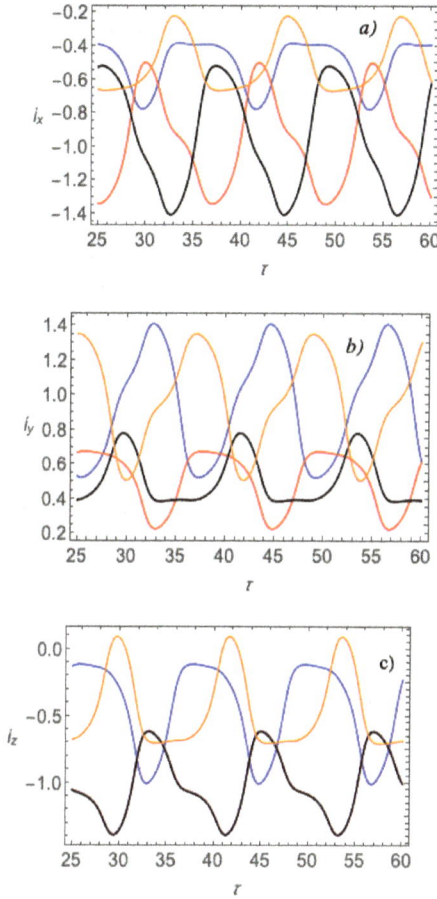

Fig. 9.15. Branch currents $i_\xi(\vec{r}, \tau)$ flowing through the junctions in a current-biased cubic network with parameters $\beta = 1.0$ and $m = 0.1$, for $\vec{\psi}_{ex} = 0.2\,\hat{z}$ and $i_B = 2.93$. Blue, red, black, and orange curves represent the quantities $i_\xi(\vec{r}, \tau)$ ($\xi = x, y, z$) in increasing hierarchical order. For branch currents $i_x(\vec{r})$ in panel a) the order is the following: $i_x(0)$, $i_x(a\hat{y})$, $i_x(a\hat{z})$, $i_x(a\hat{y} + a\hat{z})$. For branch currents $i_y(\vec{r})$ in panel b) the order is the following: $i_y(0)$, $i_y(a\hat{x})$, $i_y(a\hat{z})$, $i_y(a\hat{x} + a\hat{z})$. Finally, for branch currents $i_z(\vec{r})$ in panel c) the order is the following: $i_z(0)$, $i_z(a\hat{x})$, $i_z(a\hat{y})$, $i_y(a\hat{x} + a\hat{y})$. In the latter panel the red curve coincides with the black curve and they both appear in the lowest part of the graph. The period can be estimated to be $T \approx 12.0$.

We have seen that, in the absence of a bias current, the magnetic flux transitions in two-dimensional or three-dimensional networks were temporary, lasting only the time for flux quanta to penetrate the system and relieve the magnetic flux gradient felt by the network. The consequent lowering of the absolute values of the branch currents was then sufficient to restore the ZVS in all junctions.

In the case a bias current is present, however, we might briefly recall the properties of the equivalent single-junction model of a symmetric two-junction quantum interferometer, given in Section 3.3. By this model, a magnetic field applied to a two-junction interferometer generates a distortion in the effective washboard potential as seen in Section 3.4. For values of the bias current i_B different from zero, a tilting of the undulated potential is observed. A metastable magnetic state of the system is characterized by a local minimum in this tilted effective washboard potential. Both identical junctions, in this case, are in a ZVS state. A change in shape of the same potential, due to an external magnetic field, could thus break the ZVS of the two identical junctions by destroying the local minimum. When this occurs, both junctions go into a resistive state and a train of fluxons enters the superconducting loop through one JJ and leaves the system through the second JJ. Therefore, we expect that an increase of the magnetic field amplitude could give rise to a flux dynamics scenario as the one described above. However, when dealing with current-biased three-dimensional cubic networks, some junctions may not be affected by the external field variation, so that they will continue to stay in the ZVS. In this way, when trying to interpret the numerical data, we need to distinguish between junctions suffering phase-slip processes and those anchored to a ZVS.

In Fig. 9.14a–c we show the superconducting phase differences $\phi_\xi(\vec{r}, \tau)$ across the junctions in a current-biased cubic network with parameters $\beta = 1.0$ and $m = 0.1$, for $\psi_{ex} = 0.2$ and $i_B = 2.93$. In particular, in Fig. 9.14a, in Fig. 9.14b, and in Fig. 9.14c, the quantities $\phi_x(\vec{r}, \tau)$, $\phi_y(\vec{r}, \tau)$, and $\phi_z(\vec{r}, \tau)$ are reported, respectively. In Fig. 9.15a–c the branch currents $i_\xi(\vec{r}, \tau)$ flowing through the junctions in a current-biased cubic network are shown for the same choice of parameters as in Fig. 9.14a–c and for the same value of the forcing terms: $\psi_{ex} = 0.2$, $i_B = 2.93$. In Fig. 9.15a, in Fig. 9.15b, and in Fig. 9.15c, the quantities

$i_x(\vec{r}, \tau)$, $i_y(\vec{r}, \tau)$, and $i_z(\vec{r}, \tau)$ are reported, respectively. Whenever not explicitly reported the time-dependence of the quantities $\phi_\xi(\vec{r}, \tau)$ and $i_\xi(\vec{r}, \tau)$ in the text or in the captions to the figures, it has been omitted for brevity.

From Fig. 9.14a and Fig. 9.15a, we may argue that a first pair of junctions lying along the x −axis, one located at $\vec{r} = a\hat{y}$, the other at $\vec{r} = a\hat{z}$, is in the resistive state. A second pair, consisting of one JJ located at $\vec{r} = 0$ and one at $\vec{r} = a\hat{y} + a\hat{z}$, is in an effective ZVS. In fact, the step-like decrease of the superconducting phase differences $\phi_x(a\hat{y}, \tau)$ and $\phi_x(a\hat{z}, \tau)$ and the absolute value of the corresponding branch currents $i_x(a\hat{y}, \tau)$ and $i_x(a\hat{z}, \tau)$, temporarily greater than one, are indicative of phase-slip processes taking place in the first pair of junctions. On the other hand, the oscillations of the phase differences $\phi_x(0, \tau)$ and $\phi_x(a\hat{y} + a\hat{z}, \tau)$ about a null value and the absolute values of the corresponding branch currents $i_x(0, \tau)$ and $i_x(a\hat{z}, \tau)$ less than one tell us that an effective ZVS is realized in the second pair of junctions. By the same type of reasoning, considering now Fig. 9.14b and Fig. 9.15b we may establish that the pair of junctions lying along the y −axis, one JJ located at $\vec{r} = a\hat{x}$ and the other at $\vec{r} = a\hat{z}$, is in the resistive state, while a second pair, the first JJ located at $\vec{r} = 0$ and the second at $\vec{r} = a\hat{x} + a\hat{z}$, is in an effective ZVS. Again by a similar reasoning, considering Fig. 9.14c and Fig. 9.15c we may argue that the pair of junctions lying along the z −axis, one JJ located at $\vec{r} = a\hat{x}$ and the other at $\vec{r} = a\hat{y}$ is in the resistive state, while a second pair, the first JJ located at $\vec{r} = 0$ and the second at $\vec{r} = a\hat{x} + a\hat{y}$, is in an effective ZVS. Notice that junctions lying on a branch which starts or ends with node a or node b are in the resistive state.

In order to keep track of the variation of flux inside the cubic structure, in Fig. 9.16 we report the time dependence of the flux variables $\psi_\xi(\vec{r}, \tau)$. We notice that a train of pulses in the variables $\psi_\xi(\vec{r}, \tau)$ give a behavior similar to what seen in a two-junction quantum interferometer. Notice also that the sums $\psi_x(0) + \psi_y(0)$ and $\psi_x(a\hat{x}) + \psi_y(a\hat{y})$ are exactly zero at all times, while only a net negative flux survives in the z −direction, meaning that the system presents and overall diamagnetic behavior. Therefore, in response to the normalized externally applied magnetic field $\vec{\psi}_{ex}$, currents circulate in the network in such a way to generate a magnetic field which opposes $\vec{\psi}_{ex}$.

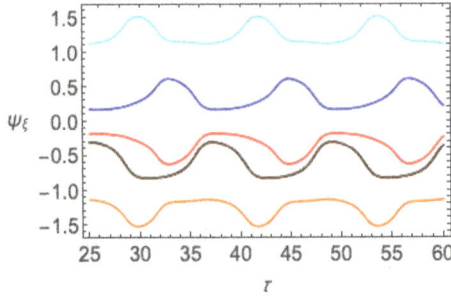

Fig. 9.16. Time dependence of the magnetic flux variables $\psi_\xi(\vec{r}, \tau)$ threading the loops of a current-biased cubic network with parameters $\beta = 1.0$ and $m = 0.1$, for $\vec{\psi}_{ex} = 0.2\,\hat{z}$ and $i_B = 2.93$. Blue, cyan, red, and orange curves represent $\psi_x(0)$, $\psi_x(a\hat{x})$, $\psi_y(0)$, $\psi_y(a\hat{y})$, respectively. The quantities $\psi_z(0)$ and $\psi_z(a\hat{z})$ are both represented by the brown curve, being $\psi_z(0) = \psi_z(a\hat{z})$. The period can be estimated to be $T \approx 12.0$.

Let us now proceed in calculating the solutions for an external magnetic field applied at an angle $\theta = \pi/4$ with respect to the z −axis.

We take the vector Γ to have the following form:

$$\Gamma = (\sin\theta,\ \sin\theta,\ 0,\ 0,\ \cos\theta,\ \cos\theta)^T. \qquad (9.91)$$

By solving Eq. (9.88) for $\beta = 1.0$, $m = 0.1$, $\vec{\psi}_{ex} = 0.2(\sqrt{2}\hat{x}/2 + \sqrt{2}\hat{z}/2)$, and $i_B = 2.90$, we obtain solutions for $\phi_\xi(\vec{r}, \tau)$ ($\xi = x, y, z$), which are reported in Fig. 9.17a–b. In Fig. 9.18a–b the branch currents $i_\xi(\vec{r}, \tau)$ for the same choice of parameters are shown. In Fig. 9.19a–b, 9.20a–b, 9.21a–b the dynamics of the voltages $v_x(\vec{r}, \tau)$, $v_y(\vec{r}, \tau)$, $v_z(\vec{r}, \tau)$, respectively, is exhibited. Finally, in Fig. 9.22 the fluxes $\psi_\xi(\vec{r}, \tau)$ threading the six loops in the cubic structure are shown.

In this case it is rather difficult to envision the magnetic and electrodynamic processes occurring in the network by a simple mental scheme. In fact, the inclination of the magnetic field may greatly complicate the scenario. Let us then choose a bias current value $i_B = 2.90$ less than the one adopted in the case $\vec{\psi}_{ex} = 0.2\,\hat{z}$ ($i_B = 2.93$). We still find a partial resistive state for the junctions in the network for $\vec{\psi}_{ex} = 0.2(\sqrt{2}\hat{x}/2 + \sqrt{2}\hat{z}/2)$.

In this respect, consider the superconducting phase differences $\phi_\xi(\vec{r}, \tau)$ in Fig. 9.17a–b. In Fig. 9.17b we detect a step-like decrease of $\phi_x(a\hat{y}, \tau)$, a symmetric increase of $\phi_y(0, \tau)$, with an additional step-like decrease of $\phi_z(a\hat{x}, \tau)$.

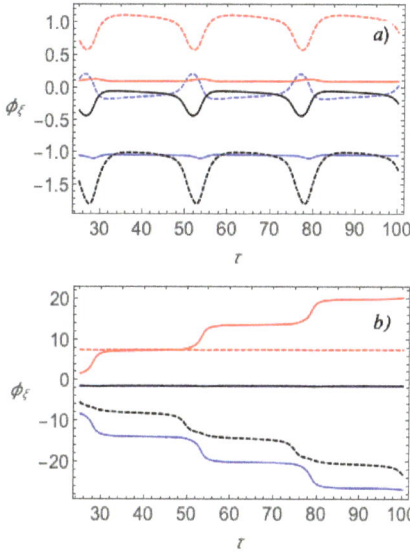

Fig. 9.17. Superconducting phase differences $\phi_\xi(\vec{r}, \tau)$ across the junctions in a current-biased cubic network with parameters $\beta = 1.0$ and $m = 0.1$, for $\vec{\psi}_{ex} = 0.2(\sqrt{2}\hat{x}/2 + \sqrt{2}\hat{z}/2)$ and $i_B = 2.90$. In panel a) we have: $\phi_x(0)$, $\phi_x(a\hat{y} + a\hat{z})$ (blue full and dotted lines, respectively), $\phi_y(a\hat{x})$, $\phi_y(a\hat{z})$ (red full and dotted lines, respectively), $\phi_z(0)$, $\phi_z(a\hat{x} + a\hat{y})$ (black full and dotted lines, respectively). In panel b): $\phi_x(a\hat{y})$, $\phi_x(a\hat{z})$ (blue full and dotted lines, respectively, the second not visible, since coincident with the full-line black curve), $\phi_y(0)$, $\phi_y(a\hat{x} + a\hat{z})$ (red full and dotted lines, respectively), $\phi_z(a\hat{x})$, $\phi_z(a\hat{y})$ (black full and dotted lines, respectively).

This is indicative of the fact that three junctions, the first in the x −direction at $\vec{r} = a\hat{y}$, the second in the y −direction at $\vec{r} = 0$, and the third in the z −direction at $\vec{r} = a\hat{x}$ are in the resistive state and allow flux transitions in and out the system. The remaining superconducting phase in Fig. 9.17a–b are either constant or only temporarily perturbed by the phase-slip processes occurring in the three junctions mentioned above.

The branch currents $i_\xi(\vec{r}, \tau)$ in Fig. 9.18a–b do confirm that the three junctions mentioned above are in the running state. In fact, by looking at Fig. 9.18b, we may notice that the absolute values of the branch currents $i_x(a\hat{y}, \tau)$, $i_y(0, \tau)$, and $i_z(a\hat{x}, \tau)$ temporarily exceeds unity.

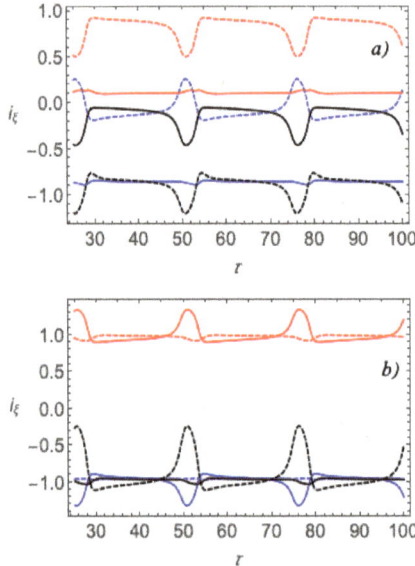

Fig. 9.18. Branch currents $i_\xi(\vec{r}, \tau)$ flowing in the junctions of a current-biased cubic network with parameters $\beta = 1.0$ and $m = 0.1$, for $\vec{\psi}_{ex} = 0.2(\sqrt{2}\hat{x}/2 + \sqrt{2}\hat{z}/2)$ and $i_B = 2.90$. In panel a) we have: $i_x(0)$, $i_x(a\hat{y} + a\hat{z})$ (blue full and dotted lines, respectively), $i_y(a\hat{x})$, $i_y(a\hat{z})$ (red full and dotted lines, respectively), $i_z(0)$, $i_z(a\hat{x} + a\hat{y})$ (black full and dotted lines, respectively). In panel b) we have: $i_x(a\hat{y})$, $i_x(a\hat{z})$ (blue full and dotted lines, respectively), $i_y(0)$, $i_y(a\hat{x} + a\hat{z})$ (red full and dotted lines, respectively), $i_z(a\hat{x})$, $i_z(a\hat{y})$ (black full and dotted lines, respectively). The period can be estimated to be $T \approx 25.5$.

A peculiarity is present in the graphs shown in Fig. 9.18a. In fact, while all other branch current values are less than one in absolute value, this does not happen for the branch current $i_z(a\hat{x} + a\hat{y})$, whose absolute value temporarily exceeds unity. However, as we shall also see by considering the voltage curves, a temporary access to the resistive state of the junction in which the branch current $i_z(a\hat{x} + a\hat{y})$ flow does not allow phase-slip processes and, therefore, flux transitions. In fact, in this case, the temporary crossing of the maximum Josephson current only gives a reentrant phase variation in the black dotted curve in Fig. 9.17a.

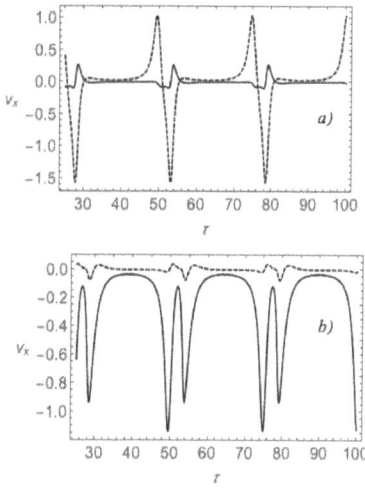

Fig. 9.19. Voltages $v_x(\vec{r}, \tau)$ across the branches of a current-biased cubic network with parameters $\beta = 1.0$ and $m = 0.1$, for $\vec{\psi}_{ex} = 0.2(\sqrt{2}\hat{x}/2 + \sqrt{2}\hat{z}/2)$ and $i_B = 2.90$. In panel a) $v_x(0)$ (full line), $v_x(a\hat{y} + a\hat{z})$ (dashed line) are shown. In panel b) $v_x(a\hat{y})$ (full line), $v_x(a\hat{z})$ (dashed line) are shown.

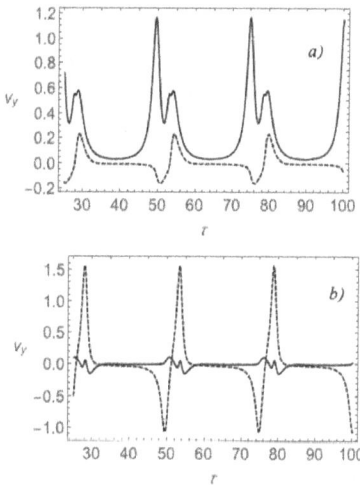

Fig. 9.20. Voltages $v_y(\vec{r}, \tau)$ across the branches of a current-biased cubic network with parameters $\beta = 1.0$ and $m = 0.1$, for $\vec{\psi}_{ex} = 0.2(\sqrt{2}\hat{x}/2 + \sqrt{2}\hat{z}/2)$ and $i_B = 2.90$. In panel a) $v_y(0)$ (full line), $v_y(a\hat{x} + a\hat{z})$ (dashed line) are shown. In panel b) $v_y(a\hat{x})$ (full line), $v_y(a\hat{z})$ (dashed line) are shown.

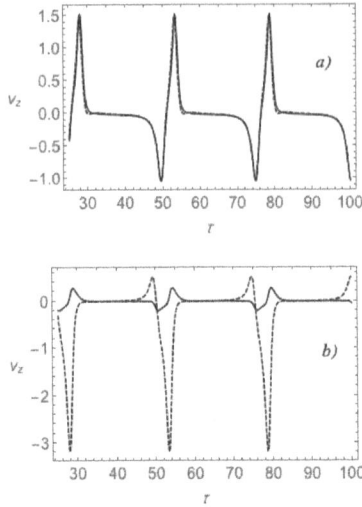

Fig. 9.21. Voltages $v_z(\vec{r}, \tau)$ in a current-biased cubic network with parameters $\beta = 1.0$ and $m = 0.1$, for $\vec{\psi}_{ex} = 0.2(\sqrt{2}\hat{x}/2 + \sqrt{2}\hat{z}/2)$ and $i_B = 2.90$. In panel a) $v_z(0)$ (full line), $v_z(a\hat{x} + a\hat{y})$ (dashed line) are shown. In panel b) the quantities $v_z(a\hat{x})$ (full line) and $v_z(a\hat{y})$ (dashed line) are shown. Notice that the values of $v_z(0)$ and $v_z(a\hat{x} + a\hat{y})$ are lmost juxtaposed in panel *a*).

The same type of behavior can be noted when considering the branch currents $i_x(a\hat{z})$, $i_y(a\hat{x} + a\hat{z})$, $i_z(a\hat{x})$ in Fig. 9.18b. These currents, although temporarily presenting absolute values greater or equal to 1, do not allow phase-slip processes in the junctions through which they flow, generating only a perturbation of the superconducting phase state.

The partial resistive state in the system for $\vec{\psi}_{ex} = 0.2(\sqrt{2}\hat{x}/2 + \sqrt{2}\hat{z}/2)$ and $i_B = 2.90$ is also confirmed by the voltage dynamics shown in Fig. 9.19a–b for $v_x(\vec{r}, \tau)$, in Fig. 9.20a–b for $v_y(\vec{r}, \tau)$, and in Fig. 9.21a–b for $v_z(\vec{r}, \tau)$. In Fig. 9.19a we notice that both voltages, $v_x(0)$ and $v_x(a\hat{y} + a\hat{z})$, have null average value, so that an effective ZVS is realized. In Fig. 9.19b, on the other hand, we may see that the voltage $v_x(a\hat{y})$ has a non-null average value, while $v_x(a\hat{z})$ presents oscillations about zero, thus meaning that the junction is in an effective ZVS. In Fig. 9.20a the voltage $v_y(0)$ has a non-null average value, confirming the resistive state of the junction lying on the same branch, while $v_y(a\hat{x} + a\hat{z})$ has a null average value. In Fig. 9.20b both voltages

$v_y(a\hat{x})$ and $v_y(a\hat{z})$ have null average value. In Fig. 9.21a, the curves of the voltages $v_z(0)$ and $v_z(a\hat{x} + a\hat{y})$ are juxtaposed and show null average value. By considering, in particular, the voltage $v_z(a\hat{x} + a\hat{y})$, we may confirm the argument set forth before; i.e. that the branch current $i_z(a\hat{x} + a\hat{y})$ does not generate phase-slip processes in the junction in which it flows. Finally, in Fig. 9. 21b, the voltage $v_z(a\hat{x})$ shows null average value. On the other hand, the voltage $v_z(a\hat{y})$ has non-null average value, confirming that the current $i_z(a\hat{y})$ generates phase slip processes in the junction in which it flows and, thus, flux transitions in the cubic structure.

As for the time dependence of the magnetic flux inside the network, in Fig. 9.22 we report the six fluxes $\psi_\xi(\vec{r}, \tau)$. In this figure we see that the fluxes $\psi_x(0, \tau)$ and $\psi_y(a\hat{y}, \tau)$ are the most affected by the phase-slip processes in the junctions.

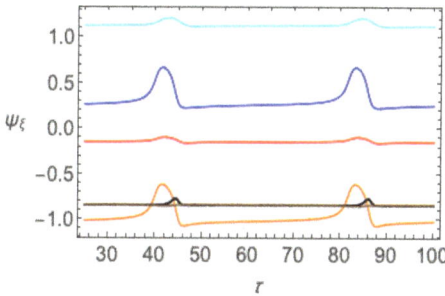

Fig. 9.22. Time dependence of the magnetic flux variables $\psi_\xi(\vec{r}, \tau)$ threading the loops of a current-biased cubic network with parameters $\beta = 1.0$ and $m = 0.1$, for $\vec{\psi}_{ex} = 0.2(\sqrt{2}\hat{x}/2 + \sqrt{2}\hat{z}/2)$ and $i_B = 2.90$. Blue, cyan, red, and orange curves represent $\psi_x(0)$, $\psi_x(a\hat{x})$, $\psi_y(0)$, $\psi_y(a\hat{y})$, respectively. The quantities $\psi_z(0)$ and $\psi_z(a\hat{z})$ are represented by the black and brown curves, respectively, and are almost coincident on this scale.

The remaining fluxes, in fact, do not appreciably vary with respect to specific constant values. Notice, finally, that the relation $\psi_x(0) + \psi_y(0) + \psi_z(0) = \psi_x(a\hat{x}) + \psi_y(a\hat{y}) + \psi_z(a\hat{z})$ can be rather easily proven true by getting data from Fig. 9.22.

Let us now calculate the solutions for an external magnetic field applied at an angle $\theta = \pi/4$ with respect to the z −axis and an angle

$\varphi = \pi/4$ with respect to x −axis. In this case we take the vector Γ to have the following form:

$$\Gamma = \frac{1}{2}\left(1, 1, 1, 1, \sqrt{2}, \sqrt{2}\right)^T. \tag{9.92}$$

The solutions $\phi_\xi(\vec{r},\tau)$, with $\xi = x, y, z$, coming from the numerical analysis for $\beta = 1.0$, $m = 0.1$, $\vec{\psi}_{ex} = 0.2(\hat{x}/2 + \hat{y}/2 + \sqrt{2}\hat{z}/2)$, and $i_B = 2.90$, are shown in Fig. 9.23a–b. In Fig. 9.24a–b the branch currents $i_\xi(\vec{r},\tau)$ for the same choice of parameters are reported. In Fig. 9.25a–b, 9.26a–b, 9.27a–b the dynamics of the voltages $v_x(\vec{r},\tau)$, $v_y(\vec{r},\tau)$, $v_z(\vec{r},\tau)$, respectively, is exhibited. Finally, in Fig. 9.28 the fluxes $\psi_\xi(\vec{r},\tau)$ threading the six loops in the cubic structure are shown.

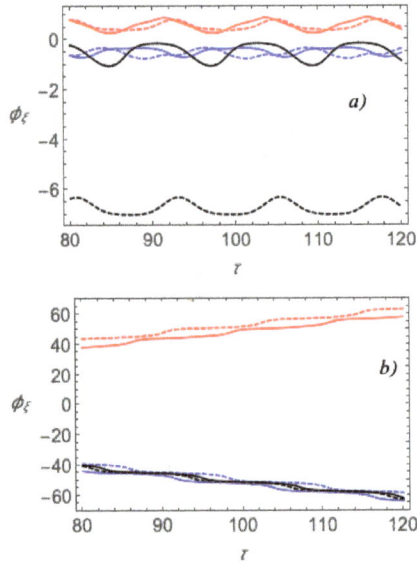

Fig. 9.23. Superconducting phase differences $\phi_\xi(\vec{r},\tau)$ across the junctions in a current-biased cubic network with parameters $\beta = 1.0$ and $m = 0.1$, for $\vec{\psi}_{ex} = 0.2(\hat{x}/2 + \hat{y}/2 + \sqrt{2}\hat{z}/2)$ and $i_B = 2.90$. In panel a) we have: $\phi_x(0)$, $\phi_x(a\hat{y} + a\hat{z})$ (blue full and dotted lines, respectively), $\phi_y(a\hat{x})$, $\phi_y(a\hat{z})$ (red full and dotted lines, respectively), $\phi_z(0)$, $\phi_z(a\hat{x} + a\hat{y})$ (black full and dotted lines, respectively). In panel b) we have: $\phi_x(a\hat{y})$, $\phi_x(a\hat{z})$ (blue full and dotted lines, respectively), $\phi_y(0)$, $\phi_y(a\hat{x} + a\hat{z})$ (red full and dotted lines, respectively), $\phi_z(a\hat{x})$, $\phi_z(a\hat{y})$ (black full and dotted lines, respectively, almost coincident).

In this case, we chose a bias current value $i_B = 2.90$ as done in the previous example. By reasoning in a similar way, we can argue that all junctions whose phase differences are reported in Fig. 9.23a are in an effective ZVS. In fact, the superconducting phases $\phi_x(0)$, $\phi_x(a\hat{y} + a\hat{z})$, $\phi_y(a\hat{x})$, $\phi_y(a\hat{z})$, $\phi_z(0)$, and $\phi_z(a\hat{x} + a\hat{y})$ do not show a step-like increment or decrement with time, but oscillate about a fixed value, as seen in panel a) of Fig. 9.23. On the other hand, all junctions whose phase differences are reported in Fig. 9.23b are in the RS. This can be seen by the step-like increase or decrease of the superconducting $\phi_x(a\hat{y})$, $\phi_x(a\hat{z})$, $\phi_y(0)$, $\phi_y(a\hat{x} + a\hat{z})$, $\phi_z(a\hat{x})$, and $\phi_z(a\hat{y})$ in panel b) of Fig. 9.23. The network branches, on which these junctions lie, converge either in node a or in node b.

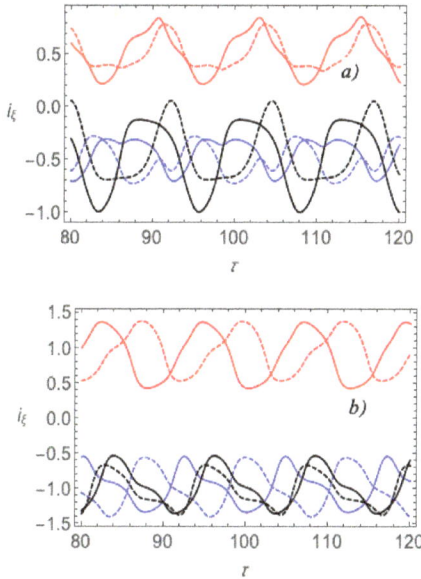

Fig. 9.24. Branch currents $i_\xi(\vec{r}, \tau)$ flowing in the junctions of a current-biased cubic network with parameters $\beta = 1.0$ and $m = 0.1$, for $\vec{\psi}_{ex} = 0.2(\hat{x}/2 + \hat{y}/2 + \sqrt{2}\hat{z}/2)$ and $i_B = 2.90$. In panel a) we have: $i_x(0)$, $i_x(a\hat{y} + a\hat{z})$ (blue full and dotted lines, respectively), $i_y(a\hat{x})$, $i_y(a\hat{z})$ (red full and dotted lines, respectively), $i_z(0)$, $i_z(a\hat{x} + a\hat{y})$ (black full and dotted lines, respectively). In panel b) we have: $i_x(a\hat{y})$, $i_x(a\hat{z})$ (blue full and dotted lines, respectively), $i_y(0)$, $i_y(a\hat{x} + a\hat{z})$ (red full and dotted lines, respectively), $i_z(a\hat{x})$, $i_z(a\hat{y})$ (black full and dotted lines, respectively). The period can be estimated to be $T \approx 12.0$.

We can confirm this argument by looking at Fig. 9.24a–b. In fact, in panel a) of Fig. 9.24 we may notice that the absolute values of all branch currents are contained within the interval $[0, 1]$. On the other hand, in panel b) of Fig. 9.24 we see that portions of each i_ξ vs. τ curve lie outside the interval $[-1, 1]$. Another type of validation of the argument is provided by the voltage curves in Fig. 9.25a–b, 9. 26a–b, 9.27a–b.

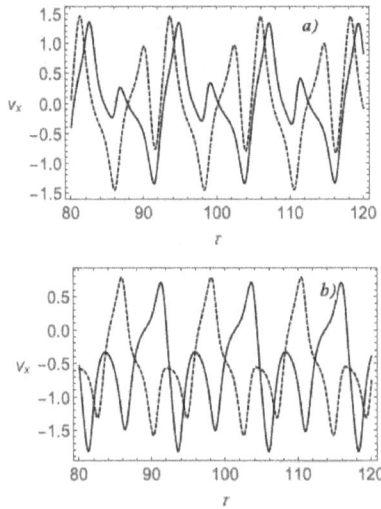

Fig. 9.25. Voltages $v_x(\vec{r}, \tau)$ across the branches of a current-biased cubic network with parameters $\beta = 1.0$ and $m = 0.1$, for $\vec{\psi}_{ex} = 0.2(\hat{x}/2 + \hat{y}/2 + \sqrt{2}\hat{z}/2)$ and $i_B = 2.90$. In panel a) $v_x(0)$ (full line), $v_x(a\hat{y} + a\hat{z})$ (dashed line) are shown. In panel b) $v_x(a\hat{y})$ (full line), $v_x(a\hat{z})$ (dashed line) are shown.

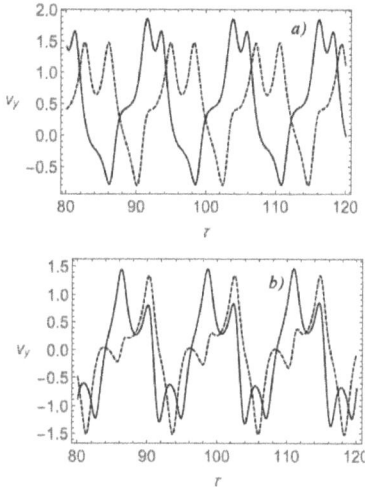

Fig. 9.26. Voltages $v_y(\vec{r}, \tau)$ across the branches of a current-biased cubic network with parameters $\beta = 1.0$ and $m = 0.1$, for $\vec{\psi}_{ex} = 0.2(\hat{x}/2 + \hat{y}/2 + \sqrt{2}\hat{z}/2)$ and $i_B = 2.90$. In panel a) $v_y(0)$ (full line), $v_y(a\hat{x} + a\hat{z})$ (dashed line) are shown. In panel b) $v_y(a\hat{x})$ (full line), $v_y(a\hat{z})$ (dashed line) are shown.

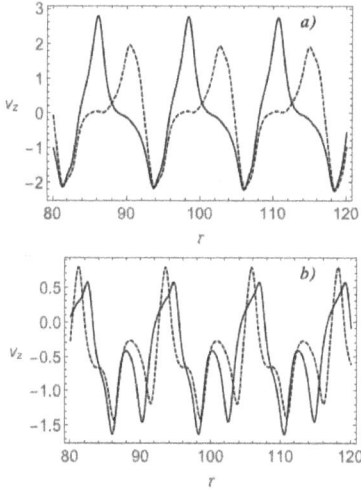

Fig. 9.27. Voltages $v_z(\vec{r}, \tau)$ in a current-biased cubic network with parameters $\beta = 1.0$ and $m = 0.1$, for $\vec{\psi}_{ex} = 0.2(\hat{x}/2 + \hat{y}/2 + \sqrt{2}\hat{z}/2)$ and $i_B = 2.90$. In panel a) $v_z(0)$ (full line), $v_z(a\hat{x} + a\hat{y})$ (dashed line) are shown. In panel b) the quantities $v_z(a\hat{x})$ (full line) and $v_z(a\hat{y})$ (dashed line) are shown. Notice that the values of $v_z(0)$ and $v_z(a\hat{x} + a\hat{y})$ are juxtaposed in panel a).

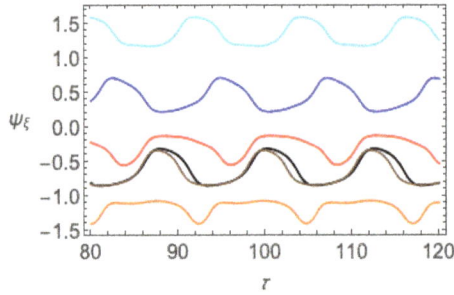

Fig. 9.28. Time dependence of the magnetic flux variables $\psi_\xi(\vec{r}, \tau)$ threading the loops of a current-biased cubic network with parameters $\beta = 1.0$ and $m = 0.1$, for $\vec{\psi}_{ex} = 0.2(\hat{x}/2 + \hat{y}/2 + \sqrt{2}\hat{z}/2)$ and $i_B = 2.90$. Blue, cyan, red, and orange curves represent $\psi_x(0)$, $\psi_x(a\hat{x})$, $\psi_y(0)$, $\psi_y(a\hat{y})$, respectively. The quantities $\psi_z(0)$ and $\psi_z(a\hat{z})$ are represented by the black and brown curves, respectively. The period can be estimated to be $T \approx 12.0$, coherently to what seen in fig. 9.24a-b.

In fact, from Fig. 9.25a it can be argued that the junctions, whose phase differences are $\phi_x(0)$ and $\phi_x(a\hat{y} + a\hat{z})$, are in an effective ZVS, because they oscillate evenly about zero. On the other hand, from Fig. 9.25b we notice that the junctions, whose phase differences are $\phi_x(a\hat{y})$ and $\phi_x(a\hat{z})$, are in the RS.

Conversely, from Fig. 9.26a it can be argued that the junctions, whose phase differences are $\phi_y(0)$ and $\phi_y(a\hat{x} + a\hat{z})$, are in the RS, while in Fig. 9.26b the junctions, whose phase differences are $\phi_y(a\hat{x})$ and $\phi_y(a\hat{z})$, are in an effective ZVS. Finally, in Fig. 9.27a we see that the junctions, whose phase differences are $\phi_z(0)$ and $\phi_z(a\hat{x} + a\hat{y})$, are in an effective ZVS, while in Fig. 9.27b the junctions, whose phase differences are $\phi_z(a\hat{x})$ and $\phi_z(a\hat{y})$, are in the RS.

The dynamics of the magnetic flux variables $\psi_\xi(\vec{r}, \tau)$ threading the loops of the cubic network are shown in Fig. 9.28. We notice that, for the values of the forcing terms, namely, $\vec{\psi}_{ex} = 0.2(\hat{x}/2 + \hat{y}/2 + \sqrt{2}\hat{z}/2)$ and $i_B = 2.90$, the fluxes are all appreciably affected by the flux transition processes in the network. A characteristic feature present in Fig. 9.28 is the time offset of the curves, which makes it difficult to verify, for any normalized time τ, the condition of null total flux trough the network given by Eq. 9.21. In this case, a numerical validation of Eq. 9.21 can be successfully performed.

9.4.2. The flux-voltage curves

In the present subsection we dedicate our attention to verify some of the general features, described in Section 9.3, of the current-biased cubic network of Josephson junctions. In particular, we shall focus on the periodicities of the time-averaged voltages $v_\xi(\vec{r}, \tau)$ across the branches of a current-biased cubic network with parameters $\beta = 1.0$ and $m = 0.1$ in three cases: a magnetic field applied along one the coordinate axis, so that, for example, $\vec{\psi}_{ex} = \psi_{ex}\hat{z}$; a magnetic field applied on a plane orthogonal to one coordinate axis, so that, for example, $\vec{\psi}_{ex} = \psi_{ex}(\sqrt{2}\hat{x}/2 + \sqrt{2}\hat{z}/2)$; a magnetic field with equal components along the three axes, so that $\vec{\psi}_{ex} = \psi_{ex}(\sqrt{3}\hat{x}/3 + \sqrt{3}\hat{y}/3 + \sqrt{3}\hat{z}/3)$. We would like to prove, by numerical analysis, the existence of the periodicities predicted in Section 9.3 for these cases. In the first case, for $\vec{\psi}_{ex} = \psi_{ex}\hat{z}$, we shall look for appearance of the periodicities $\Delta\psi_{ex} = |n|$, where n is an integer. In the second case, for $\vec{\psi}_{ex} = \psi_{ex}(\sqrt{2}\hat{x}/2 + \sqrt{2}\hat{z}/2)$, proof of the existence of periodicities $\Delta\psi_{ex} = \sqrt{2}|n|$ will be given. Finally, in the third case, for $\vec{\psi}_{ex} = \psi_{ex}(\sqrt{3}\hat{x}/3 + \sqrt{3}\hat{y}/3 + \sqrt{3}\hat{z}/3)$, periodicities of the type $\Delta\psi_{ex} = \sqrt{3}|n|$ will be sought.

Let us start by applying a magnetic field in the z −direction in such a way that $\vec{\psi}_{ex} = \psi_{ex}\hat{z}$. By means of the solutions to Eq. (9.88) for various values of the normalized externally applied flux amplitude ψ_{ex}, we determine the voltages $v_\xi(\vec{r}, \tau)$ across the branches of the network by means of Eq. (9.89). Time averaging is performed by summing, over a fixed interval of time, sample values $v_\xi(\vec{r}, \tau_k)$, for $k = 1, \ldots, N$, and by dividing by N. This procedure, although not very accurate and susceptible of numerical noise, as we shall see, allows us to avoid finding the exact value of the period T case by case.

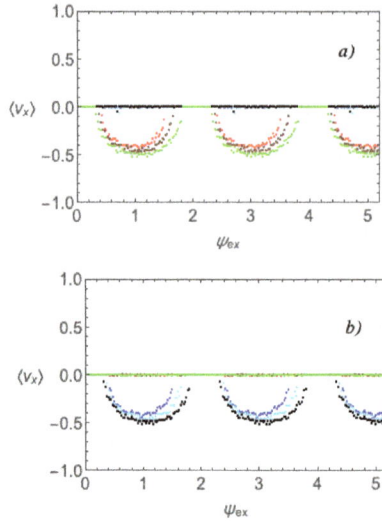

Fig. 9.29. Time-averaged voltages $\langle v_x(\vec{r})\rangle$ vs. ψ_{ex} curves for a cubic network with $\beta = 1.0$ and $m = 0.1$ for $\vec{\psi}_{ex} = \psi_{ex}\hat{z}$. In panel a) $\langle v_x(0)\rangle$ and $\langle v_x(a\hat{y})\rangle$ are shown. In panel b) $\langle v_x(a\hat{z})\rangle$ and $\langle v_x(a\hat{y} + a\hat{z})\rangle$ are represented. In both panels, the first quantities, $\langle v_x(0)\rangle$ and $\langle v_x(a\hat{z})\rangle$, are in blue for $i_B = 2.65$, cyan for $i_B = 2.75$, and black for $i_B = 2.85$; the second, $\langle v_x(a\hat{y})\rangle$ and $\langle v_x(a\hat{y} + a\hat{z})\rangle$, are in red for $i_B = 2.65$, brown for $i_B = 2.75$, green for $i_B = 2.85$.

In Fig. 9.29a–b, 9.30a–b, and 9.31a–b, time-averaged voltages $\langle v_x(\vec{r})\rangle$, $\langle v_y(\vec{r})\rangle$, $\langle v_z(\vec{r})\rangle$, respectively, are shown as functions of the normalized quantity ψ_{ex} for values of $i_B = 2.65, 2.75, 2.85$. In all figures we notice a periodicity $\Delta\psi_{ex} = 2.0$ and a behavior rather similar to the well-known curves, represented in Fig. 3.2a, for a two-junction quantum interferometer. Furthermore, we notice that the six branches connected to nodes a and b show non-null average voltages, while for the remaining branches the average voltages are zero.

This means that the junctions lying on the latter branches are in an effective ZVS, while the former, for field intervals in which the voltages $\langle v_\xi(\vec{r})\rangle$ acquire non-null values, are in the RS. In particular, in Fig. 9.29a–b, the quantities $\langle v_x(0)\rangle$ and $\langle v_x(a\hat{y} + a\hat{z})\rangle$ are zero for all field values, while the voltages $\langle v_x(a\hat{y})\rangle$ and $\langle v_x(a\hat{z})\rangle$, in specific field intervals, are negative and equal. In Fig. 9.30a–b, the quantities $\langle v_y(a\hat{x})\rangle$ and $\langle v_y(a\hat{z})\rangle$ represent an effective ZVS for the junctions lying along these branches, while the voltages $\langle v_y(0)\rangle$ and $\langle v_x(a\hat{y} + a\hat{z})\rangle$ are positive and equal in the same field intervals for which the voltages $\langle v_x(a\hat{y})\rangle$ and $\langle v_x(a\hat{z})\rangle$ are non-zero. Finally, in Fig. 9.31a–b, the quantities $\langle v_z(0)\rangle$ and $\langle v_z(a\hat{x} + a\hat{y})\rangle$ are zero for all field values, while the voltages $\langle v_z(a\hat{x})\rangle$ and $\langle v_z(a\hat{y})\rangle$ are negative and equal in the same field intervals where the voltages $\langle v_x(a\hat{y})\rangle$, $\langle v_x(a\hat{z})\rangle$, $\langle v_x(a\hat{y})\rangle$, and $\langle v_x(a\hat{z})\rangle$ present a non-null value. Therefore, there exist, for values of the bias current i_B in Fig. 9.29a–b, 9.30a–b, 9.31a–b, intervals of ψ_{ex} for which a total effective ZVS is realized in the network.

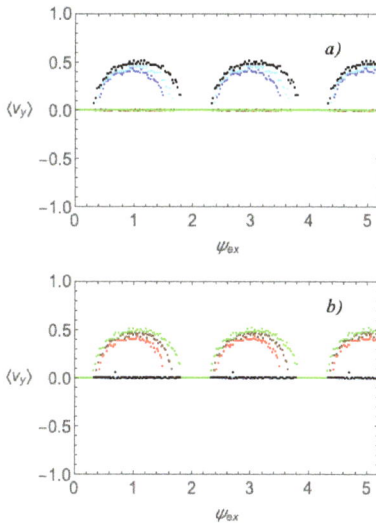

Fig. 9.30. Time-averaged voltages $\langle v_y(\vec{r})\rangle$ vs. ψ_{ex} curves for a cubic network with $\beta = 1.0$ and $m = 0.1$ for $\vec{\psi}_{ex} = \psi_{ex}\hat{z}$. In panel a) $\langle v_y(0)\rangle$ and $\langle v_y(a\hat{x})\rangle$ are shown. In panel b) $\langle v_y(a\hat{z})\rangle$ and $\langle v_y(a\hat{x} + a\hat{z})\rangle$ are represented. In both panels, the first quantities, $\langle v_y(0)\rangle$ and $\langle v_y(a\hat{z})\rangle$, are shown in blue for $i_B = 2.65$, cyan for $i_B = 2.75$, and black for $i_B = 2.85$; the second, $\langle v_y(a\hat{x})\rangle$ and $\langle v_y(a\hat{x} + a\hat{z})\rangle$, in red for $i_B = 2.65$, brown for $i_B = 2.75$, green for $i_B = 2.85$.

Up to this point we have seen that the $\langle v_\xi(\vec{r})\rangle$ vs. ψ_{ex} curves shown in Fig. 9.29a–b, 9.30a–b, 9.31a–b have a periodicity $\Delta\psi_{ex} = 2.0$ which, of course, confirms the analytic properties of the system seen in Section 9.3. However, let us take a look at what happens in the system for higher value of the bias current. Therefore, in Fig. 9.32a–b, 9.33a–b, and

9.34a–b we show the time-averaged voltages $\langle v_x(\vec{r})\rangle$, $\langle v_y(\vec{r})\rangle$, $\langle v_z(\vec{r})\rangle$, respectively, for the following values of the bias current i_B: 2.95, 3.05, 3.15. In these curves we may see that some of the $\langle v_\xi(\vec{r})\rangle$ vs. ψ_{ex} curves present unitary periodicity, thus confirming once more the analytic properties studied in Section 9.3. In particular, in Fig. 9.32a–b the quantities $\langle v_x(0)\rangle$ and $\langle v_x(a\hat{y}+a\hat{z})\rangle$ are still zero for all field values, while the voltages $\langle v_x(a\hat{y})\rangle$ and $\langle v_x(a\hat{z})\rangle$, in specific field intervals, are negative and equal, as in Fig. 9.29a–b.

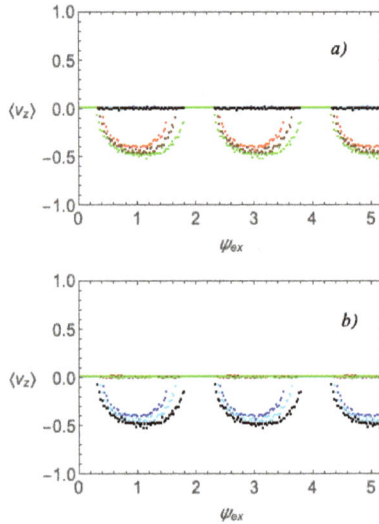

Fig. 9.31. Time-averaged voltages $\langle v_z(\vec{r})\rangle$ vs. ψ_{ex} curves for a cubic network with $\beta = 1.0$ and $m = 0.1$ for $\vec{\psi}_{ex} = \psi_{ex}\hat{z}$. In panel a) $\langle v_z(0)\rangle$ and $\langle v_z(a\hat{x})\rangle$ are shown. In panel b) $\langle v_z(a\hat{y})\rangle$ and $\langle v_z(a\hat{x}+a\hat{y})\rangle$ are represented. In both panels, the first quantities, $\langle v_z(0)\rangle$ and $\langle v_z(a\hat{y})\rangle$, are shown in blue for $i_B = 2.65$, cyan for $i_B = 2.75$, and black for $i_B = 2.85$; the second, $\langle v_z(a\hat{x})\rangle$ and $\langle v_z(a\hat{x}+a\hat{y})\rangle$, in red for $i_B = 2.65$, brown for $i_B = 2.75$, green for $i_B = 2.85$.

We notice that both periodicities $\Delta\psi_{ex} = 1.0$ and $\Delta\psi_{ex} = 2.0$ occur in these curves. In fact, while for $i_B = 2.95$ and $i_B = 3.05$ the voltages $\langle v_x(a\hat{y})\rangle$ and $\langle v_x(a\hat{z})\rangle$ present a periodicity $\Delta\psi_{ex} = 1.0$, for $i_B = 3.15$ a periodicity $\Delta\psi_{ex} = 2.0$ appears in the green curve for $\langle v_x(a\hat{y})\rangle$ in Fig. 9.32a and in the black curve for $\langle v_x(a\hat{z})\rangle$ in Fig. 9.32b. The same features can be detected in Fig. 9.33a–b, where the quantities $\langle v_y(a\hat{x})\rangle$

and $\langle v_y(a\hat{z})\rangle$ represent an effective ZVS for the junctions lying along these branches, and the voltages $\langle v_y(0)\rangle$ and $\langle v_x(a\hat{y}+a\hat{z})\rangle$ are positive and equal. These curves, as for the ones in Fig. 9.32a–b, show unitary periodicity for $i_B = 2.95$ and $i_B = 3.05$, and periodicity $\Delta\psi_{ex} = 2.0$ for $i_B = 3.15$. Finally, in Fig. 9.34a–b, the quantities $\langle v_z(0)\rangle$ and $\langle v_z(a\hat{x}+a\hat{y})\rangle$ are zero for all field values, while the voltages $\langle v_z(a\hat{x})\rangle$ and $\langle v_z(a\hat{y})\rangle$ are negative and equal.

In all these cases, therefore, in which the field is applied along the z −direction and the values of the bias currents are greater than $i_B = 2.65$, there do not exist intervals of ψ_{ex} for which a total effective ZVS is realized in the network.

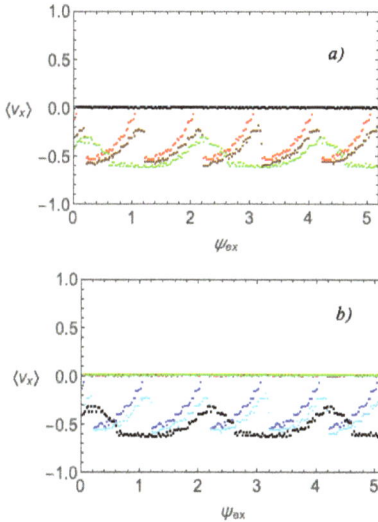

Fig. 9.32. Time-averaged voltage $\langle v_x(\vec{r})\rangle$ vs. ψ_{ex} curves for $\beta = 1.0$, $m = 0.1$, and $\vec{\psi}_{ex} = \psi_{ex}\hat{z}$. In panel a) $\langle v_x(0)\rangle$ and $\langle v_x(a\hat{y})\rangle$ and in panel b) $\langle v_x(a\hat{z})\rangle$ and $\langle v_x(a\hat{y}+a\hat{z})\rangle$ are represented. In both panels, the first quantities, $\langle v_x(0)\rangle$ and $\langle v_x(a\hat{z})\rangle$, are in blue for $i_B = 2.95$, cyan for $i_B = 3.05$, and black for $i_B = 3.15$; the second, $\langle v_x(a\hat{y})\rangle$ and $\langle v_x(a\hat{y}+a\hat{z})\rangle$, are in red for $i_B = 2.95$, brown for $i_D = 3.05$, green for $i_B = 3.15$.

In order to give an example of a magnetic field applied in a direction lying on a plane orthogonal to one coordinate axis, we set $\vec{\psi}_{ex} = \psi_{ex}(\sqrt{2}\hat{x}/2 + \sqrt{2}\hat{z}/2)$. We thus proceed as in the previous case. Therefore, we first solve Eq. (9.88) for the superconducting phase variables $\phi_\xi(\vec{r}, \tau)$ for various values of ψ_{ex}. Successively, we find the voltages $v_\xi(\vec{r}, \tau)$ according to Eq. (9.89). Finally, by the same averaging process described in the previous case, the $\langle v_\xi(\vec{r}) \rangle$ vs. ψ_{ex} curves are obtained. These curves are shown in Fig. 9.35a–b, 9.36a–b, 9.37a–b for $\langle v_x(\vec{r}) \rangle$, $\langle v_y(\vec{r}) \rangle$, $\langle v_z(\vec{r}) \rangle$, respectively. The bias current values have been chosen to be the following: $i_B = 2.85, 2.95, 3.05$.

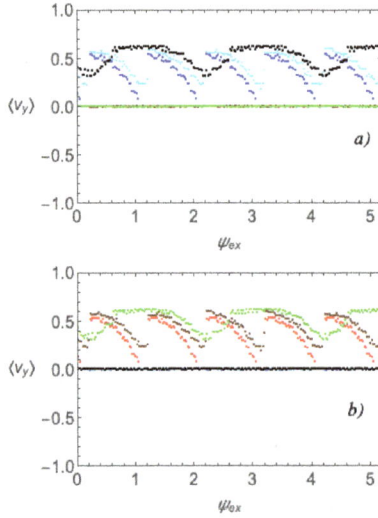

Fig. 9.33. Time-averaged voltage $\langle v_y(\vec{r}) \rangle$ vs. ψ_{ex} curves for $\beta = 1.0$, $m = 0.1$, and $\vec{\psi}_{ex} = \psi_{ex}\hat{z}$. In panel a) $\langle v_y(0) \rangle$ and $\langle v_y(a\hat{x}) \rangle$ and in panel b) $\langle v_y(a\hat{z}) \rangle$ and $\langle v_y(a\hat{x} + a\hat{z}) \rangle$ are represented. In both panels, the first quantities, $\langle v_y(0) \rangle$ and $\langle v_y(a\hat{z}) \rangle$, are shown in blue for $i_B = 2.95$, cyan for $i_B = 3.05$, and black for $i_B = 3.15$; the second, $\langle v_y(a\hat{x}) \rangle$ and $\langle v_y(a\hat{x} + a\hat{z}) \rangle$, in red for $i_B = 2.95$, brown for $i_B = 3.05$, green for $i_B = 3.15$.

In particular, in Fig. 9.35a the average voltages $\langle v_x(0) \rangle$ and $\langle v_x(a\hat{y}) \rangle$ are shown. The first voltage, for the values of the bias current chosen, is zero for all values of ψ_{ex}. Therefore, the junction lying along the branch is in an effective ZVS. The second quantity, $\langle v_x(a\hat{y}) \rangle$, attains non-zero values and is represented by a red curve with periodicity $\Delta\psi_{ex} = \sqrt{2}$ for $i_B = 2.85$, a brown curve with same periodicity for $i_B = 2.95$, and a green curve with periodicity $\Delta\psi_{ex} = 1.0$ for $i_B = 3.05$. In Fig. 9.35b the average voltages $\langle v_x(a\hat{z}) \rangle$ and $\langle v_x(a\hat{y} + a\hat{z}) \rangle$ are shown. The first voltage, $\langle v_x(a\hat{z})\rangle$, attains non-zero values and is represented by a blue curve with periodicity $\Delta\psi_{ex} = \sqrt{2}$ for $i_B = 2.85$, a cyan curve with same periodicity for $i_B = 2.95$, and a black curve with periodicity $\Delta\psi_{ex} = 1.0$ for $i_B = 3.05$. In the same panel *b*) of Fig. 9.35 the voltage $\langle v_x(a\hat{y} + a\hat{z})\rangle$ is seen to be zero for the values of the bias current chosen.

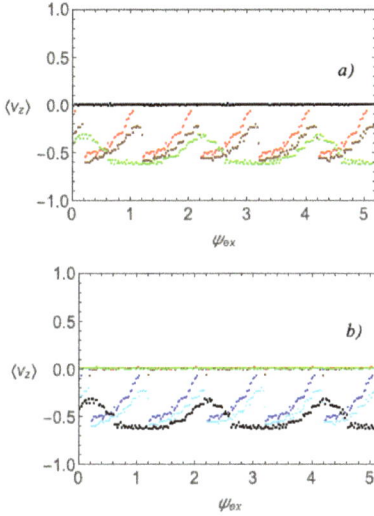

Fig. 9.34. Time-averaged voltage $\langle v_z(\vec{r})\rangle$ vs. ψ_{ex} curves for $\beta = 1.0$, $m = 0.1$, and $\vec{\psi}_{ex} = \psi_{ex}\hat{z}$. In panel a) $\langle v_z(0)\rangle$ and $\langle v_z(a\hat{x})\rangle$ and in panel b) $\langle v_z(a\hat{y})\rangle$ and $\langle v_z(a\hat{x} + a\hat{y})\rangle$ are represented. In both panels, the first quantities, $\langle v_z(0)\rangle$ and $\langle v_z(a\hat{y})\rangle$, are shown in blue for $i_B = 2.95$, cyan for $i_B = 3.05$, and black for $i_B = 3.15$; the second, $\langle v_z(a\hat{x})\rangle$ and $\langle v_z(a\hat{x} + a\hat{y})\rangle$, in red for $i_B = 2.95$, brown for $i_B = 3.05$, green for $i_B = 3.15$.

In Fig. 9.36a the average voltages $\langle v_y(0)\rangle$ and $\langle v_y(a\hat{x})\rangle$ are shown. The first voltage, $\langle v_y(0)\rangle$, attains non-zero values and is represented by a blue curve with periodicity $\Delta\psi_{ex} = \sqrt{2}$ for $i_B = 2.85$, a cyan curve with same periodicity for $i_B = 2.95$, and a black curve with periodicity $\Delta\psi_{ex} = 1.0$ for $i_B = 3.05$. The second voltage is zero for all values of i_B. In Fig. 9.36b the average voltages $\langle v_y(a\hat{z})\rangle$ and $\langle v_y(a\hat{x} + a\hat{z})\rangle$ are shown. The first voltage, for the values of the bias current chosen, is zero for all values of ψ_{ex}. The second quantity, $\langle v_y(a\hat{x} + a\hat{z})\rangle$, attains non-zero values and is represented by a red curve with periodicity $\Delta\psi_{ex} = \sqrt{2}$ for

$i_B = 2.85$, a brown curve with same periodicity for $i_B = 2.95$ and a green curve with periodicity $\Delta\psi_{ex} = 1.0$ for $i_B = 3.05$.

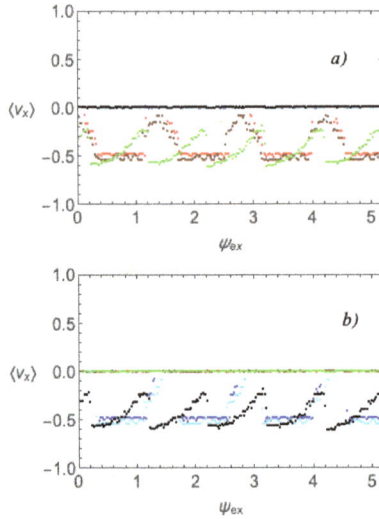

Fig. 9.35. Time-averaged voltage $\langle v_x(\vec{r}) \rangle$ vs. ψ_{ex} curves for $\beta = 1.0$, $m = 0.1$, and $\vec{\psi}_{ex} = \psi_{ex}(\sqrt{2}\hat{x}/2 + \sqrt{2}\hat{z}/2)$. In panel a) $\langle v_x(0) \rangle$ and $\langle v_x(a\hat{y}) \rangle$ and in panel b) $\langle v_x(a\hat{z}) \rangle$ and $\langle v_x(a\hat{y} + a\hat{z}) \rangle$ are represented. In both panels, the first quantities, $\langle v_x(0) \rangle$ and $\langle v_x(a\hat{z}) \rangle$, are in blue for $i_B = 2.85$, cyan for $i_B = 2.95$, and black for $i_B = 3.05$; the second, $\langle v_x(a\hat{y}) \rangle$ and $\langle v_x(a\hat{y} + a\hat{z}) \rangle$, are in red for $i_B = 2.85$, brown for $i_B = 2.95$, green for $i_B = 3.05$.

In Fig. 9.37a the average voltages $\langle v_z(0) \rangle$ and $\langle v_z(a\hat{x}) \rangle$ are shown. The first voltage, for the values of the bias current chosen, is zero for all values of ψ_{ex}. The second quantity, $\langle v_z(a\hat{x}) \rangle$, attains non-zero values and is represented by a red curve with periodicity $\Delta\psi_{ex} = \sqrt{2}$ for $i_B = 2.85$, a brown curve with same periodicity for $i_B = 2.95$, and a green curve with periodicity $\Delta\psi_{ex} = 1.0$ for $i_B = 3.05$. In Fig. 9.37b the average voltages $\langle v_z(a\hat{y}) \rangle$ and $\langle v_z(a\hat{x} + a\hat{y}) \rangle$ are shown. The first voltage, $\langle v_z(a\hat{y}) \rangle$, attains non-zero values and is represented by a blue curve with periodicity $\Delta\psi_{ex} = \sqrt{2}$ for $i_B = 2.85$, a cyan curve with same periodicity for $i_B = 2.95$, and a black curve with periodicity $\Delta\psi_{ex} = 1.0$ for $i_B = 3.05$. In the same panel b) of Fig. 9.37 the voltage $\langle v_x(a\hat{y} + a\hat{z}) \rangle$ is seen to be zero for all values of the bias current chosen.

In the cases considered for fields orthogonal to the y-axis, taking $\vec{\psi}_{ex} = \psi_{ex}(\sqrt{2}\hat{x}/2 + \sqrt{2}\hat{z}/2)$, we have therefore detected the periodicity $\Delta\psi_{ex} = \sqrt{2}$ along with periodicity $\Delta\psi_{ex} = 1.0$. We notice that the first periodicity can be predicted by means of Eq. (9.76). Contextual appearance of the second periodicity, on the other hand, is a result of a more complex dynamics of the system, by which the assumption of the existence of a single periodicity in Eq. (9.63) does not hold.

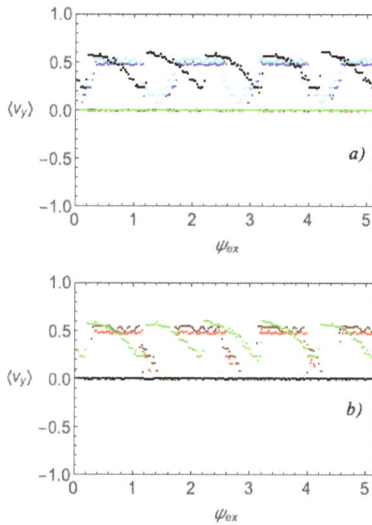

Fig. 9.36. Time-averaged voltage $\langle v_y(\vec{r})\rangle$ vs. ψ_{ex} curves for $\beta = 1.0$, $m = 0.1$, and $\vec{\psi}_{ex} = \psi_{ex}(\sqrt{2}\hat{x}/2 + \sqrt{2}\hat{z}/2)$. In panel a) $\langle v_y(0)\rangle$ and $\langle v_y(a\hat{x})\rangle$ and in panel b) $\langle v_y(a\hat{z})\rangle$ and $\langle v_y(a\hat{x} + a\hat{z})\rangle$ are represented. In both panels, the first quantities, $\langle v_y(0)\rangle$ and $\langle v_y(a\hat{z})\rangle$, are shown in blue for $i_B = 2.85$, cyan for $i_B = 2.95$, and black for $i_B = 3.05$; the second, $\langle v_y(a\hat{x})\rangle$ and $\langle v_y(a\hat{x} + a\hat{z})\rangle$, in red for $i_B = 2.85$, brown for $i_B = 2.95$, green for $i_B = 3.05$.

As for the example in which a magnetic field applied in such a way that all components are equal, giving $\vec{\psi}_{ex} = \psi_{ex}(\sqrt{3}\hat{x}/3 + \sqrt{3}\hat{x}/3 + \sqrt{3}\hat{z}/3)$, we proceed as in previous cases by first solving Eq. (9.88) for the superconducting phase variables $\phi_\xi(\vec{r}, \tau)$ for various values of ψ_{ex}. Successively, the voltages $v_\xi(\vec{r}, \tau)$ are found according to Eq. (9.89). Finally, by the same averaging process described in the first case, the $\langle v_\xi(\vec{r})\rangle$ vs. ψ_{ex} curves are obtained.

In Fig. 9.38a–b, 9.39a–b, 9.40a–b the $\langle v_x(\vec{r}) \rangle$, $\langle v_y(\vec{r}) \rangle$, $\langle v_z(\vec{r}) \rangle$ curves are reported, respectively. In Fig. 9.38a the average voltages $\langle v_x(0) \rangle$ and $\langle v_x(a\hat{y}) \rangle$ are shown. The first voltage, for the values of the bias current chosen, is zero for all values of ψ_{ex}. The second quantity, $\langle v_x(a\hat{y}) \rangle$, attains non-zero values and is represented by a red curve with periodicity $\Delta\psi_{ex} = 1$ for $i_B = 2.95$, a brown curve with periodicity $\Delta\psi_{ex} = \sqrt{3}$ for $i_B = 3.05$, and a green curve with periodicity $\Delta\psi_{ex} = 2.0$ for $i_B = 3.15$. Therefore, in these curves, the periodicity $\Delta\psi_{ex} = \sqrt{3}$ appears together with integer values of $\Delta\psi_{ex}$.

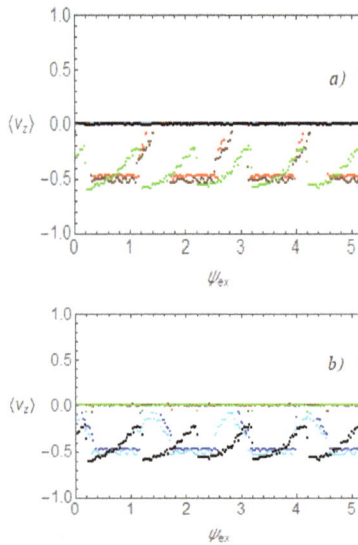

Fig. 9.37. Time-averaged voltage $\langle v_z(\vec{r}) \rangle$ vs. ψ_{ex} curves for $\beta = 1.0$, $m = 0.1$, and $\vec{\psi}_{ex} = \psi_{ex}(\sqrt{2}\hat{x}/2 + \sqrt{2}\hat{z}/2)$. In panel a) $\langle v_z(0) \rangle$ and $\langle v_z(a\hat{x}) \rangle$ and in panel b) $\langle v_z(a\hat{y}) \rangle$ and $\langle v_z(a\hat{x} + a\hat{y}) \rangle$ are represented. In both panels, the first quantities, $\langle v_z(0) \rangle$ and $\langle v_z(a\hat{y}) \rangle$, are shown in blue for $i_B = 2.85$, cyan for $i_B = 2.95$, and black for $i_B = 3.05$; the second, $\langle v_z(a\hat{x}) \rangle$ and $\langle v_z(a\hat{x} + a\hat{y}) \rangle$, in red for $i_B = 2.85$, brown for $i_B = 2.95$, green for $i_B = 3.05$.

In Fig. 9.38b the average voltages $\langle v_x(a\hat{z}) \rangle$ and $\langle v_x(a\hat{y} + a\hat{z}) \rangle$ are shown. The first voltage, $\langle v_x(a\hat{z}) \rangle$, attains non-zero values and is represented by a blue curve with periodicity $\Delta\psi_{ex} = 1$ for $i_B = 2.95$, a cyan curve with periodicity $\Delta\psi_{ex} = \sqrt{3}$ for $i_B = 3.05$, and a black curve with periodicity $\Delta\psi_{ex} = 2.0$ for $i_B = 3.15$. In the same panel b) of

Fig. 9.38 the voltage $\langle v_x(a\hat{y} + a\hat{z})\rangle$ is seen to be zero for all values of the bias current chosen.

In Fig. 9.39a the average voltages $\langle v_y(0)\rangle$ and $\langle v_y(a\hat{x})\rangle$ are shown. The first voltage, $\langle v_y(0)\rangle$, attains non-zero values and is represented by a blue curve with periodicity $\Delta\psi_{ex} = 1$ for $i_B = 2.95$, a cyan curve with periodicity $\Delta\psi_{ex} = \sqrt{3}$ for $i_B = 3.05$, and a black curve with periodicity $\Delta\psi_{ex} = 2.0$ for $i_B = 3.15$. In the same panel b) of Fig. 9.39 the voltage $\langle v_y(a\hat{x})\rangle$ is seen to be zero for all values of the bias current chosen.

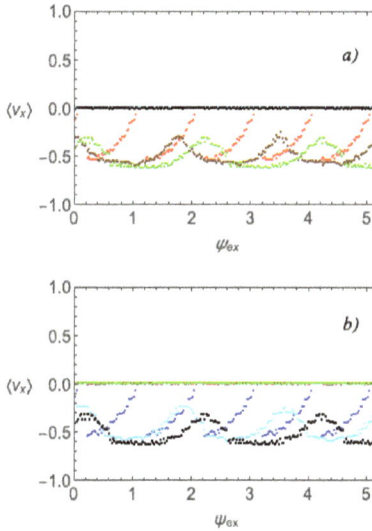

Fig. 9.38. Time-averaged voltage $\langle v_x(\vec{r})\rangle$ vs. ψ_{ex} curves for $\beta = 1.0$, $m = 0.1$, and $\vec{\psi}_{ex} = \psi_{ex}(\sqrt{3}\hat{x}/3 + \sqrt{3}\hat{y}/3 + \sqrt{3}\hat{z}/3)$. In panel a) $\langle v_x(0)\rangle$ and $\langle v_x(a\hat{y})\rangle$ and in panel b) $\langle v_x(a\hat{z})\rangle$ and $\langle v_x(a\hat{y} + a\hat{z})\rangle$ are represented. In both panels, the first quantities, $\langle v_x(0)\rangle$ and $\langle v_x(a\hat{z})\rangle$, are in blue for $i_B = 2.95$, cyan for $i_B = 3.05$, and black for $i_B = 3.15$; the second, $\langle v_x(a\hat{y})\rangle$ and $\langle v_x(a\hat{y} + a\hat{z})\rangle$, are in red for $i_B = 2.95$, brown for $i_B = 3.05$, green for $i_B = 3.15$.

In Fig. 9.39b the average voltages $\langle v_y(a\hat{z})\rangle$ and $\langle v_y(a\hat{x} + a\hat{z})\rangle$ are shown. The first voltage, for the values of the bias current chosen, is zero for all values of ψ_{ex}. The second quantity, $\langle v_y(a\hat{x} + a\hat{z})\rangle$, attains non-zero values and is represented by a red curve with periodicity $\Delta\psi_{ex} = 1$ for $i_B = 2.95$, a brown curve with periodicity $\Delta\psi_{ex} = \sqrt{3}$ for $i_B = 3.05$, and a green curve with periodicity $\Delta\psi_{ex} = 2.0$ for $i_B = 3.15$.

In Fig. 9.40a the average voltages $\langle v_z(0) \rangle$ and $\langle v_z(a\hat{x}) \rangle$ are shown. The first voltage, for the values of the bias current chosen, is zero for all values of ψ_{ex}. The second quantity, $\langle v_z(a\hat{x}) \rangle$, attains non-zero values and is represented by a red curve with periodicity $\Delta\psi_{ex} = 1$ for $i_B = 2.95$, a brown curve with periodicity $\Delta\psi_{ex} = \sqrt{3}$ for $i_B = 3.05$, and a green curve with periodicity $\Delta\psi_{ex} = 2.0$ for $i_B = 3.15$.

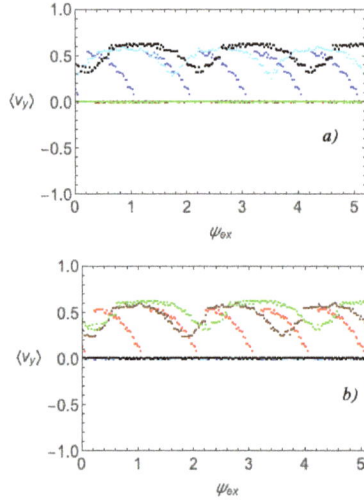

Fig. 9.39. Time-averaged voltage $\langle v_y(\vec{r}) \rangle$ vs. ψ_{ex} curves for $\beta = 1.0$, $m = 0.1$, and $\vec{\psi}_{ex} = \psi_{ex}(\sqrt{3}\hat{x}/3 + \sqrt{3}\hat{y}/3 + \sqrt{3}\hat{z}/3)$. In panel a) $\langle v_y(0) \rangle$ and $\langle v_y(\hat{x}) \rangle$ and in panel b) $\langle v_y(a\hat{z}) \rangle$ and $\langle v_y(a\hat{x} + a\hat{z}) \rangle$ are represented. In both panels, the first quantities, $\langle v_y(0) \rangle$ and $\langle v_y(a\hat{z}) \rangle$, are shown in blue for $i_B = 2.95$, cyan for $i_B = 3.05$, and black for $i_B = 3.15$; the second, $\langle v_y(a\hat{x}) \rangle$ and $\langle v_y(a\hat{x} + a\hat{z}) \rangle$, in red for $i_B = 2.95$, brown for $i_B = 3.05$, green for $i_B = 3.15$.

In Fig. 9.40b the average voltages $\langle v_z(a\hat{y}) \rangle$ and $\langle v_z(a\hat{x} + a\hat{y}) \rangle$ are shown. The first voltage, $\langle v_z(a\hat{y}) \rangle$, attains non-zero values and is represented by a blue curve with periodicity $\Delta\psi_{ex} = 1$ for $i_B = 2.95$, a cyan curve with periodicity $\Delta\psi_{ex} = \sqrt{3}$ for $i_B = 3.05$, and a black curve with periodicity $\Delta\psi_{ex} = 2.0$ for $i_B = 3.15$. In the same panel b) of Fig. 9.38 the voltage $\langle v_z(a\hat{x} + a\hat{y}) \rangle$ is seen to be zero for all values of the bias current chosen. In the cases considered for fields $\vec{\psi}_{ex} = \psi_{ex}(\sqrt{3}\hat{x}/3 + \sqrt{3}\hat{y}/3 + \sqrt{3}\hat{z}/3)$, we did detect the periodicity $\Delta\psi_{ex} = \sqrt{3}$ along with periodicities $\Delta\psi_{ex} = 1.0$ and $\Delta\psi_{ex} = 2.0$. The first periodicity could be predicted by means of Eq. (9.83). As for integer periodicities, we again notice that their appearance is a result of a more complex dynamics of the system, by which the assumption of the existence of a single periodicity claimed in Eq. (9.63) does not hold.

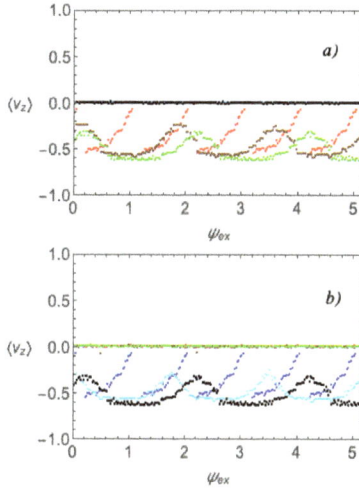

Fig. 9.40. Time-averaged voltage $\langle v_z(\vec{r}) \rangle$ vs. ψ_{ex} curves for $\beta = 1.0$, $m = 0.1$, and $\vec{\psi}_{ex} = \psi_{ex}(\sqrt{3}\hat{x}/3 + \sqrt{3}\hat{y}/3 + \sqrt{3}\hat{z}/3)$. In panel a) $\langle v_z(0) \rangle$ and $\langle v_z(a\hat{x}) \rangle$ and in panel b) $\langle v_z(a\hat{y}) \rangle$ and $\langle v_z(a\hat{x} + a\hat{y}) \rangle$ are represented. In both panels, the first quantities, $\langle v_z(0) \rangle$ and $\langle v_z(a\hat{y}) \rangle$, are shown in blue for $i_B = 2.95$, cyan for $i_B = 3.05$, and black for $i_B = 3.15$; the second, $\langle v_z(a\hat{x}) \rangle$ and $\langle v_z(a\hat{x} + a\hat{y}) \rangle$, in red for $i_B = 2.95$, brown for $i_B = 3.05$, green for $i_B = 3.15$.

9.4.3. *The voltage-current curves*

In the present subsection we shall study the way the average voltages $\langle v_\xi(\vec{r}) \rangle$ depends upon the bias current i_B when the magnetic field is kept constant. Therefore, we first solve Eq. (9.88) for the superconducting phase variables $\phi_\xi(\vec{r}, \tau)$ for various values of i_B for fixed values of ψ_{ex}, setting the parameter β equal to 1.0 and m equal to 0.1. Successively, we find the voltages $v_\xi(\vec{r}, \tau)$ according to Eq. (9.89). Finally, by the same averaging process described in the previous subsection, the $\langle v_\xi(\vec{r}) \rangle$ vs. i_B curves are obtained. By the knowledge of the voltages across the single branches, we shall determine the voltage difference $V_{ab} = V_b - V_a$

between node b, in which the bias current is extracted from the network, and node a, in which the bias current is injected. The numerical analysis will be completely shown only for the case in which the magnetic field is directed along the z −axis. The influence of the orientation of the field upon the voltage-current curves will be shown only by means of the V_{ab} vs. i_B curves.

Therefore, starting from the case in which the applied field is directed along the z −direction, i.e., in which $\vec{\psi}_{ex} = \psi_{ex}\hat{z}$, in Fig. 9.41a–b, 9.42a–b, and 9.43a–b we show the average voltages $\langle v_x(\vec{r})\rangle$, $\langle v_y(\vec{r})\rangle$, $\langle v_z(\vec{r})\rangle$, respectively. The chosen values of ψ_{ex} are the following: $\psi_{ex} = 0.0, 0.4, 0.8$.

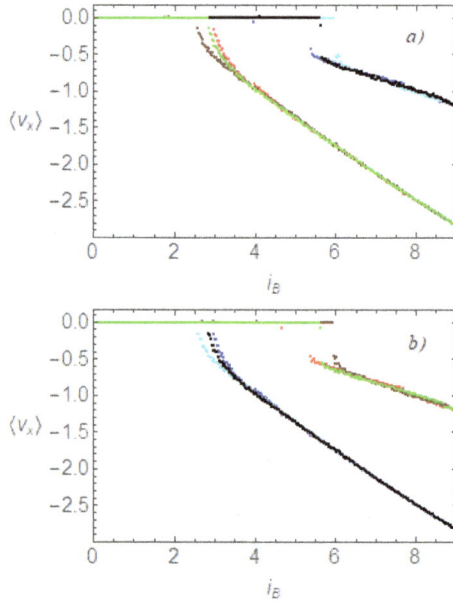

Fig. 9.41. Time-averaged voltage $\langle v_x(\vec{r})\rangle$ vs. i_B curves for $\beta = 1.0$, $m = 0.1$, and $\vec{\psi}_{ex} = \psi_{ex}\hat{z}$. In panel a) $\langle v_x(0)\rangle$ and $\langle v_x(a\hat{y})\rangle$ and in panel b) $\langle v_x(a\hat{z})\rangle$ and $\langle v_x(a\hat{y} + a\hat{z})\rangle$ are represented. In both panels, the first quantities, $\langle v_x(0)\rangle$ and $\langle v_x(a\hat{z})\rangle$, are in blue for $\psi_{ex} = 0.0$, cyan for $\psi_{ex} = 0.4$, and black for $\psi_{ex} = 0.8$; the second, $\langle v_x(a\hat{y})\rangle$ and $\langle v_x(a\hat{y} + a\hat{z})\rangle$, are in red for $\psi_{ex} = 0.0$, brown for $\psi_{ex} = 0.4$, green for $\psi_{ex} = 0.8$.

In particular, in Fig. 9.41a the average voltages $\langle v_x(0) \rangle$ and $\langle v_x(a\hat{y}) \rangle$ are shown as a function of the bias current. The first voltage, $\langle v_x(0) \rangle$, is reported in blue for $\psi_{ex} = 0.0$, in cyan for $\psi_{ex} = 0.4$, and in black for $\psi_{ex} = 0.8$, while the second voltage, $\langle v_x(a\hat{y}) \rangle$, is reported in red for $\psi_{ex} = 0.0$, in brown for $\psi_{ex} = 0.4$, and in green for $\psi_{ex} = 0.8$.

In Fig. 9.41b, the average voltages $\langle v_x(a\hat{z}) \rangle$ and $\langle v_x(a\hat{y} + a\hat{z}) \rangle$ are shown as a function of the bias current. The first voltage, $\langle v_x(a\hat{z}) \rangle$, is reported in blue for $\psi_{ex} = 0.0$, in cyan for $\psi_{ex} = 0.4$, and in black for $\psi_{ex} = 0.8$, while the second voltage, $\langle v_x(a\hat{y} + a\hat{z}) \rangle$, is reported in red for $\psi_{ex} = 0.0$, in brown for $\psi_{ex} = 0.4$, and in green for $\psi_{ex} = 0.8$. By looking at Fig. 41a–b we may soon notice that all voltages are negative.

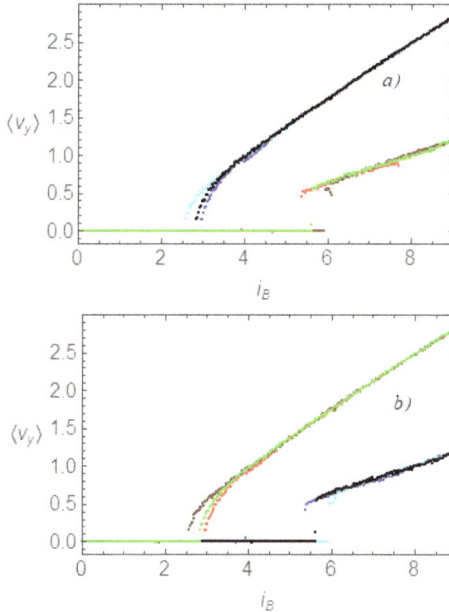

Fig. 9.42. Time-averaged voltage $\langle v_y(\vec{r}) \rangle$ vs. i_B curves for $\beta = 1.0$, $m = 0.1$, and $\vec{\psi}_{ex} = \psi_{ex}\hat{z}$. In panel a) $\langle v_y(0) \rangle$ and $\langle v_y(a\hat{x}) \rangle$ and in panel b) $\langle v_y(a\hat{z}) \rangle$ and $\langle v_y(a\hat{x} + a\hat{z}) \rangle$ are represented. In both panels, the first voltages, $\langle v_y(0) \rangle$ and $\langle v_y(a\hat{z}) \rangle$, are in blue for $\psi_{ex} = 0.0$, cyan for $\psi_{ex} = 0.4$, and black for $\psi_{ex} = 0.8$; the second, $\langle v_y(ax) \rangle$ and $\langle v_y(a\hat{x} + a\hat{z}) \rangle$, are in red for $\psi_{ex} = 0.0$, brown for $\psi_{ex} = 0.4$, green for $\psi_{ex} = 0.8$.

Moreover, the voltages across the branches converging either to node a or to node b, namely, $\langle v_x(a\hat{z})\rangle$ and $\langle v_x(a\hat{y})\rangle$, are equal. Also the voltages $\langle v_x(0)\rangle$ and $\langle v_x(a\hat{y} + a\hat{z})\rangle$ are seen to be equal.

In Fig. 9.42a the average voltages $\langle v_y(0)\rangle$ and $\langle v_y(a\hat{x})\rangle$ are shown as a function of the bias current. The first voltage, $\langle v_y(0)\rangle$, is reported in blue for $\psi_{ex} = 0.0$, in cyan for $\psi_{ex} = 0.4$, and in black for $\psi_{ex} = 0.8$, while the second voltage, $\langle v_y(a\hat{x})\rangle$, is reported in red for $\psi_{ex} = 0.0$, in brown for $\psi_{ex} = 0.4$, and in green for $\psi_{ex} = 0.8$.

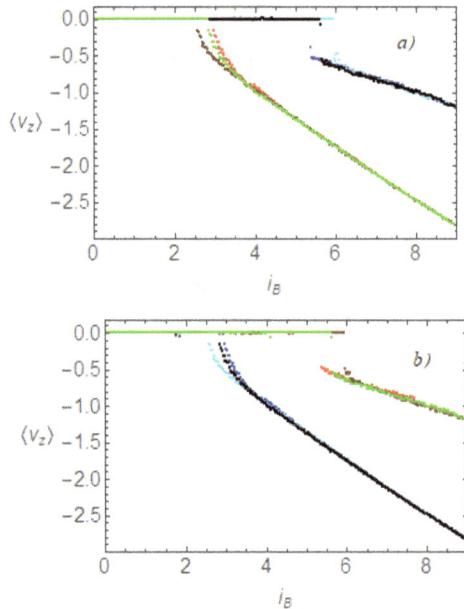

Fig. 9.43. Time-averaged voltage $\langle v_z(\vec{r})\rangle$ vs. i_B curves for $\beta = 1.0$, $m = 0.1$, and $\vec{\psi}_{ex} = \psi_{ex}\hat{z}$. In panel a) $\langle v_z(0)\rangle$ and $\langle v_z(a\hat{x})\rangle$ and in panel b) $\langle v_z(a\hat{y})\rangle$ and $\langle v_z(a\hat{x} + a\hat{y})\rangle$ are represented. In both panels, the first voltages, $\langle v_z(0)\rangle$ and $\langle v_z(a\hat{y})\rangle$, are in blue for $\psi_{ex} = 0.0$, cyan for $\psi_{ex} = 0.4$, and black for $\psi_{ex} = 0.8$; the second, $\langle v_z(a\hat{x})\rangle$ and $\langle v_z(a\hat{x} + a\hat{y})\rangle$, are in red for $\psi_{ex} = 0.0$, brown for $\psi_{ex} = 0.4$, green for $\psi_{ex} = 0.8$.

In Fig. 9.42b, on the other hand, the average voltages $\langle v_y(a\hat{z})\rangle$ and $\langle v_y(a\hat{x} + a\hat{z})\rangle$ are shown as a function of the bias current. The first voltage, $\langle v_y(a\hat{z})\rangle$, is reported in blue for $\psi_{ex} = 0.0$, in cyan for $\psi_{ex} = 0.4$, and in black for $\psi_{ex} = 0.8$, while the second voltage, $\langle v_y(a\hat{x} +$

$a\hat{z})\rangle$, is reported in red for $\psi_{ex} = 0.0$, in brown for $\psi_{ex} = 0.4$, and in green for $\psi_{ex} = 0.8$.

By looking at Fig. 9.42a–b we may soon notice that all voltages are positive. Moreover, the voltages across the branches converging either to node a or to node b, namely, $\langle v_y(0)\rangle$ and $\langle v_y(a\hat{x} + a\hat{z})\rangle$, are equal. Also the voltages $\langle v_y(a\hat{x})\rangle$ and $\langle v_y(a\hat{z})\rangle$ are seen to be equal.

In Fig. 9.43a the average voltages $\langle v_z(0)\rangle$ and $\langle v_z(a\hat{x})\rangle$ are shown as a function of the bias current. The first voltage, $\langle v_z(0)\rangle$, is reported in blue for $\psi_{ex} = 0.0$, in cyan for $\psi_{ex} = 0.4$, and in black for $\psi_{ex} = 0.8$, while the second voltage, $\langle v_z(a\hat{x})\rangle$, is reported in red for $\psi_{ex} = 0.0$, in brown for $\psi_{ex} = 0.4$, and in green for $\psi_{ex} = 0.8$. In Fig. 9.43b, on the other hand, the average voltages $\langle v_z(a\hat{y})\rangle$ and $\langle v_z(a\hat{x} + a\hat{y})\rangle$ are shown as a function of the bias current. The first voltage, $\langle v_z(a\hat{y})\rangle$, is reported in blue for $\psi_{ex} = 0.0$, in cyan for $\psi_{ex} = 0.4$, and in black for $\psi_{ex} = 0.8$, while the second voltage, $\langle v_z(a\hat{x} + a\hat{y})\rangle$, is reported in red for $\psi_{ex} = 0.0$, in brown for $\psi_{ex} = 0.4$, and in green for $\psi_{ex} = 0.8$. In these figures we still notice that voltages across the branches positioned at both ends of the two diagonals of the square faces lying on the $x - y$ plane are equal, so that $\langle v_z(a\hat{x})\rangle = \langle v_z(a\hat{y})\rangle$ and $\langle v_z(0)\rangle = \langle v_z(a\hat{x} + a\hat{y})\rangle$.

As an overall comment on the curves obtained for the average voltages across the twelve branches of the network, we may say that the main features of the $I - V$ characteristics of superconducting quantum interferometers is well reproduced in the branches having one end coincident with either with node a or node b. It can be noted that these curves show a transition to the RS for bias currents close to $i_B = 3$. The particular value of i_B, at which the first transition to the resistive state occurs, can be identified with a critical current i_c for the network. This current depends upon the applied magnetic field, as it can be argued from Fig. 9.41a–b, Fig. 9.42a–b, and Fig. 9.43a–b. As a matter of fact, the different point of inception of the resistive state obtained from these figures at different values of ψ_{ex} clearly indicates this dependence. It can be also noticed that a bias current close to $i_B = 3$ needs to flow in the network in order to have transition to the resistive state, because it splits into three parts when entering the network at node a and recombines from three different branch currents when leaving the network at node b. The remaining $\langle v_\xi(\vec{r})\rangle$ vs. i_B curves, for those branches not connected to the nodes a or b, show a pronounced discontinuity at values of the bias current close to $i_B = 6$. In fact, in order to obtain a resistive state in the junctions lying on the latter branches, the current flowing in the three branches connected to node a or in the three branches connected to node b must be close to the value of 2. The current coming from these branches will then successively split into two parts when encountering a node different from a and b.

Having obtained the voltages across all branches in the network, we may now calculate the voltage difference $V_{ab} = V_b - V_a$ from any path going from node a to node b. For example, we might express this voltage difference as follows:

$$V_{ab} = \langle v_x(a\hat{z})\rangle + \langle v_y(0)\rangle + \langle v_z(0)\rangle. \tag{9.93}$$

In this way, the voltage V_{ab} versus the bias current curves can be determined by means of the results shown in Fig. 9.41a–b, 9.42a–b, 9.43 a–b.

Therefore, in Fig. 9.44, V_{ab} vs. i_B curves are shown in red for $\psi_{ex} = 0.0$, in blue for $\psi_{ex} = 0.4$, and in black for $\psi_{ex} = 0.8$.

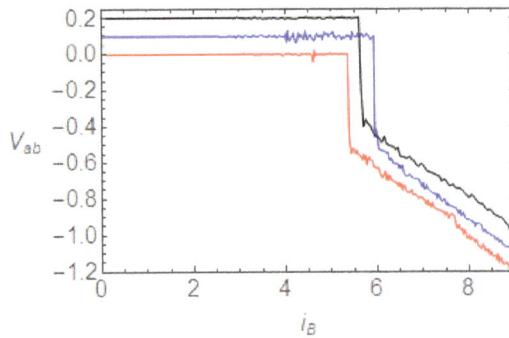

Fig. 9.44. Voltage difference V_{ab} vs. i_B curves for $\beta = 1.0$, $m = 0.1$, and $\vec{\psi}_{ex} = \psi_{ex}\hat{z}$. The red curve is for $\psi_{ex} = 0.0$, the blue curve for $\psi_{ex} = 0.4$, and the black curve for $\psi_{ex} = 0.8$. The blue and black curves are shifted of 0.1 and 0.2 on top of the red curve, respectively, for clarity reasons.

In the curves in Fig. 9.44 the main feature is the abrupt discontinuity in the vicinity of $i_B = 6$, which closely recalls the property of the voltage

$\langle v_z(0) \rangle$ across the vertical branch. This can be understood by considering that the two remaining voltages in Eq. (9.93), namely, $\langle v_x(a\hat{z}) \rangle$ and $\langle v_y(0) \rangle$, cancel each other, as it can be seen from Fig. 4.41b and Fig. 4.42a.

In this way, the voltage difference V_{ab} is dominated by the properties of voltages across the branches not directly linked to nodes a or b.

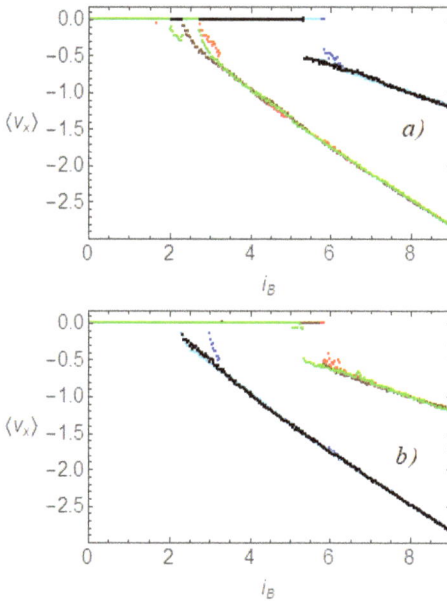

Fig. 9.45. Time-averaged voltage $\langle v_x(\vec{r}) \rangle$ vs. i_B curves for $\beta = 1.0$, $m = 0.1$, and $\vec{\psi}_{ex} = \psi_{ex}(\sqrt{2}\hat{x}/2 + \sqrt{2}\hat{z}/2)$. In panel a) $\langle v_x(0) \rangle$ and $\langle v_x(a\hat{y}) \rangle$ and in panel b) $\langle v_x(a\hat{z}) \rangle$ and $\langle v_x(a\hat{y} + a\hat{z}) \rangle$ are represented. In both panels, the first quantities, $\langle v_x(0) \rangle$ and $\langle v_x(a\hat{z}) \rangle$, are in blue for $\psi_{ex} = 0.2$, cyan for $\psi_{ex} = 0.4$, and black for $\psi_{ex} = 0.8$; the second, $\langle v_x(a\hat{y}) \rangle$ and $\langle v_x(a\hat{y} + a\hat{z}) \rangle$, are in red for $\psi_{ex} = 0.1$, brown for $\psi_{ex} = 0.4$, green for $\psi_{ex} = 0.8$.

In Fig. 9.45a–b, 9.46a–b, and 9.47a–b the $\langle v_\xi(\vec{r}) \rangle$ vs. i_B curves are shown for $\beta = 1.0$, $m = 0.1$, and for fields inclined of $45°$ with respect to the z−axis, in such a way that $\vec{\psi}_{ex} = \psi_{ex}(\sqrt{2}\hat{x}/2 + \sqrt{2}\hat{z}/2)$. The following values of the amplitude ψ_{ex} of the normalized externally applied flux have been chosen: $\psi_{ex} = 0.2$, $\psi_{ex} = 0.4$, and $\psi_{ex} = 0.8$.

In particular, in Fig. 9.45a–b the numerically evaluated $\langle v_x(\vec{r}) \rangle$ vs. i_B curves are reported. In panel $a)$ the voltages $\langle v_x(0) \rangle$ and $\langle v_x(a\hat{y}) \rangle$ are shown, while in panel $b)$ the voltages $\langle v_x(a\hat{z}) \rangle$ and $\langle v_x(a\hat{y} + a\hat{z}) \rangle$ are represented. In both panels, the first voltages, namely, $\langle v_x(0) \rangle$ and $\langle v_x(a\hat{z}) \rangle$, are in blue for $\psi_{ex} = 0.2$, cyan for $\psi_{ex} = 0.4$, and black for $\psi_{ex} = 0.8$; the second voltages, $\langle v_x(a\hat{y}) \rangle$ and $\langle v_x(a\hat{y} + a\hat{z}) \rangle$, are in red for $\psi_{ex} = 0.2$, brown for $\psi_{ex} = 0.4$, green for $\psi_{ex} = 0.8$.

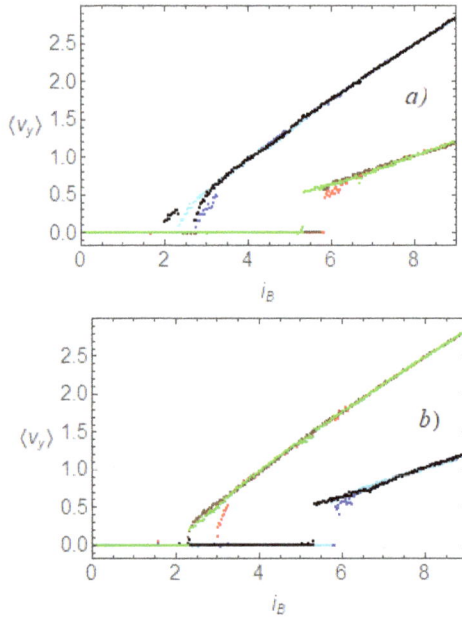

Fig. 9.47. Time-averaged voltage $\langle v_y(\vec{r}) \rangle$ vs. i_B curves for $\beta = 1.0$, $m = 0.1$, and $\vec{\psi}_{ex} = \psi_{ex}(\sqrt{2}\hat{x}/2 + \sqrt{2}\hat{z}/2)$. In panel a) $\langle v_y(0) \rangle$ and $\langle v_y(a\hat{x}) \rangle$ and in panel b) $\langle v_y(a\hat{z}) \rangle$ and $\langle v_y(a\hat{x} + a\hat{z}) \rangle$ are represented. In both panels, the first voltages, $\langle v_y(0) \rangle$ and $\langle v_y(a\hat{z}) \rangle$, are in blue for $\psi_{ex} = 0.2$, cyan for $\psi_{ex} = 0.4$, and black for $\psi_{ex} = 0.8$; the second, $\langle v_y(ax) \rangle$ and $\langle v_y(a\hat{x} + a\hat{z}) \rangle$, are in red for $\psi_{ex} = 0.2$, brown for $\psi_{ex} = 0.4$, green for $\psi_{ex} = 0.8$.

In Fig. 9.46a–b the numerically evaluated $\langle v_y(\vec{r}) \rangle$ vs. i_B curves are reported. In panel a) the voltages $\langle v_y(0) \rangle$ and $\langle v_y(a\hat{x}) \rangle$ are shown, while in panel b) the voltages $\langle v_y(a\hat{z}) \rangle$ and $\langle v_y(a\hat{x} + a\hat{z}) \rangle$ are represented. In both panels, the first voltages, namely, $\langle v_y(0) \rangle$ and $\langle v_y(a\hat{z}) \rangle$, are in blue for $\psi_{ex} = 0.2$, cyan for $\psi_{ex} = 0.4$, and black for $\psi_{ex} = 0.8$; the second voltages, $\langle v_y(a\hat{y})\rangle$ and $\langle v_y(a\hat{x}+a\hat{z})\rangle$, are in red for $\psi_{ex}=0.2$, brown for $\psi_{ex}=0.4$, green for $\psi_{ex}=0.8$. Finally, Fig. 9.47a–b the numerically evaluated $\langle v_z(\vec{r})\rangle$ vs. i_B curves are reported. In panel a) the voltages $\langle v_z(0)\rangle$ and $\langle v_z(a\hat{x})\rangle$ are shown, while in panel b) the voltages $\langle v_z(a\hat{y})\rangle$ and $\langle v_z(a\hat{x}+a\hat{y})\rangle$ are represented. In both panels, the first voltages, namely, $\langle v_z(0)\rangle$ and $\langle v_z(a\hat{y})\rangle$, are in blue for $\psi_{ex}=0.2$, cyan for $\psi_{ex}=0.4$, and black for $\psi_{ex}=0.8$; the second voltages, $\langle v_z(a\hat{y})\rangle$ and $\langle v_z(a\hat{x}+a\hat{y})\rangle$, are in red for $\psi_{ex}=0.2$, brown for $\psi_{ex}=0.4$, green for $\psi_{ex}=0.8$.

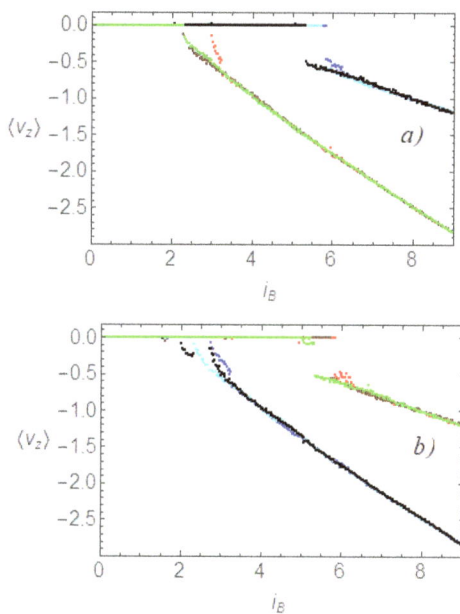

Fig. 9.47. Time-averaged voltage $\langle v_z(\vec{r})\rangle$ vs. i_B curves for $\beta=1.0$, $m=0.1$, and $\vec{\psi}_{ex}=\psi_{ex}(\sqrt{2}\hat{x}/2+\sqrt{2}\hat{z}/2)$. In panel a) $\langle v_z(0)\rangle$ and $\langle v_z(a\hat{x})\rangle$ and in panel b) $\langle v_z(a\hat{y})\rangle$ and $\langle v_z(a\hat{x}+a\hat{y})\rangle$ are represented. In both panels, the first voltages, $\langle v_z(0)\rangle$ and $\langle v_z(a\hat{y})\rangle$, are in blue for $\psi_{ex}=0.2$, cyan for $\psi_{ex}=0.4$, and black for $\psi_{ex}=0.8$; the second, $\langle v_z(a\hat{x})\rangle$ and $\langle v_z(a\hat{x}+a\hat{y})\rangle$, are in red for $\psi_{ex}=0.2$, brown for $\psi_{ex}=0.4$, green for $\psi_{ex}=0.8$..

Finally, in Fig. 9.47a–b, the numerically evaluated $\langle v_z(\vec{r})\rangle$ vs. i_B curves are reported. In panel a) the voltages $\langle v_z(0)\rangle$ and $\langle v_z(a\hat{x})\rangle$ are shown, while in panel b) the voltages $\langle v_z(a\hat{y})\rangle$ and $\langle v_z(a\hat{x}+a\hat{y})\rangle$ are represented. In both panels, the first voltages, namely, $\langle v_z(0) \rangle$ and $\langle v_z(a\hat{y}) \rangle$, are in blue for $\psi_{ex} = 0.2$, cyan for $\psi_{ex} = 0.4$, and black for $\psi_{ex} = 0.8$; the second voltages, $\langle v_z(a\hat{y}) \rangle$ and $\langle v_z(a\hat{x} + a\hat{y}) \rangle$, are in red for $\psi_{ex} = 0.2$, brown for $\psi_{ex} = 0.4$, green for $\psi_{ex} = 0.8$.

The curves in Fig. 9.45a–b, 9.46a–b, 9.47a–b present similar features to those in Fig. 9.41a–b, 9.42a–b, 9.43a–b. However, for $\psi_{ex} = 0.8$ and around $i_B = 2.0$, we notice a reentrant behavior to a ZVS of the junctions across which a voltage $\langle v_x(a\hat{y}) \rangle$, $\langle v_y(a\hat{x}) \rangle$, and $\langle v_z(a\hat{x} + a\hat{y}) \rangle$ can be measured. In fact, the green curve in Fig. 9.45a, the black curve in Fig. 9.46a and the black curve in Fig. 9.47b show a little bump in the vicinity of $i_B = 2.0$, which identifies a transition to a resistive state in a small interval of i_B. The definitive transition to the resistive state, after reentrance in the ZVS of these junctions, occurs at $i_B \approx 2.7$.

This feature is visible in Fig. 9.48, where V_{ab} vs. i_B curves are shown in red for $\psi_{ex} = 0.2$, in blue for $\psi_{ex} = 0.4$, and in black for $\psi_{ex} = 0.8$. These curves are shifted of a quantity $\Delta V = 0.1$ for clarity reasons.

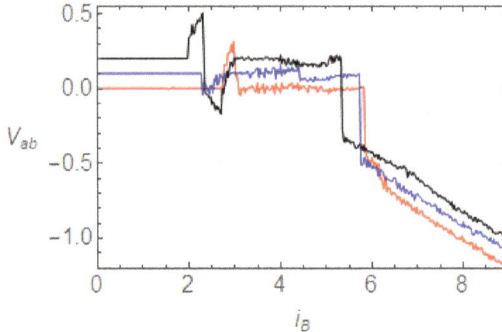

Fig. 9.48. Voltage difference V_{ab} vs. i_B curves for $\beta = 1.0$, $m = 0.1$, and $\vec{\psi}_{ex} = \psi_{ex}(\sqrt{2}\hat{x}/2 + \sqrt{2}\hat{z}/2)$. The red curve is for $\psi_{ex} = 0.2$, the blue curve for $\psi_{ex} = 0.4$, and the black curve for $\psi_{ex} = 0.8$. The blue and black curves are shifted of 0.1 and 0.2 on top of the red curve, respectively, for clarity reasons.

Apart from small deviations from $V_{ab} = 0$, clear bumps in the red, blue, and black curves are detected around $i_B = 2.5$, where some junctions in the network go from the zero-voltage state to the resistive state. In this way, even if the voltage difference V_{ab} is still dominated by the properties of voltages across the branches not directly linked to nodes a

or b, traces of the transition to the resistive states of junctions lying on the branches linked to these nodes are visible form the V_{ab} vs. i_B curves. In Fig. 9.49a–b, 9.50a–b, and 9.51a–b the $\langle v_\xi(\vec{r})\rangle$ vs. i_B curves are shown for $\beta = 1.0$, $m = 0.1$, and for fields whose components are such that $\vec{\psi}_{ex} = \psi_{ex}(\sqrt{3}\hat{x}/3 + \sqrt{3}\hat{y}/3 + \sqrt{3}\hat{z}/3)$. The values of the amplitude ψ_{ex} of the normalized externally applied flux are still the following: $\psi_{ex} = 0.2$, $\psi_{ex} = 0.4$, and $\psi_{ex} = 0.8$.

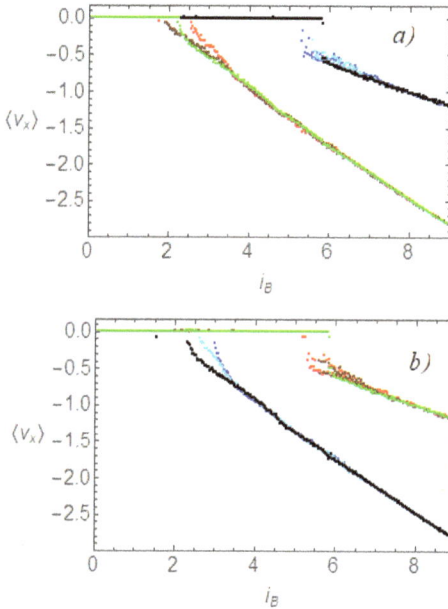

Fig. 9.49. Time-averaged voltage $\langle v_x(\vec{r})\rangle$ vs. i_B curves for $\beta = 1.0$, $m = 0.1$, and $\vec{\psi}_{ex} = \psi_{ex}(\sqrt{3}\hat{x}/3 + \sqrt{3}\hat{y}/3 + \sqrt{3}\hat{z}/3)$. In panel a) $\langle v_x(0)\rangle$ and $\langle v_x(a\hat{y})\rangle$ and in panel b) $\langle v_x(a\hat{z})\rangle$ and $\langle v_x(a\hat{y} + a\hat{z})\rangle$ are represented. In both panels, the first quantities, $\langle v_x(0)\rangle$ and $\langle v_x(a\hat{z})\rangle$, are in blue for $\psi_{ex} = 0.2$, cyan for $\psi_{ex} = 0.4$, and black for $\psi_{ex} = 0.8$; the second, $\langle v_x(a\hat{y})\rangle$ and $\langle v_x(a\hat{y} + a\hat{z})\rangle$, are in red for $\psi_{ex} = 0.1$, brown for $\psi_{ex} = 0.4$, green for $\psi_{ex} = 0.8$.

In particular, in Fig. 9.49a–b the numerically evaluated $\langle v_x(\vec{r})\rangle$ vs. i_B curves are reported. In panel a) the voltages $\langle v_x(0)\rangle$ and $\langle v_x(a\hat{y})\rangle$ are shown, while in panel b) the voltages $\langle v_x(a\hat{z})\rangle$ and $\langle v_x(a\hat{y} + a\hat{z})\rangle$ are represented. In both panels, the first voltages, namely, $\langle v_x(0)\rangle$ and $\langle v_x(a\hat{z})\rangle$, are in blue for $\psi_{ex} = 0.2$, cyan for $\psi_{ex} = 0.4$, and black for

$\psi_{ex} = 0.8$; the second voltages, $\langle v_x(a\hat{y}) \rangle$ and $\langle v_x(a\hat{y} + a\hat{z}) \rangle$, are in red for $\psi_{ex} = 0.2$, brown for $\psi_{ex} = 0.4$, green for $\psi_{ex} = 0.8$.

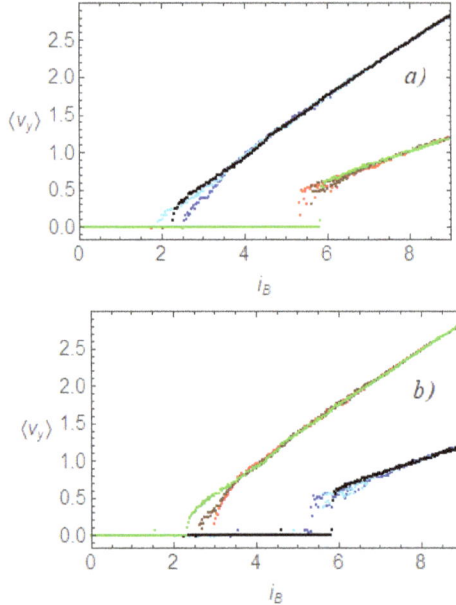

Fig. 9.50. Time-averaged voltage $\langle v_y(\vec{r}) \rangle$ vs. i_B curves for $\beta = 1.0$, $m = 0.1$, and $\vec{\psi}_{ex} = \psi_{ex}(\sqrt{3}\hat{x}/3 + \sqrt{3}\hat{y}/3 + \sqrt{3}\hat{z}/3)$. In panel a) $\langle v_y(0) \rangle$ and $\langle v_y(a\hat{x}) \rangle$ and in panel b) $\langle v_y(a\hat{z}) \rangle$ and $\langle v_y(a\hat{x} + a\hat{z}) \rangle$ are represented. In both panels, the first voltages, $\langle v_y(0) \rangle$ and $\langle v_y(a\hat{z}) \rangle$, are in blue for $\psi_{ex} = 0.2$, cyan for $\psi_{ex} = 0.4$, and black for $\psi_{ex} = 0.8$; the second, $\langle v_y(ax) \rangle$ and $\langle v_y(a\hat{x} + a\hat{z}) \rangle$, are in red for $\psi_{ex} = 0.2$, brown for $\psi_{ex} = 0.4$, green for $\psi_{ex} = 0.8$.

In Fig. 9.50a–b the numerically evaluated $\langle v_y(\vec{r}) \rangle$ vs. i_B curves are reported. In panel a) the voltages $\langle v_y(0) \rangle$ and $\langle v_y(a\hat{x}) \rangle$ are shown, while in panel b) the voltages $\langle v_y(a\hat{z}) \rangle$ and $\langle v_y(a\hat{x} + a\hat{z}) \rangle$ are represented. In both panels, the first voltages, namely, $\langle v_y(0) \rangle$ and $\langle v_y(a\hat{z}) \rangle$, are in blue for $\psi_{ex} = 0.2$, cyan for $\psi_{ex} = 0.4$, and black for $\psi_{ex} = 0.8$; the second voltages, $\langle v_y(a\hat{y}) \rangle$ and $\langle v_y(a\hat{x} + a\hat{z}) \rangle$, are in red for $\psi_{ex} = 0.2$, brown for $\psi_{ex} = 0.4$, green for $\psi_{ex} = 0.8$.

Finally, in Fig. 9.51a–b, the numerically evaluated $\langle v_z(\vec{r}) \rangle$ vs. i_B curves are reported. In panel a) the voltages $\langle v_z(0) \rangle$ and $\langle v_z(a\hat{x}) \rangle$ are shown, while in panel b) the voltages $\langle v_z(a\hat{y}) \rangle$ and $\langle v_z(a\hat{x} + a\hat{y}) \rangle$ are represented. In both panels, the first voltages, namely, $\langle v_z(0)\rangle$ and $\langle v_z(a\hat{y})\rangle$, are in blue for $\psi_{ex} = 0.2$, cyan for $\psi_{ex} = 0.4$, and black for $\psi_{ex} = 0.8$; the second voltages, $\langle v_z(a\hat{y})\rangle$ and $\langle v_z(a\hat{x} + a\hat{y})\rangle$, are in red for $\psi_{ex} = 0.2$, brown for $\psi_{ex} = 0.4$, green for $\psi_{ex} = 0.8$.

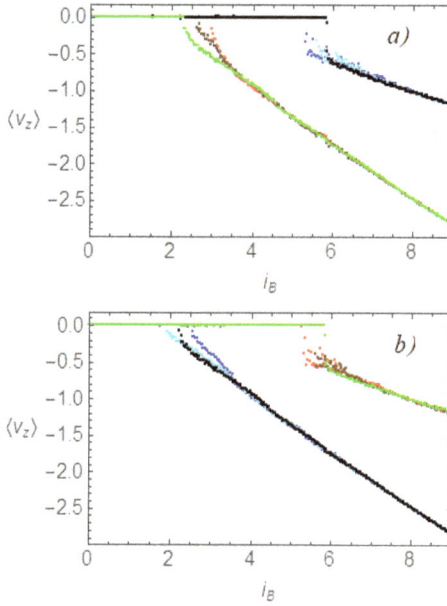

Fig. 9.51. Time-averaged voltage $\langle v_z(\vec{r})\rangle$ vs. i_B curves for $\beta = 1.0$, $m = 0.1$, and $\vec{\psi}_{ex} = \psi_{ex}(\sqrt{3}\hat{x}/3 + \sqrt{3}\hat{y}/3 + \sqrt{3}\hat{z}/3)$. In panel a) $\langle v_z(0)\rangle$ and $\langle v_z(a\hat{x})\rangle$ and in panel b) $\langle v_z(a\hat{y})\rangle$ and $\langle v_z(a\hat{x} + a\hat{y})\rangle$ are represented. In both panels, the first voltages, $\langle v_z(0)\rangle$ and $\langle v_z(a\hat{y})\rangle$, are in blue for $\psi_{ex} = 0.2$, cyan for $\psi_{ex} = 0.4$, and black for $\psi_{ex} = 0.8$; the second, $\langle v_z(a\hat{x})\rangle$ and $\langle v_z(a\hat{x} + a\hat{y})\rangle$, are in red for $\psi_{ex} = 0.2$, brown for $\psi_{ex} = 0.4$, green for $\psi_{ex} = 0.8$..

Finally, Fig. 9.51a–b the numerically evaluated $\langle v_z(\vec{r})\rangle$ vs. i_B curves are reported. In panel a) the voltages $\langle v_z(0)\rangle$ and $\langle v_z(a\hat{x})\rangle$ are shown, while in panel b) the voltages $\langle v_z(a\hat{y})\rangle$ and $\langle v_z(a\hat{x} + a\hat{y})\rangle$ are represented. In both panels, the first voltages, namely, $\langle v_z(0)\rangle$ and $\langle v_z(a\hat{y})\rangle$, are in blue for $\psi_{ex} = 0.2$, cyan for $\psi_{ex} = 0.4$, and black for $\psi_{ex} = 0.8$; the second voltages, $\langle v_z(a\hat{y})\rangle$ and $\langle v_z(a\hat{x} + a\hat{y})\rangle$, are in red for $\psi_{ex} = 0.2$, brown for $\psi_{ex} = 0.4$, green for $\psi_{ex} = 0.8$.

The curves in Fig. 9.49a–b, 9.50a–b, 9.51a–b present similar features when compared to those in Fig. 9.41a–b, 9.42a–b, 9.43a–b. However, for $\psi_{ex} = 0.8$ and around $i_B = 2.0$, we notice a reentrant behavior to a ZVS of the junctions across which a voltage $\langle v_x(a\hat{y})\rangle$, $\langle v_y(a\hat{x})\rangle$, and $\langle v_z(a\hat{x} + a\hat{y})\rangle$ can be measured. In fact, the green curve in Fig. 9.45a, the black curve in Fig. 9.46a and the black curve in Fig. 9.47b show a little bump in the vicinity of $i_B = 2.0$, which identifies a transition to a resistive state in a small interval of i_B. The definitive transition to the resistive state, after reentrance in the ZVS of these junctions, occurs at $i_B \approx 2.7$.

This feature is visible in Fig. 9.52, where V_{ab} vs. i_B curves are shown in red for $\psi_{ex} = 0.2$, in blue for $\psi_{ex} = 0.4$, and in black for $\psi_{ex} = 0.8$. These curves are shifted of a quantity $\Delta V = 0.1$ for clarity reasons.

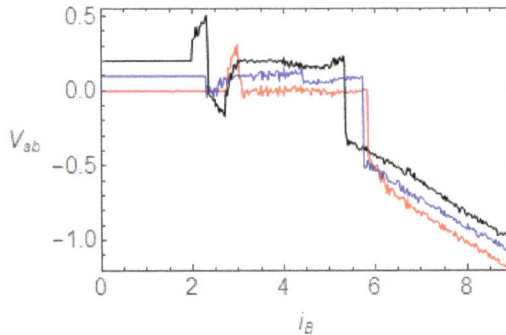

Fig. 9.52. Voltage difference V_{ab} vs. i_B curves for $\beta = 1.0$, $m = 0.1$, and $\vec{\psi}_{ex} = \psi_{ex}(\sqrt{3}\hat{x}/3 + \sqrt{3}\hat{y}/3 + \sqrt{3}\hat{z}/3)$. The red curve is for $\psi_{ex} = 0.2$, the blue curve for $\psi_{ex} = 0.4$, and the black curve for $\psi_{ex} = 0.8$. The blue and black curves are shifted of 0.1 and 0.2 on top of the red curve, respectively, for clarity reasons.

Apart from small deviations from $V_{ab} = 0$, clear bumps in the red, blue, and black curves are detected around $i_B = 2.5$, where the junctions lying on the branches connected to nodes a and b go from the zero-voltage state to the resistive state.

After having investigated, in three different case, the behavior of the network under a gradual increase of the bias current in the presence of an applied magnetic field not necessarily oriented along the axial direction, we shall now try to detect some differences between these three cases.

Let us then consider the transition to the resistive state of the junctions lying on the branches connected to nodes a and b. The voltages across the junctions close to node a, belonging to a first group, are the following: $\langle v_x(a\hat{z})\rangle$, $\langle v_y(a\hat{x} + a\hat{z})\rangle$, and $\langle v_z(a\hat{x})\rangle$. On the other hand, the voltages across the junctions close to node b, belonging to a second group, are: $\langle v_x(a\hat{y})\rangle$, $\langle v_y(0)\rangle$, and $\langle v_z(a\hat{y})\rangle$. From the collected $\langle v_\xi(\vec{r})\rangle$ vs. i_B curves we may notice that both groups of voltages present the same type of behavior for a magnetic field oriented in the z −direction. The $\langle v_\xi(\vec{r})\rangle$ vs. i_B curves of the junctions lying on the branches connected to nodes a or b is different, depending on the particular group to which the junction belongs. As an example, from Fig. 9.45a, 9.46a, and 9.47b, we may notice that the junctions in the second group show a bump around the value of the bias current $i_B = 2.5$ in their $\langle v_\xi(\vec{r})\rangle$ vs. i_B curves, obtained for $\vec{\psi}_{ex} = \psi_{ex}(\sqrt{2}\hat{x}/2 + \sqrt{2}\hat{z}/2)$ with $\psi_{ex} = 0.4$. Contrarily, the junctions belonging to the first group do not show this feature in their $\langle v_\xi(\vec{r})\rangle$ vs. i_B curves obtained for the same magnetic field value. We recall that this feature is indicative of the inception of the resistive state, at relatively low values of i_B, in the most exposed junctions in the network.

Bibliography

Abramowitz, M., & Stegun, I. A. (1965). *Handbook of Mathematical Functions*. New York: Dover.

Abrikosov, A. A. (1957). On the magnetic properties of superconductors of the second group. *Sov. Phys. JETP, 5*, 1174.

Ambegaokar, V., Halperin, B. I. (1969). Voltage due to thermal noise in the dc Josephson effect. *Phys. Rev. Lett.*, 1364.

Ambegaokar, V., & Baratoff, A. (1963). Tunneling between superconductors. *Phys. Rev. Lett., 10*, 486-489.

Amin, M. H. S., Smirnov, A. Yu., Zagoskin, A. M., Lindström, T., Charlebois, S. A., Claeson, T., Tzalenchuk, A. Ya. (2005). Silent phase qubit based on d-wave Josephson junctions, *Phys. Rev. B, 71*, 064516.

Arfken, G. W. (2012). *Mathematical Methods for Physicists*. Oxford, UK: Academic Press.

Barone, A., & Paternò, G. (1982). *Physics and Applications of the Josephson Effect*. New York: Wiley.

Baselmans, J. J. A., Morpurgo, A. F., Van Wees, B. J., and Klapwijk T. M. (1999). Reversing the direction of the supercurrent in a controllable Josephson junction. *Nature, 397*, 43.

Bednorz, J. G., & Muller, K. A. (1986). Possible highTc superconductivity in the Ba−La−Cu−O system. *Z. Physik B - Condensed Matter, 64*, 189-193.

Blatter, G., Geshkenbein, V. B., Ioffe, L. B. (2001). Design aspects of superconducting-phase quantum bits. *Phys. Rev. B, 63*, 174511.

Bocko, M. F., Herr, A. M., Feldman, M. J. (1997). Prospect for Quantum Coherence Computation Using Superconducting Electronics. *IEEE Transactions on Applied Superconductivity, 7*, 3638.

Boulaevskii, L.N., Kuzii, V. V., Sobyanin, A. A. (1977). Superconducting system with weak coupling to the current in the ground state. *JEPT Lett., 25*, 290.

Boyce, W. E., DiPrima R. C. (1976). *Elementary differential equations and boundary value problems*. New York: John Wiley & Sons.

Braunish, W., Knauf, N., Bauer, G., Kock, A., Becker, A., Freitag, B., Grütz, A., Kataev V, V., Neuhausen, S., Roden, B., Khomskii, D., Wohlleben, D., Bock, J., Preisler, E. (1993). Paramegnetic Meissner effect in high-temperature superconductors. *Phys. Rev. B*, 4030-4042.

Chen, D. X. Sanchez, A. Hernando, A. (1994). Magnetic properties of slablike Josephson-junction arrays. *Phys. Rev. B, 50*, 13735.

Chesca, B. (1998). Analytical theory of DC SQUIDS operating in the presence of thermal fluctuations. *Journal of Low Temparature Physics, 112*, 165.

Chesca, B. (1999). Magnetic field dependencies of the critical current and of the resonant modes of dc SQUIDs fabricated from superconductors with $s + id_{(x^2-y^2)}$ order-parameter symmetries. *Ann. Phys. (Leipzig), 8*, 511.

Clarke, J., Braginsky, A. I. (2004). *The SQUID Handbook, Vol. I.* Weinheim: Wiley-VCH.

Clem, J. R. (1988). Granular and superconducting-glass properties of the high-temperature superconductors. *Physica C, 153-155*, 50-55.

Crankshaw, D. S., Orlando, T. P. (2001). Inductance Effect in the Persistent Current Qubit. *IEEE Transactions on Applied Superconductivity, 11*, 1006.

De Leo, C., Rotoli, G., Barbara, P., Nielsen, A. P., Lobb, C. J. (2001). Mutual-inductance route to the paramagnetic Meissner effect in two-dimensional Josephson-junction arrays. *Phys. Rev. B, 64*, 1445181-1445185.

De Luca, R. (2000). Magnetic response of field-cooled type-I superconducting hollow cylinders. *Physica B*, 801-802.

De Luca, R. (2001). A general analytic approach to the study of the electrodynamic properties of d.c. SQUIDs. *Phys. Lett. A, 280*, 209-214.

De Luca, R., Di Matteo, T., Tuohimaa, A., Paasi, J. (1998). Three-dimensional Josephson-junction arrays: Static magnetic response. *Phys. Rev. B, 57*, 1173-1180.

De Luca, R., Di Matteo, T., Tuohimaa, A. , Paasi, J. (2000). Electrodynamics of current-biased inductive cubic networks of Josephson junctions. *Phylosophical Magazine B, 80*, 897-905.

De Luca, R., Fedullo, A., Gasanenko A. V. (2010). Equivalent single junction model for two-junction quantum interferometers with small inductance in the presence of noise. *Physica C, 470*, 134.

De Luca, R., Romeo, F. (2002). Electrodynamic response of current-biased elementary cubic networks of Josephson junctions. *Phys. Rev. B, 66*, 0245091-0245097.

De Luca, R., & Romeo, F. (2009). Sawtooth current-phase relation of a superconducting trilayer system described using Ohta's model. *Phys. Rev. B, 79*, 0945161-6.

De Luca, R., Pace, S., & Raiconi, G. (1993). Critical state in Josephson junction arrays as models of a bi-dimensional superconducting granular system. *Phys. Lett. A, 172*, 391-398.

Deutscher, G. and Muller, K. A. (1987). Origin of superconductive glassy state and extrinsic critical currents in high-Tc oxides. *Phys. Rev. Lett., 59*, 1745-1747.

Ebner, C. and Stroud, D. (1985). Diamagnetic susceptibility of superconducting clusters: Spin-glass behavior. *Phys. Rev. B, 31*, 165-171.

Feynman, R. P., Leighton, R. B., & Sands, M. (1965). *The Feynman's Lectures on Physics* (Vol. III). Reading, MA: Addison-Wesley.

Geshkenbein, V. B., Larkin, A. I., Barone, A. (1987). Vortices with half magnetic flux quanta in "heavy-fermion" superconductors. *Phys. Rev. B, 36*, 235.

Ghosh, S. P., Chakraborty, A. K. (2010). *Network Analyisi and Synthesis.* New Delhi: McGraw Hill.

Ginzburg, S. L., Khavronin, V. P., Logvinova, G. Yu., Luzyanin, I. D., Herrmann, J. Lippold, B., Borner, H., and Schmiedel, H. (1991). Low-field electrodynamics of high-Tc superconductors — theory and experiment. *Physica C, 174*, 109-116.

Goodman, W. L., & Deaver, B. S. (1970). Detailed measurements of the quantized flux states of hollow superconducting cylinders. *Phys. Rev. Lett.*, 870-873.

Gradshteyn, I. S., Ryzhik, I. M. (1979). *Table of integrals, series, and products.* San Diego: Academic Press.

Greenberg, Y. S. (2001). Theory of the voltage–current characteristic of high Tc DC SQUIDs. *Physica C, 371*, 156.

Grønbech-Jensen, N., Thompson, D. B., Cirillo, M., Cosmelli, C. (2003). Thermal escape from zero-voltage states in hysteretic superconducting interferometers. *Phys. Rev. B, 67*, 224505.

Halliday, D., Resnick, R., Walker, J. (2005). *Fundamentals of Physics, Seventh Edition.* New York: John Wiley & Sons.

Hardy, G. G., Wright, E. M. 1979. *An introduction to the theory of numbers.* Oxford University Press.

Ioffe, L. B., Geshkenbein, V. B., Feigel'man, M. V., Fauchère, A. L., Blatter, G. (1999). *Nature, 398*, 679.

Jackson, J. D. (1975). *Classical Electrodynamics.* New York: John Wiley and Sons.

Josephson, B. D. (1962). Possible new effects in superconducting tunnelling. *Phys. Lett., 1*, 251-253.

Kramers, H. A. (1940). Brownian motion in a field of force and the diffusion model of chemical reactions. *Physica, 284*.

Lang, S. (1987). *Linear Algebra.* New York: Springer-Verlag.

Lewandowski, S. J. (1991). dc properties of series-parallel arrays of Josephson junctions in an external magnetic field. *Phys. Rev. B, 43*, 7776.

Li, M. S. (2003). Paramagnetic Meissner effect and related dynamical phenomena. *Physics Reports, 376*, 133-223.

Likharev, K. K. (1986). *Dynamics of Josepshon Junctions and Circuits.* Amsterdam: Gordon and Breach.

Meissner, W., & Ochsenfeld, R. (1933). Ein neuer Effekt bei Eintritt der Supraleitfähigkeit. *Naturwiss*, 787-788.

Mooij, J. E., Orlando, T. P., Levitov, L. S., Tian, L., C. H. Van der Wal, C. H., Lloyd, S. 1999. *Science, 285*, 1036.

Müller, K. A., Tagashike, M., & Bednorz, J. G. (1987). Flux trapping and superconductive glass state in La_2CuO_4-y:Ba. *Phys. Rev. Lett.*, 1143-1146.

Muller, K. H., Matthews, D. N., Driver, R. (1992). Critical current density of ceramic high-temperature superconductors in a low magnetic field. *Physica C, 191*, 339-346.

Newrock, R. S., Lobb, C. J., Geigenmüller, U., Octavio, M. (2000). The two-dimensional physics of Josephson junction arrays. *Solid State Physics, 54*, 263-512.

Ohta, H. (1977). A Self-consinstent Model of the Josephson Junction. In H. D. Hahlbonm, & H. Lubbig, *Superconducting Quantum Interference Devices and their Applications* (pp. 35-49). Berlin: de Gruyter.

Oppenländer, J., Häussler, Ch., Schopohl, N. (1999). Dynamic electromagnetic response of three-dimensional Josephson junction arrays. *J. Appl. Phys., 86*, 5775-5779.

Oppenländer, J., Häussler, Ch., and Schopohl, N. (2000). Non-Φ_0-periodic quantum interference in one-dimensional parallel Josephson junction arrays with unconventional grating structure. *Phys. Rev. B, 63*, 024511.

Oppenländer, J., Häussler, Ch., Friesch, A., Tomes, J., Caputo, P., Traeuble, T., and Schopohl, N. (2005). Superconducting Quantum Interference Filters operated in commercial miniature cryocoolers. *IEEE Trams. Appl. Superc., 15*, 936.

Ryazanov, V. V., Oboznov, V. A., Rusanov, A- Yu., Veretennikov, A. V., Golubov, A. A. and Aarts, J. (2001). Coupling of Two Superconductors through a Ferromagnet: Evidence for a π Junction. *Phys. Rev. Lett., 86,* 2427.

Sakurai, J. J. (2004). *Modern Quantum Mechanics.* Singapore: Pearson Education.

Scharinger, S., Gürlich, C., Mints, R. G., Weides, M., Kohlstedt, H., Goldobin, E., Koelle, D. and Kleiner, R. (2010). Interference patterns of multifacet 20×(0−π) Josephson junctions with ferromagnetic barrier. *Phys. Rev. B, 81,* 174535.

Schmidt, V. V. (1997). *The Physics of Superconductors* (1st ed.). (P. Muller, & A. V. Ustinov, Eds.) Berlin: Springer.

Schultz, R. R., Chesca,B., Goetz, B., Schneider, C. W., Schmehl, A., Bielefeldt, H., Hilgenkamp, H., Mannhart, J. and Tsuei, C. C. (2000). Design and realization of an all d-wave dc π-superconducting quantum interference device . *Appl. Phys. Lett., 76,* 912.

Sears, F. W., Zemansky, M. W., & Young, H. D. (1977). *University Physics.* Reading, MA: Addison-Wesley.

Shapiro, S., Janus, A. R., & Holly, S. (1964). Effects of microwaves on Josephson current in superconducting tunneling. *Rev. Mod. Phys., 36,* 223-225.

Smilde, H. J. H., Ariando, Blank, D. H. A., Hilgenkamp, H., and Rogalla, H. (2004). pi-SQUIDs based on Josephson contacts between high-Tc and low-Tc superconductors. *Phys. Rev. B, 70,* 024519.

Tinkham, M.and Lobb, C. J. (1989). Physical Properties of the New Superconductors. *Solid State Physics, 42,* 91-134.

Weides, M., Kemmler, M., Kohlstedt, H., Waser, R., Koelle, D., Kleiner, R. and Goldobin E. (2006). 0-π Josephson tunnel junctions with ferromagnetic barrier. *Phys. Rev. Lett., 97,* 247001.

Wolf, T. and Majhofer, A. (1993). ac susceptibilities of a network of resistively shunted Josephson junctions with self-inductances. *Phys. Rev. B, 47,* 5383--5389.

Wollman, D. A., Van Harlingen, D. J., Lee, W. C., Ginsberg, D. M. and Leggett, A. J. (1993). Experimental determination of the superconducting pairing state in YBCO from the phase oherenece of YBCO-Pb dc SQUIDs. *Phys. Rev. Lett., 71,* 2134.

Yamashita, T., Tanikawa, K., Takahashi, S., Maekawa, S. (2005). *Phys. Rev. Lett., 95,* 097001.

Index

www.ingramcontent.com/pod-product-compliance
Lightning Source LLC
Chambersburg PA
CBHW050534190326
41458CB00007B/1785